QUANTUM FIELD THEORY

An Introduction for Chemical Physicists

QUANTUM FIELD THEORY

An Introduction for Chemical Physicists

RICHARD N PORTER

Stony Brook University, USA

World Scientific

EW JERSEY • LONDON • SINGAPORE • BEIJING • SHANGHAI • HONG KONG • TAIPEI • CHENNAI • TOKYO

Published by

World Scientific Publishing Co. Pte. Ltd.

5 Toh Tuck Link, Singapore 596224

USA office: 27 Warren Street, Suite 401-402, Hackensack, NJ 07601

UK office: 57 Shelton Street, Covent Garden, London WC2H 9HE

Library of Congress Cataloging-in-Publication Data

Names: Porter, Richard N. (Richard Needham), 1932– author.
Title: Quantum field theory : an introduction for chemical physicists /
Richard N. Porter, Stony Brook University, USA.
Description: New Jersey : World Scientific, [2022] |
Includes bibliographical references and index.
Identifiers: LCCN 2021017845 | ISBN 9789811239885 (hardcover) |
ISBN 9789811239892 (ebook) | ISBN 9789811239908 (ebook other)
Subjects: LCSH: Quantum field theory.
Classification: LCC QC174.45 .P68 2022 | DDC 530.14/3--dc23
LC record available at https://lccn.loc.gov/2021017845

British Library Cataloguing-in-Publication Data

A catalogue record for this book is available from the British Library.

For any available supplementary material, please visit
https://www.worldscientific.com/worldscibooks/10.1142/12359#t=suppl

Desk Editor: Shaun Tan Yi Jie

Printed in Singapore

Preface

How did a person who wanted to do organic chemistry get involved with quantum field theory? After completing my bachelors degree at Texas A&M I felt that organic chemistry was in line for a sea change if it were better informed by physical chemistry. In turn, the heart of physical chemistry is physics and mathematics. So I concentrated on chemical physics and mathematics in graduate school at the University of Illinois, Urbana, under the tutelage of Frederick Wall, Charles Schlichter, William Wyld, Aron Kuppermann, and Martin Karplus, the mathematics being largely self-taught from Kaplan's *Advanced Calculus and Courant* and Hilbert's *Methods of Mathematical Physics*, with Gradshteyn and Ryzhik close at hand.

As I learned quantum chemistry from Martin's lectures, I began to question the consistency of formulating the Schrödinger equation in terms of the positions and momenta of the electrons in atoms and molecules when the resulting wave function presented probability "clouds" for these variables. When I raised the point with Martin, he suggested that I pursue it in quantum field theory, which I began to do.

My quest became sharply focused at Stony Brook when my colleague Phillip Johnson and his student Li Le-ping presented me with unusual results from their observation of the multiphoton ionization of argon. Modeling photons as excitations of a quantized electromagnetic field was well-known, but what about the ensemble of atoms and molecules with which they interact? I began wrestling with quantum-mechanical modeling of an ideal gas at equilibrium, and after going down several blind alleys, I discovered the isodasic ensemble by randomizing the phases of the molecules. This discovery is the subject of Chapter 6; the extension to non-ideal gases and gas-liquid phase transitions through use of a simple equation satisfied by the isodasic ensemble operator is the subject of Chapter 7. To complete the book, I added mathematical tools, in particular an introduction to Fock algebra and its application to spinless particles and electrons, and finished with a quantum-field treatment of photons and their interaction with molecular ensembles.

Come join me in the joy of discovery!

Richard N. Porter

Setauket, New York, April 21, 2021

About the Author

Dr. Richard N. Porter is Professor Emeritus at the Department of Chemistry, Stony Brook University, USA. His research interests are in the theory of chemical reactions resulting from molecular collisions, the theory of nuclear motion in small stable molecules, and in many-body and field-theoretic formulation of the properties of molecular ensembles and their interaction with light. He has 60 years of experience in teaching undergraduate and graduate students. He has published over 80 papers in top reviewed journals such as *Physical Review Letters* and *Journal of Chemical Physics*. He holds a PhD from the University of Illinois, Urbana, USA.

Contents

Chapter 1

Introduction

1.1. Why this Book?

With so many books on quantum field theory available both in print and on the internet,[1] a valid question is "why another?" We offer two answers. First, few of these treatises are accessible to an untapped pool of potential users, namely, graduate students in chemical physics. This book, although no claim is made that it is easy reading, starts where most chemistry courses in quantum theory leave off and develops the necessary mathematical methods either in the text or in appendices. Second, some of the physicist authors have skirted over features of quantum field theory that have relevance to the study of ensembles of interacting molecules, including the modeling of phase transitions. This last subject is left open-ended in hopes that this book will whet the curiosity of its readers to the point that they will want to extend the theory of the molecular ensemble operator beyond what is presented here. At the very least, the power of the quantum field methodology, and in particular the Fock operators, should fascinate the reader as it has us over the course of many years.

1.2. What is Required of the Reader?

We assume the reader to have taken a graduate course in quantum theory and came away with a desire to learn more about the

[1] An excellent detailed introduction is Tong, D., *Lectures on Quantum Field Theory*, Cambridge University (2006).

subject. A prerequisite for quantum theory is a good grounding in multivariant differential and integral calculus. The reader would do well to review those subjects before tackling this book, or have a good text or class notes handy to review as s/he goes along. A prior knowledge of perturbation theory as applied in quantum theory would be useful, though not necessary, as we develop it here in some detail. But a facility with series expansions and their summation is a definite advantage. However, the advice given above should not keep the interested reader away. Most of the intended audience will do just fine, although multiple readings with pencil and paper at hand will most likely be required.

1.3. What is the Subject Matter of this Book?

1.3.1. *Chapter 2: The Harmonic Oscillator*

We begin with the Schrödinger equation for the harmonic oscillator, which will be familiar to all readers. But instead of solving it by analytical methods, we introduce the Fock operators, whose algebra is essential to all the rest of the book. Many readers will be familiar with the Fock technique, but we take it farther than some authors. We use Fock operators to define the Hermite polynomials and their recursion relations and normalization integrals. Dirac's *bra-* and *ket-* notation is introduced here and used to elucidate the notions of superposition of states and completeness. These concepts are in turn used to define a minimal wave packet and its time evolution is derived in both the Schrödinger and Heisenberg pictures.

1.3.2. *Chapter 3: Time-dependent Perturbation Expansions*

The Dirac (interaction) picture is defined and used to develop the Dyson expansion from which the time-evolution operator can be obtained by iteration. Evaluation of the terms in the Dyson expansion is facilitated by the use of Wick's theorem together with graphical analysis to keep up with the various contraction patterns that arise. The graphical technique is used to sum the series to infinite order in a special application. In the course of this analysis the notion of connected and disconnected graphs appears. The concepts introduced in Chapter 3 will be used in later applications of quantum field theory.

1.3.3. *Chapter 4: Spinless Particles*

This chapter treats the simplest case of the field theory of particles, in which the particles have no spin. Thus the Fock algebra remains bosonic in this case. First we use the formalism of Lagrangian classical mechanics to solve the classical harmonic oscillator problem, using Noether's theorem to identify the constants of motion. The oscillator is quantized by applying the uncertainty principle to the position and momentum variables. We introduce classical relativistic field theory of the oscillator by giving the Lagrangian *density* in the four-dimensional space of special relativity; that is, Klein-Gordon Lagrangian. The constants of the motion are found for the Klein-Gordon Lagrangian as well as the canonical coordinate and momentum densities. The Klein-Gordon equation is then obtained from the identification of the Hamiltonian density with the energy density. Quantization is then achieved by applying the coordinate-momentum commutation rule to their Klein-Gordon densities. The result is that the Fock operators are generalized to creation and annihilation operators for spinless particles, which are interpreted as excitations of the field.

1.3.4. *Chapter 5: Charge and Spin*

Here is where things become interesting, although complicated. The goal of this chapter is to obtain Dirac's beautiful equation describing electrons and positrons and put it into field-theoretic form. The quantum field theory of charged particles is developed and its coupling to a classical electromagnetic field is done with the introduction of the charge-current density tensor. We then proceed to Pauli's inserting spin into the Schrödinger equation and coupling spin to the electromagnetic field, which requires the wave function to become a 2-vector and the use of the 2×2 Pauli spin matrices in the wave equation. The resulting "Pauli equation" is unsatisfactory for treating high-speed electrons, however, since it is non-relativistic. This is where P. A. M. Dirac's great discovery of the 4-vector wave-function and the 4×4 Dirac matrices are required. In the field-theoretic form, the Dirac theory is applicable to positrons as well as electrons. Our treatment of these developments is as concise as we think possible, since our goal is to apply the *non-relativistic* theory to molecules in the next chapter.

1.3.5. *Chapter 6: The Perfect Molecular Gas*

How does one describe a perfect gas in quantum field theory? The answer is the subject of this chapter. It turns out that a generalization of the density operator is more useful in this regard than a wave function if the gas is assumed to be at thermodynamic equilibrium and has uniform density in coördinate space. We have shown that creation and annihilation operators can be used for both elementary bosons (Chapter 4) and fermions (Chapter 5). Here, we use an extension of the Fock operator technique to apply quantum field theory to an ensemble of non-interacting molecules. In this case, the field quanta are *molecules*, which may be either bosons or fermions, with momentum $\boldsymbol{p}$ and in internal state α. In the "classical" limit of high temperature and large volume V and number of molecules N, both types of molecules obey Boltzmann statistics. But even in this limit, quantum field theory remains a useful tool for deriving properties of perfect gases and provides a basis for extension of the theory to imperfect gases and liquids, as we shall see in Chapter 7. To obtain the ensemble operator for a perfect gas, Glauber theory, introduced in Chapter 3, is first used to obtain a *coherent* ensemble of molecules. When the molecular phases are included and then randomized, the resulting ensemble operator becomes *isodasic*[2]; *i.e.*, it describes a gas with uniform density filling a volume V.

When the entropy of the ensemble is maximized at constant N (the ensemble average of the number of molecules) and energy E

[2]From the Greek *iso* = same plus *dasys* = dense.

(the average total energy), the ensemble describes a perfect gas at thermal equilibrium.

The trace of the isodasic ensemble operator is the grand canonical partition function. Although the explicit form of the isodasic ensemble operator is somewhat complicated, it obeys a very simple equation which facilitates the calculations. The thermodynamic properties of the gas derived directly from the isodasic ensemble operator by the application of quantum field theory are identical to those derived from the grand canonical partition function in the traditional way. The ensemble operator in the Fock operator form also provides a means of deriving statistical properties of Bose, Fermi, and Boltzmann gases.

It is intended that Chapters 6 and 7 on the quantum field theory of perfect gases and of imperfect gases be self-contained. We therefore review briefly in Section 6.2 the Fock operators introduced in Chapter 2. We show how the finite volume enters the present theory through a delta function with zero argument. A succinct form of Wick's theorem for bosons and fermions, required for calculating the trace of the ensemble operator and average values of properties of the gas, is given in Section 6.3. In Section 6.4 we present Glauber's theory of coherent states, generalized to apply to molecules, and obtain the essential properties of coherent ensembles of non-interacting molecules. We randomize the phases of the molecules in Section 6.5 and show that the resulting ensemble operator obeys a simple equation which is key to the extension of the theory to imperfect gases and provides a novel approach to gas-liquid phase transitions given in Chapter 7. We determine the trace of the isodasic operator in Section 6.6 by a linked-cluster expansion and derive the distribution function for thermal equilibrium. In Section 6.7 we compare molecular statistics for coherent and isodasic ensembles.

1.3.6. *Chapter 7: Real Gases; Phase Transitions*

Quantum field theory, shown in Chapter 6 to provide a systematic tool for ensembles of molecules, is used in Chapter 7 to treat the effects of interaction of the molecules. A diagrammatic perturbation theory is developed for this purpose. Feynman-like graphs corresponding to long- and short-range interactions are summed to infinite order, yielding a model of the pair-correlation function and a new avenue for the study of liquid-gas transitions. In the case of ionic intermolecular forces, first-order perturbation theory gives the result obtained heuristically by Peter Debye as a first approach to the study of liquid solutions. An outline of suggestions for future work is included at the end of the chapter.

1.3.7. *Chapter 8: Photons*

In this chapter we describe the quantum-field ensemble operator for photons for both coherent and homogeneous light, the latter being analogous to the isodasic ensemble operator for molecules. This is accomplished by generalizing Glauber's theory to describe a focused light beam with a finite band width and a Gaussian profile. The $\boldsymbol{k}$ vectors (analogous to the $\boldsymbol{p}$ vectors for material particles) are parameterized by characteristics of the light intensity in the beam. Ensemble averaging is illustrated by calculating the photon density in $\boldsymbol{k}$-space (a Poisson distribution for coherent ensembles, the Bose-Einstein distribution for homogeneous and incoherent ensembles) and the spatial dependence of the light intensity in the beam. For homogeneous light (*e.g.*, incandescence), the ensemble operator gives the Planck distribution law in the limit of thermal equilibrium. The distinguishing mathematical and physical properties of ensemble operators for coherent and incoherent beams and for homogeneous light are presented, and the importance of phase in the representation of grand canonical ensembles of photons is illustrated.

In Section 8.2, we outline briefly the main features of quantum electrodynamics and in Section 8.3 we establish our notation by reviewing the relevant parts of the theory of coherence. In Section 8.4 we give the general form and some mathematical properties of the normalized ensemble operator for a coherent system of photons and derive the Poisson distribution for photons in a given sub-volume of $\boldsymbol{k}$-space. We obtain in Section 8.5 the explicit form for the photon number density in a focused coherent Gaussian pulse of light and calculate ensemble averages for the vector potential, energy density, and Poynting vector. We show how phase randomization leads to an ensemble operator for homogeneous radiation (*i.e.*, uniform energy density) and for a focused incoherent light pulse. A summary is given in Section 8.6 in which the distinguishing properties of coherent and incoherent light pulses and homogeneous light are reviewed, and their implications for experimental design are discussed.

1.3.8. *Chapter 9: Light-molecule Interactions*

In order to explore some of the phenomena that can be observed when light interacts with molecules, we first determine the contribution to the Hamiltonian from the interaction; that is, H_{I}. For guidance, we look back at Chapter 5, in which the invariance of the Lagrangian to local phase changes led to the result, described in Section 5.3.2.5 of that chapter, in which the momentum $\boldsymbol{p}$ is replaced by $\boldsymbol{p} - (q/c)\mathbf{A}$, where q is the charge, c the speed of light, and $\mathbf{A}$ the vector potential. In classical mechanics the kinetic energy becomes

$$\frac{(\boldsymbol{p} - \frac{q}{c}\mathbf{A})^2}{2m} - \frac{1}{2m}\left[p^2 - 2\frac{q}{c}\boldsymbol{p}\cdot\mathbf{A} + \left(\frac{q}{c}\right)^2\right],$$

from which we obtain

$$H_{\mathrm{I}} = -\frac{q}{mc}\boldsymbol{p}\cdot\mathbf{A} + \left(\frac{q}{c}\right)^2\mathbf{A}^2.$$

We then proceed to express this Hamiltonian in second-quantized form.

From there we go on to treat multipolar interaction, stimulated and spontaneous photon emission, photon absorption, the Einstein coefficients, and Planck's black body radiation law in the context of this chapter.

Finally, there is a brief discussion of multi-photon, multi-molecule interactions.

1.3.9. *Chapter 10: Conclusions, Acknowledgments, and Notes*

A feature of this brief chapter is a list of reference notes together with the pages on which they appear in the previous chapters as footnotes.

Chapter 2

The Harmonic Oscillator: A Treatment by Fock Operators*

2.1. Introduction

In Chapter 2, we introduce some of the elementary concepts and techniques that will be useful to us in later applications. To keep the development relatively simple, we confine ourselves to a familiar problem: the harmonic oscillator. Not only is this choice of starting point historical, since Fock's original work on the use of "raising" and "lowering" operators was to solve the oscillator problem, but it is a quantum-mechanical system with which both physicists and chemists have become familiar early in their studies. Indeed, the Fock operator technique is treated in many courses at the undergraduate as well as the graduate level.

Here, as throughout this volume, we try to demystify the derivations by showing most of them in full detail. Occasionally, when a derivation is so similar to another that the reader should have no trouble working it out from a cited example, we give just the results, as a slight bow to space constraints and the reader's patience. If there is too much detail for the reader's taste, or if the reader sees short cuts or other simplifications, the derivations appearing here might be ignored. But in doing so, the reader may miss out on practicing mathematical drills implicit in the details that help to provide the facility that will make the later chapters

*Avery, J., *Creation and Annihilation Operators*, McGraw-Hill (1976).

easier to follow. Where derivations are only sketched out, the reader is invited to work out the details as exercises so as to obtain greater familiarity with commutation relations, in particular with those of the Fock operators. We have chosen what we hope are some interesting questions that can be answered with these methods – the calculation of the eigenergies, the generation of the explicit energy eigenfunctions and their normalization, details of the *ket*, *bra* shorthand (with which some chemists are often even now not totally facile), and its application to expansions in an appropriate basis set and the calculation of operator matrix elements, and finally the dynamics of a minimal wave packet in both the Schrödinger and Heisenberg pictures.

Readers who work through this chapter thoroughly will be well prepared for Chapter 3, which deals with time-dependent perturbation theory and presents an introduction to diagrammatic techniques in an application to wave-packet dynamics, and for Chapter 4, where we apply the theory to spinless particles.

2.2. The Schrödinger Equation

The Schrödinger equation for a harmonic oscillator with reduced mass μ, force constant k, and energy E is

$$\hat{H}\psi(x) = E\psi(x), \tag{2.1a}$$

where the Hamiltonian operator is

$$\hat{H} = \frac{\hat{p}^2}{2\mu} + \frac{1}{2}kx^2, \tag{2.1b}$$

x is the displacement of the oscillator from equilibrium, and the momentum operator $\hat{p}$ is given by

$$\hat{p} = \frac{\hbar}{i}\frac{d}{dx}, \tag{2.2}$$

where $\hbar$ is the (reduced) Planck's constant $\hbar = h/2\pi$.

A natural dimensionless variable is

$$\xi = \left(\frac{\mu\omega}{\hbar}\right)^{\frac{1}{2}} x, \tag{2.3}$$

where ω is the classical frequency[1] given by

$$\omega = \left(\frac{k}{\mu}\right)^{\frac{1}{2}}. \tag{2.4}$$

In terms of ξ we have for $\hat{p}$ and x

$$\hat{p} = -i(\mu\hbar\omega)^{\frac{1}{2}}\frac{d}{d\xi}, \quad x = \left(\frac{\hbar}{\mu\omega}\right)^{\frac{1}{2}}\xi \tag{2.5}$$

and Eq. (2.1b) becomes

$$\frac{1}{2}\hbar\omega\left(-\frac{d^2}{d\xi^2} + \xi^2\right)\varphi(\xi) = E\varphi(\xi), \tag{2.6}$$

where we write $\varphi(\xi)$ for the wave function $\psi\left[\left(\frac{\hbar}{\mu\omega}\right)^{\frac{1}{2}}\xi\right]$.

2.3. Fock Operators

We now introduce the Fock operator

$$\hat{a} = \frac{1}{\sqrt{2}}\left(\frac{d}{d\xi} + \xi\right) \tag{2.7a}$$

and its Hermitian conjugate

$$\hat{a}^\dagger = \frac{1}{\sqrt{2}}\left(-\frac{d}{d\xi} + \xi\right). \tag{2.7b}$$

[1]The property ω is often called angular velocity, while frequency is $\omega/2\pi$. For simplicity, we call ω the "frequency".

It is easily verified that

$$\hat{p} = i\left(\frac{\mu\hbar\omega}{2}\right)^{\frac{1}{2}}(\hat{a}^\dagger - \hat{a}) \tag{2.8}$$

$$\hat{x} = \left(\frac{\hbar}{2\mu\omega}\right)^{\frac{1}{2}}(\hat{a}^\dagger + \hat{a}). \tag{2.9}$$

The commutation relation for the Fock operators is simply unity, or the identity operator, as seen by

$$\begin{aligned}
[\hat{a}, \hat{a}^\dagger]_- &= \hat{a}\hat{a}^\dagger - \hat{a}^\dagger\hat{a} \\
&= \frac{1}{2}\left[\left(\frac{d}{d\xi} + \xi\right)\left(-\frac{d}{d\xi} + \xi\right) - \left(-\frac{d}{d\xi} + \xi\right)\left(\frac{d}{d\xi} + \xi\right)\right] \\
&= \frac{1}{2}\left[\left(-\frac{d^2}{d\xi^2} - \xi\frac{d}{d\xi} + \xi\frac{d}{d\xi} + \xi^2 + 1\right)\right. \\
&\qquad \left. - \left(-\frac{d^2}{d\xi^2} + \xi\frac{d}{d\xi} - \xi\frac{d}{d\xi} + \xi^2 - 1\right)\right] \\
&= 1.
\end{aligned} \tag{2.10}$$

From Eqs. (2.8)–(2.10) it follows that

$$[\hat{p}, x]_- = \frac{i\hbar}{2}[(\hat{a}^\dagger - \hat{a}), (\hat{a}^\dagger + \hat{a})]_- = \frac{\hbar}{i}[\hat{a}, \hat{a}^\dagger]_- = \frac{\hbar}{i}, \tag{2.11}$$

the well-known relation that is the basis of the uncertainty relation for position and momentum.

To write the Schrödinger equation in terms of the Fock operators, we use Eqs. (2.8)–(2.10) to obtain

$$\hat{p}^2 = -\frac{1}{2}\mu\hbar\omega(\hat{a}^\dagger - \hat{a})^2 = -\frac{1}{2}\mu\hbar\omega[(\hat{a}^\dagger)^2 - \hat{a}^\dagger\hat{a} - \hat{a}\hat{a}^\dagger + \hat{a}^2] \tag{2.12}$$

$$\hat{x}^2 = \frac{\hbar}{2\mu\omega}(\hat{a}^\dagger + \hat{a})^2 = \frac{\hbar}{2\mu\omega}[(\hat{a}^\dagger)^2 + \hat{a}^\dagger\hat{a} + \hat{a}\hat{a}^\dagger + \hat{a}^2]. \tag{2.13}$$

Then the Hamiltonian takes a simple form in terms of the Fock operators:

$$\begin{aligned}
\hat{H} &= \frac{\hat{p}^2}{2\mu} + \frac{1}{2}kx^2 \\
&= \frac{1}{4}\hbar\omega\left(-[(\hat{a}^\dagger)^2 - \hat{a}^\dagger\hat{a} - \hat{a}\hat{a}^\dagger + \hat{a}^2] + [(\hat{a}^\dagger)^2 + \hat{a}^\dagger\hat{a} + \hat{a}\hat{a}^\dagger + \hat{a}^2]\right) \\
&= \frac{1}{2}\hbar\omega(\hat{a}^\dagger\hat{a} + \hat{a}\hat{a}^\dagger) = \frac{1}{2}\hbar\omega[\hat{a}^\dagger\hat{a} + \hat{a}^\dagger\hat{a} + (\hat{a}\hat{a}^\dagger - \hat{a}^\dagger\hat{a})] \\
&= \hbar\omega\left(\hat{a}^\dagger\hat{a} + \frac{1}{2}\right),
\end{aligned} \tag{2.14}$$

where we have used the commutation relation of Eq. (2.10) in the last step. Thus, the Schrödinger equation, Eq. (2.1b), becomes

$$\hbar\omega(\hat{a}^\dagger\hat{a} + \frac{1}{2}\varphi(\xi) = E\varphi(\xi). \tag{2.15}$$

If we divide both sides of Eq. (2.15) by $\hbar\omega$ and write the dimensionless energy ε for $E/\hbar\omega$, we have

$$\hat{h}\varphi(\xi) = \varepsilon\varphi(\xi), \tag{2.16a}$$

where $\hat{h}$ is the reduced (dimensionless) Hamiltonian operator in Fock notation, namely

$$\hat{h} = \frac{1}{2}\left(-\frac{d^2}{d\xi^2} + \xi^2\right) = \hat{a}^\dagger\hat{a} + \frac{1}{2}. \tag{2.16b}$$

2.4. Solution of the Harmonic Oscillator Problem with Fock Algebra

Not only is the Fock representation for $\hat{h}$ very simple, but it affords an elegant algebraic solution to Eq. (2.16). We begin by working out the commutator

$$[\hat{h}, \hat{a}^\dagger]_- = [\hat{a}^\dagger\hat{a}, \hat{a}^\dagger]_- = \hat{a}^\dagger\hat{a}\hat{a}^\dagger - (\hat{a}^\dagger)^2\hat{a} = \hat{a}^\dagger[\hat{a}, \hat{a}^\dagger]_- = \hat{a}^\dagger. \tag{2.17}$$

Eq. (2.17) allows us to evaluate

$$\hat{h}\hat{a}^\dagger\varphi = \left([\hat{h}, \hat{a}^\dagger]_- + \hat{a}^\dagger\hat{h}\right)\varphi = (\hat{a}^\dagger + \hat{a}^\dagger\hat{h})\varphi. \tag{2.18}$$

If φ is an eigenfunction of $\hat{h}$; *i.e.*, if Eq. (2.16a) holds, then Eq. (2.18) shows that

$$\hat{h}\hat{a}^\dagger\varphi = (\varepsilon + 1)\hat{a}^\dagger\varphi. \tag{2.19}$$

Thus, if φ is an eigenfunction of $\hat{h}$ with eigenvalue ε, then $\hat{a}^\dagger\varphi$ is an eigenfunction of $\hat{h}$ with eigenvalue $\varepsilon + 1$.

$$\begin{aligned}[\hat{h}, (\hat{a}^\dagger)^2]_- &= \hat{h}(\hat{a}^\dagger)^2 - (\hat{a}^\dagger)^2\hat{h} \\ &= \left(\hat{h}\hat{a}^\dagger - \hat{a}^\dagger\hat{h}\hat{a}^\dagger + \hat{a}^\dagger\ (\hat{h}\hat{a}^\dagger - \hat{a}^\dagger\hat{h}\right) \\ &= (\hat{a}^\dagger)\hat{a}^\dagger + \hat{a}^\dagger(\hat{a}^\dagger) = 2(\hat{a}^\dagger)^2. \end{aligned} \tag{2.20a}$$

From Eq. (2.17), we see that

$$\begin{aligned}[\hat{h}, (\hat{a}^\dagger)^3]_- &= \hat{h}(\hat{a}^\dagger)^3 - (\hat{a}^\dagger)^3\hat{h} \\ &= \left[\hat{h}(\hat{a}^\dagger)^2 - (\hat{a}^\dagger)^2\hat{h}\right]\hat{a}^\dagger + (\hat{a}^\dagger)^2(\hat{h}\hat{a}^\dagger - \hat{a}^\dagger\hat{h}) \\ &= 2(\hat{a}^\dagger)^2\hat{a}^\dagger + (\hat{a}^\dagger)^2\hat{a}^\dagger = 3(\hat{a}^\dagger)^3 \end{aligned} \tag{2.20b}$$

and from Eq. (2.20a) we obtain

$$[\hat{h}, (\hat{a}^\dagger)^n]_- = n(\hat{a}^\dagger)^n. \tag{2.21}$$

It appears that if we continue as in Eq. (2.20), we will finally obtain

$$\begin{aligned}[\hat{h}, (\hat{a}^\dagger)^{n+1}]_- &= [\hat{h}, (\hat{a}^\dagger)^n]_-\hat{a}^\dagger + (\hat{a}^\dagger)^n\hat{h}\hat{a}^\dagger - (\hat{a}^\dagger)^{n+1}\hat{h} \\ &= [\hat{h}, (\hat{a}^\dagger)^n]_-\hat{a}^\dagger + (\hat{a}^\dagger)^n[\hat{h}, \hat{a}^\dagger]_- \\ &= n(\hat{a}^\dagger)^{n+1} + (\hat{a}^\dagger)^{n+1} \\ &= (n+1)(\hat{a}^\dagger)^{n+1}. \end{aligned} \tag{2.22}$$

That Eq. (2.21) is correct can be proved by induction: Eqs. (2.17) and (2.20a) show that Eq. (2.21) is true for $n = 1$ and $n = 2$. Then if Eq. (2.21) is true for some n, we have

$$\begin{aligned}\hat{h}(\hat{a}^\dagger)^n\varphi &= \left([\hat{h},(\hat{a}^\dagger)^n]_- + (\hat{a}^\dagger)^n\hat{h}\right)\varphi \\ &= [n(\hat{a}^\dagger)^n + (\hat{a}^\dagger)^n\varepsilon]\varphi = (\varepsilon+n)(\hat{a}^\dagger)^n\varphi, \qquad (2.23)\end{aligned}$$

which completes the induction proof.

From Eq. (2.23) we now have the result that if φ is an eigenfunction of $\hat{h}$ with eigenvalue ε, then $(\hat{a}^\dagger)^n\varphi$ is an eigenfunction of $\hat{h}$ with eigenvalue $\varepsilon + n$.

Similarly, from Eq. (2.21) and the rule for taking the Hermitian conjugate of a commutator, namely,

$$[\hat{A},\hat{B}]_-^\dagger = [\hat{B}^\dagger,\hat{A}^\dagger]_-, \qquad (2.24)$$

and the fact that $\hat{h}$ is Hermitian; *i.e.*, $\hat{h}^\dagger = \hat{h}$, we have

$$[\hat{h},\hat{a}^n]_- = [(\hat{a}^\dagger)^n,\hat{h}]_-^\dagger = [\hat{h},(\hat{a}^\dagger)^n]_-^\dagger = -n\hat{a}^n. \qquad (2.25)$$

Eq. (2.25) allows us to obtain

$$\begin{aligned}\hat{h}\hat{a}^n\varphi &= \left([\hat{h},\hat{a}^n]_\mp \hat{a}^n\hat{h}\right)\varphi \\ &= (-n\hat{a}^n + \hat{a}^n\varepsilon)\varphi = (\varepsilon - n)\hat{a}^n\varphi, \qquad (2.26)\end{aligned}$$

and $(\hat{a})^n\varphi$ is seen to be an eigenfunction of $\hat{h}$ with eigenvalue $\varepsilon - n$. For these reasons $\hat{a}^\dagger$ and $\hat{a}$ are often called the raising operator and the lowering operator, respectively.

Unless there is a lower limit to ε, we can obtain eigenfunctions corresponding to ever lower eigenvalues by operating on φ repeatedly with $\hat{a}$. It follows that we can eventually obtain an eigenfunction φ_0 with corresponding (positive) eigenvalue ε_0 such that one additional operation with $\hat{a}$ yields

$$\hat{h}\hat{a}\varphi_0 = (\varepsilon_0 - 1)\hat{a}\varphi_0, \qquad (2.27)$$

for which

$$(\varepsilon_0 - 1) < 0. \qquad (2.28)$$

But Eq. (2.28) cannot be true, since both the kinetic and potential energies for a harmonic oscillator are positive; this requires that all the energy levels must be positive. It follows that in order for ε to cut off at the least positive value, we must have

$$\hat{a}\varphi_0 = 0. \tag{2.29}$$

The property of $\hat{a}$ given in Eq. (2.29) leads to the designation of $\hat{a}$ as the *annihilation* operator. The operator $\hat{a}^\dagger$ is often called the *creation* operator. We may find the function $\varphi_0(\xi)$ by writing out Eq. (2.29) explicitly. From Eq. (2.7a) we have

$$\varphi_0'(\xi) + \xi\varphi_0(\xi) = 0, \tag{2.30}$$

the solution to which is

$$\varphi_0(\xi) = e^{-\frac{1}{2}\xi^2}, \tag{2.31}$$

where we have arbitrarily taken the normalizing constant to be unity. Insertion of $\varphi_0(\xi)$ from Eq. (2.31) into Eq. (2.16) gives

$$\hat{h}\varphi_0 = \frac{1}{2}\varphi_0, \tag{2.32a}$$

from which we conclude that the minimum energy is

$$\varepsilon_0 = \frac{1}{2}. \tag{2.32b}$$

Thus, in the general case, writing $\varphi_n(\xi)$ for $(\hat{a}^\dagger)^n\varphi_0$, we have from Eqs. (2.32b) and (2.23)

$$\hat{h}\varphi_n(\xi) = \left(n + \frac{1}{2}\right)\varphi_n(\xi). \tag{2.32c}$$

From Eqs. (2.32c) and (2.16b) we see that

$$\hat{a}^\dagger\hat{a}\varphi_n = n\varphi_n. \tag{2.33}$$

For this reason $\hat{a}^\dagger\hat{a}$ is often called the *number operator.* The next eigenfunction in order of increasing energy is found from $\varphi_0(\xi)$ by operating upon it with $\hat{a}^\dagger$. Since the eigenfunctions will be normalized later, for simplicity we can ignore the factor $1/\sqrt{2}$ in Eq. (2.7b) to obtain

$$\varphi_1(\xi) = -\varphi_0'(\xi) + \xi\varphi_0(\xi) = 2\xi e^{-\frac{1}{2}\xi^2}. \tag{2.34a}$$

Similarly we have

$$\varphi_2(\xi) = \varphi_1'(\xi) + \xi\varphi_1(\xi) = (4\xi^2 - 2)e^{-\frac{1}{2}\xi^2} \tag{2.34b}$$

$$\varphi_3(\xi) = \varphi_2(\xi) + \xi\varphi_2(\xi) = (8\xi^3 - 12\xi)e^{-\frac{1}{2}\xi^2} \tag{2.34c}$$

$$\varphi_4(\xi) = -\varphi_3'(\xi) + \xi\varphi_3(\xi) = (16\xi^4 - 48\xi^2 + 12)e^{-\frac{1}{2}\xi^2} \tag{2.34d}$$

$$\varphi_5(\xi) = (32\xi^5 - 160\xi^3 + 120\xi)e^{-\frac{1}{2}\xi^2}, \tag{2.34e}$$

or more generally,

$$\varphi_{(n+1)}(\xi) = \left(-\frac{d}{d\xi} + \xi\right)\varphi_n(\xi) \tag{2.34f}$$

$$\varphi_n(\xi) = \left(-\frac{d}{d\xi} + \xi\right)^n \varphi_0(\xi). \tag{2.34g}$$

Eqs. (2.34a)–(2.34g) show that the form of the eigenfunctions is

$$\varphi_n(\xi) = H_n(\xi)e^{-\frac{1}{2}\xi^2}, \tag{2.34h}$$

where $H_n(\xi)$ is a polynomial in ξ of degree n. These polynomials, called the *Hermite polynomials*, have well-known properties, several of which we now derive.

2.5. Some Properties of the Hermite Polynomials

From Eq. (2.34f) we have

$$\begin{aligned} H_{(n+1)}(\xi)e^{-\frac{1}{2}\xi^2} &= \left(-\frac{d}{d\xi} + \xi\right) H_n(\xi)e^{-\frac{1}{2}\xi^2} \\ &= \left[-H_n'(\xi) + 2\xi H_n(\xi)\right] e^{-\frac{1}{2}\xi^2} \end{aligned} \tag{2.35}$$

Left-multiplying both sides of Eq. (2.35) by $e^{\frac{1}{2}\xi^2}$ gives the *recursion relation*

$$H_{(n+1)}(\xi) = 2\xi H_n(\xi) - H_n'(\xi). \tag{2.36}$$

Comparing Eq. (2.7b) with Eq. (2.34f), we obtain

$$\hat{a}^\dagger \varphi_{n-1} = \frac{1}{\sqrt{2}}\varphi_n. \tag{2.37}$$

If we now apply $\hat{a}$ to the left-hand side of Eq. (2.37), we may use Eq. (2.7a) to write

$$\hat{a}\hat{a}^\dagger \varphi_{(n-1)}(\xi) = \frac{1}{2}\left(\frac{d}{d\xi} + \xi\right)\varphi_n(\xi). \tag{2.38}$$

But from the commutation relation of Eq. (2.10) we have

$$\hat{a}\hat{a}^\dagger = \hat{a}^\dagger\hat{a} + 1, \tag{2.39}$$

so that

$$\hat{a}\hat{a}^\dagger \varphi_{(n-1)} = (\hat{a}^\dagger\hat{a} + 1)\varphi_{(n-1)} = [(n-1)+1]\,\varphi_{(n-1)} = n\varphi_{(n-1)}. \tag{2.40}$$

Thus, Eqs. (2.38) and (2.40) give

$$n\varphi_{n-1}(\xi) = \frac{1}{2}[\varphi_n'(\xi) + \xi\varphi_n(\xi)]. \tag{2.41}$$

From the form of $\varphi_n(\xi)$ given by Eq. (2.34k), we have

$$\varphi_n'(\xi) = \left[H_n'(\xi) - \xi H_n(\xi)\right] e^{-\frac{1}{2}\xi^2}, \tag{2.42}$$

so that Eqs. (2.41) and (2.42) give the useful result

$$H_n'(\xi) = 2nH_{n-1}(\xi). \tag{2.43}$$

Insertion of Eq. (2.43) into Eq. (2.36) gives the additional recursion relation

$$H_{(n+1)}(\xi) = 2\xi H_n(\xi) - 2nH_{(n-1)}(\xi). \tag{2.44}$$

2.6. Evaluation of the Normalization Constant

We now evaluate the ortho-normalization integral for $\varphi_n(\xi)$, which is defined by

$$I_{nm} = \int_{-\infty}^{\infty} \varphi_n(\xi)\varphi_m(\xi)d\xi. \tag{2.45}$$

Using Eqs. (2.34f), we can write I_{nm} as

$$I_{nm} = \int_{-\infty}^{\infty} \left[\left(-\frac{d}{d\xi} + \xi\right) \varphi_{(n-1)}(\xi)\right] \varphi_m(\xi)d\xi. \tag{2.46a}$$

We can integrate once by parts to obtain

$$\begin{aligned} I_{nm} &= \int_{-\infty}^{\infty} [\varphi_{(n-1)}(\xi)] \left[\left(\frac{d}{d\xi} + \xi\right) \varphi_m(\xi)\right] d\xi \\ &= \sqrt{2} \int_{-\infty}^{\infty} \varphi_{(n-1)}(\xi)\hat{a}\varphi_m(\xi)d\xi. \end{aligned} \tag{2.46b}$$

Repetition of this process gives eventually

$$\begin{aligned} I_{nm} &= \left(\sqrt{2}\right)^n \int_{-\infty}^{\infty} \varphi_0(\xi)\hat{a}^n\varphi_m(\xi)d\xi \\ &= \left(\sqrt{2}\right)^{(n+m)} \int_{-\infty}^{\infty} \varphi_0(\xi)\hat{a}^n(\hat{a}^\dagger)^m\varphi_0(\xi)d\xi. \end{aligned} \tag{2.47}$$

One can prove by induction that

$$\hat{a}^n(\hat{a}^\dagger)^m = \sum_{k=0}^{n_<} \frac{n!m!}{k!(m-k)!(n-k)!}(\hat{a}^\dagger)^{(m-k)}\hat{a}^{(n-k)}. \tag{2.48}$$

If Eq. (2.48) holds for some n and m, it can be shown that it holds also for $\hat{a}^{(n+1)}(\hat{a}^\dagger)^m$ and $\hat{a}^n(\hat{a}^\dagger)^{(m+1)}$, to complete the induction proof. We defer the details until our treatment of Wick's theorem in Chapter 3.

Applying Eq. (2.48) to Eq. (2.47), we see from Eq. (2.29) that only the term in which $m - k$ and $n - k$ are both equal to zero will yield a non-zero contribution to I_{nm}. This is because integration by parts allows us to write

$$\int_{-\infty}^{\infty} \varphi_0 (\hat{a}^\dagger)^k \varphi_0 d\xi = \int_{-\infty}^{\infty} (\hat{a}^k \varphi_0)\varphi_0 d\xi = 0. \tag{2.49}$$

Only integrals of the form of Eq. (2.47) with $m = n$ are non-zero. For $m = n$, Eq. (2.48) gives

$$\hat{a}^n (\hat{a}^\dagger)^n = n! + \sum_{k=0}^{n-1} \frac{(n!)^2}{k![(n-k)!]^2} (\hat{a}^\dagger)^{(n-k)} \hat{a}^{(n-k)}. \tag{2.50}$$

Since terms in the summation in Eq. (2.50) yield zero in the integral of Eq. (2.47), the result for the integral is

$$I_{nm} = 2^n n! \int_{-\infty}^{\infty} e^{-\xi^2} d\xi \delta_{nm} = 2^n n! \sqrt{\pi} \delta_{nm}, \tag{2.51}$$

where we have used Eq. (2.31). The symbol δ_{nm} is the Kronecker delta, defined by:

$$\delta_{nm} = \begin{cases} 1; & n = m \\ 0; & n \neq m \end{cases}. \tag{2.52}$$

If we use the physical displacement x for the integration variable, we require an additional factor of $(\hbar\mu/\omega)^{1/2}$ since Eq. (2.3) shows that

$$dx = \left(\frac{\hbar}{\mu\omega}\right)^{\frac{1}{2}} d\xi. \tag{2.53}$$

Thus, writing the harmonic oscillator energy eigenfunction in the form

$$\psi_n(x) = N_n H_n \left[\left(\frac{\mu\omega}{\hbar}\right)^{\frac{1}{2}} x \right] e^{-\frac{1}{2}\frac{\mu\omega}{\hbar}x^2}, \tag{2.54}$$

where N_n is the normalization constant, we have from Eqs. (2.51)–(2.54)

$$\int_{-\infty}^{\infty} \psi_n(x)\psi_m(x)dx = N_n N_m \left(\frac{\hbar}{\mu\omega}\right)^{\frac{1}{2}} \int_{-\infty}^{\infty} H_n(\xi)H_m(\xi)e^{-\xi^2}d\xi$$

$$= N_n^2 \left(\frac{\hbar}{\mu\omega}\right)^{\frac{1}{2}} 2^n n!\sqrt{\pi}\delta_{nm}. \qquad (2.55)$$

Since $[\psi(x)]^2 dx$ is the probability of finding the value of x between x and $x + dx$, we require that

$$\int_{-\infty}^{\infty} [\psi_n(x)]^2 dx = 1, \qquad (2.56)$$

and thus that

$$N_n = \left(\frac{\mu\omega}{\pi\hbar}\right)^{\frac{1}{4}} \frac{1}{\sqrt{(2^n n!)}}. \qquad (2.57)$$

2.7. *Bra* and *Ket* Notation

A standard simplified notation for integrals such as Eq. (2.55) is $\langle n|m\rangle$, called a *Dirac bracket.* Dirac used these symbols $\langle n|$ and $|m\rangle$ separately (a *bra* and a *ket*, respectively) to represent the individual wave functions; *e.g.*, $|m\rangle$ corresponds to ψ_m and $\langle n|$ to $\psi_n^\dagger$. Thus in the Dirac notation,

$$|m\rangle = (\hat{a}^\dagger)^m|0\rangle \quad \langle n| = \langle 0|\hat{a}^n. \qquad (2.58)$$

Since it is understood that *bra*'s and *ket*'s always will be combined eventually to form brackets (*i.e.*, integrals) in any calculations, we may use our results from integration by parts to write

$$\int_{-\infty}^{\infty} \psi_n(x)\psi_m(x)dx = \langle n|m\rangle, \qquad (2.59)$$

because in the integrand $\varphi_0\hat{a}^n$ may be replaced by $(\hat{a}^\dagger)^n\varphi_0$.

A word about normalization is in order. From Eqs. (2.7b), (2.34g), and (2.54) we have

$$\psi_n = N_n H_n e^{-\frac{1}{2}\xi^2} = N_n \left(-\frac{d}{d\xi} + \xi\right)^n e^{-\frac{1}{2}\xi^2} = N_n 2^{\frac{n}{2}} (\hat{a}^\dagger)^n e^{-\frac{1}{2}\xi^2}$$
$$= \left(\frac{\mu\omega}{\pi\hbar}\right)^{\frac{1}{4}} \frac{1}{\sqrt{n!}} (\hat{a}^\dagger)^n e^{-\frac{1}{2}\xi^2}. \tag{2.60}$$

But since

$$\psi_0 = N_0 e^{-\frac{1}{2}\xi^2} = \left(\frac{\mu\omega}{\pi\hbar}\right)^{\frac{1}{4}} e^{-\frac{1}{2}\xi^2}, \tag{2.61}$$

Eq. (2.60) becomes

$$\psi_n = \frac{1}{\sqrt{n!}} (\hat{a}^\dagger)^n \psi_0. \tag{2.62}$$

In the *ket*, *bra* notation

$$|n\rangle = \frac{1}{\sqrt{n!}} (\hat{a}^\dagger)^n |0\rangle, \quad \langle n| = \frac{1}{\sqrt{n!}} \langle 0|\hat{a}^n. \tag{2.63}$$

Thus we have

$$\langle n|m\rangle = \frac{1}{\sqrt{n!m!}} \langle 0|\hat{a}^n (\hat{a}^\dagger)^m |0\rangle. \tag{2.64}$$

From use of Eqs. (2.48)–(2.50) and the discussion used to obtain Eq. (2.51), then Eq. (2.59) becomes

$$\langle n|m\rangle = \frac{1}{\sqrt{n!m!}} n!\delta_{nm} \langle 0|0\rangle = \delta_{nm}, \tag{2.65}$$

as required.

2.8. Matrix Elements of the Operators $\hat{x}$ and $\hat{p}$

To calculate x_{nm} we make use of Eq. (2.9):

$$x_{nm} = \langle n|\hat{x}|m\rangle = \left(\frac{\hbar}{2\mu\omega}\right)^{\frac{1}{2}} \langle n|\hat{a}^\dagger + \hat{a}|m\rangle. \tag{2.66}$$

From Eqs. (2.63) we have

$$\hat{a}^\dagger|m\rangle = \frac{1}{\sqrt{m!}}(\hat{a}^\dagger)^{(m+1)}|0\rangle = \sqrt{\frac{(m+1)!}{m!}}|m+1\rangle = \sqrt{m+1}|m+1\rangle$$

$$\langle n|\hat{a} = \frac{1}{\sqrt{n!}}\langle 0|\hat{a}^{(n+1)} = \sqrt{\frac{(n+1)!}{n!}}\langle n+1| = \sqrt{n+1}\langle +1|. \tag{2.67}$$

Inserting Eqs. (2.67) into Eq. (2.66), we obtain

$$\begin{aligned} x_{nm} &= \left(\frac{\hbar}{2\mu\omega}\right)^{\frac{1}{2}} \langle n|(\hat{a}^\dagger + \hat{a})|m\rangle \\ &= \left(\frac{\hbar}{2\mu\omega}\right)^{\frac{1}{2}} \left[\sqrt{m+1}\langle n|m+1\rangle + \sqrt{n+1}\langle n+1|m\rangle\right] \\ &= \left(\frac{\hbar}{2\mu\omega}\right)^{\frac{1}{2}} \left[\sqrt{n}\delta_{m,n-1} + \sqrt{n+1}\delta_{m,n+1}\right]. \end{aligned} \tag{2.68}$$

In a similar way, we have from Eqs. (2.8) and (2.67)

$$\begin{aligned} p_{nm} = \langle n|\hat{p}|m\rangle &= i\left(\frac{\mu\hbar\omega}{2}\right)^{\frac{1}{2}} \langle n|(\hat{a}^\dagger - \hat{a})|m\rangle \\ &= i\left(\frac{\mu\hbar\omega}{2}\right)^{\frac{1}{2}} \left[\sqrt{n}\delta_{m,n-1} - \sqrt{n+1}\delta_{m,n+1}\right]. \end{aligned} \tag{2.69}$$

Notice that the diagonal elements x_{nn} and p_{nn} are zero, indicating that the average values of x and p are zero, in accordance with the reflection symmetry of the potential-energy function $^1/_2kx^2$. The matrices x and p have the *tri-diagonal* form

$$x = \left(\frac{\hbar}{2\mu\omega}\right)^{\frac{1}{2}} \begin{pmatrix} 0 & 1 & 0 & 0 & \dots \\ 1 & 0 & \sqrt{2} & 0 & \dots \\ 0 & \sqrt{2} & 0 & \sqrt{3} & \dots \\ 0 & 0 & \sqrt{3} & 0 & \dots \\ \vdots & \vdots & \vdots & \vdots & \end{pmatrix},$$

$$y = i\left(\frac{\mu\hbar\omega}{2}\right)^{\frac{1}{2}} \begin{pmatrix} 0 & -1 & 0 & 0 & \dots \\ 1 & 0 & -\sqrt{2} & 0 & \dots \\ 0 & \sqrt{2} & 0 & -\sqrt{3} & \dots \\ 0 & 0 & \sqrt{3} & 0 & \dots \\ \vdots & \vdots & \vdots & \vdots & \end{pmatrix}. \quad (2.70)$$

For matrices of powers of x and p, we take the powers of the matrices for x and y. Alternatively, we may find the matrix elements of powers of the operators of Eqs. (2.8) and (2.9). So with the aid of Eq. (2.10) we obtain

$$(\hat{a}^\dagger \pm \hat{a})^2 = (\hat{a}^\dagger)^2 \pm \hat{a}^\dagger\hat{a} \pm \hat{a}\hat{a}^\dagger + \hat{a}^2 = (\hat{a}^\dagger)^2 \pm (2\hat{a}^\dagger\hat{a} + 1) + \hat{a}^2. \quad (2.71)$$

Thus,

$$\begin{aligned}(x^2)_{nm} &= \frac{\hbar}{2\mu\omega}\langle n|[(\hat{a}^\dagger)^2 + 2\hat{a}^\dagger\hat{a} + 1 + \hat{a}^2]|m\rangle \\ &= \frac{\hbar}{2\mu\omega}\left[\sqrt{\frac{(m+2)!}{m!}}\delta_{m,n-2} + (2n+1)\delta_{nm} + \sqrt{\frac{(n+2)}{n!}}\delta_{m,n+2}\right]\end{aligned}$$

$$= \frac{\hbar}{2\mu\omega}\left[\sqrt{n(n-1)}\delta_{m,n-2} + (2n+1)\delta_{nm} + \sqrt{(n+2)(n+1)}\delta_{m,n+2}\right] \tag{2.72}$$

$$(p^2)_{nm} = \frac{\mu\hbar\omega}{2}\left[\sqrt{n(n-1)}\delta_{m,n-2} - (2n+1)\delta_{nm} + \sqrt{(n+2)(n+1)}\delta_{m,n+2}\right]. \tag{2.73}$$

Both x^2 and p^2 are seen to have non-zero diagonal matrix elements, namely

$$x^2_{nn} = \frac{\hbar}{\mu\omega}\left(n+\frac{1}{2}\right) \tag{2.74}$$

$$p^2_{nn} = -\mu\hbar\omega\left(n+\frac{1}{2}\right). \tag{2.75}$$

Since

$$(\hat{a}^\dagger + \hat{a})(\hat{a}^\dagger - \hat{a}) = (\hat{a}^\dagger)^2 - \hat{a}^2 + [\hat{a}, \hat{a}^\dagger]_- = (\hat{a}^\dagger)^2 - \hat{a}^2 + 1, \tag{2.76}$$

the elements of xp are

$$(xp)_{n,m} = i\frac{\hbar}{2}\left[\sqrt{n(n-1)}\delta_{m,n-2} + \delta_{nm} - \sqrt{(n+2)(n+1)}\delta_{m,n+2}\right]. \tag{2.77a}$$

Similarly,

$$(px)_{nm} = i\frac{\hbar}{2}\left[\sqrt{n(n-1)}\delta_{m,n-2} - \delta_{nm} - \sqrt{(n+2)(n+1)}\delta_{m,n+2}\right]. \tag{2.77b}$$

2.9. The Uncertainty Product

The uncertainty in the measurement of the position x is defined as the root mean square deviation; *i.e.*,

$$|\Delta x| = \left[\langle(\hat{x} - \langle\hat{x}\rangle)^2\rangle\right]^{\frac{1}{2}} = \left[\langle\hat{x}^2\rangle - 2\langle\langle\hat{x}\rangle\hat{x}\rangle + \langle\hat{x}\rangle^2\right]^{\frac{1}{2}} = \left[\langle\hat{x}^2\rangle - \langle\hat{x}\rangle^2\right]^{\frac{1}{2}}.$$

For an oscillator in the state n,

$$|\Delta x| = \left[\langle n|x^2|n\rangle - \langle n|x|n\rangle^2\right]^{\frac{1}{2}} = \left[\langle n|x^2|n\rangle\right]^{\frac{1}{2}} = \left[\left(n+\frac{1}{2}\right)\frac{\hbar}{\mu\omega}\right]^{\frac{1}{2}}. \tag{2.78}$$

Similarly,

$$|\Delta p| = \left[\langle n|\hat{p}^2|n\rangle - \langle n|\hat{p}|n\rangle^2\right]^{\frac{1}{2}} = \left[\langle n|\hat{p}^2|n\rangle\right]^{\frac{1}{2}} = \left[\left(n+\frac{1}{2}\right)\mu\hbar\omega\right]^{\frac{1}{2}}. \tag{2.79}$$

We have made use of Eqs. (2.72)–(2.75) and the fact that $\langle x\rangle = 0$ and $\langle p\rangle = 0$ to obtain Eqs. (2.78) and (2.79). The uncertainty product for an oscillator in the state n is thus

$$|\Delta x||\Delta p| = \left(n+\frac{1}{2}\right)\hbar. \tag{2.80}$$

We see that the minimum uncertainty product $\hbar/2$ is achieved for an oscillator in the ground state; *i.e.*, for $n = 0$.

$$|\Delta x||\Delta p| = \frac{\hbar}{2}. \tag{2.81}$$

2.10. Superposition of States; Completeness

A function $F(x)$ which obeys the same boundary conditions as the harmonic oscillator eigenfunctions $\psi_n(x)$ can be written as a linear combination of the set of complete functions $\psi_n(x)$. In *ket* notation,

$$|F\rangle = \sum_{n=0}^{\infty} |n\rangle C_n = \sum_{n=0}^{\infty} |n\rangle\langle n|F\rangle. \tag{2.82}$$

Multiplication of Eq. (2.81) on the left with the *bra* $\langle n|$ gives

$$\langle n|F\rangle = \sum_{m=0}^{\infty} \langle n|m\rangle C_m = \sum_{m=0}^{\infty} C_m\delta_{nm} = C_n. \tag{2.83}$$

Since the function F is *any* function that obeys the boundary conditions and is therefore any function that can be obtained by superposition of members of the set $\{\psi_n\}$, we can see that the sum $\sum_n |n\rangle\langle n|F\rangle$ gives $|F\rangle$ for all functions of interest. We thus conclude that the sum of the diagonal *ket-bra*'s is the *identity operator*

$$\sum_{n=0}^{\infty} |n\rangle\langle n| = \hat{1}. \tag{2.84}$$

Eq. (2.84) is equivalent to the statement that the set of functions $\{\psi_n\}$ is *complete.*

2.11. Minimal Wave Packet; the Translation Operator

A *minimal wave packet* is a wave function for which the uncertainty product $|\Delta x||\Delta p|$ is a minimum, namely $\hbar/2$. We have seen that ψ_0 is such a wave function. In fact, we shall see that *any* Gaussian wave function is a minimal wave packet, centered on the position of the maximum of the Gaussian function. Let us designate a minimal wave packet centered at x_0 and having momentum p_0 by

$$\psi_W(\alpha, x_0, p_0; x) = Ne^{-\frac{\alpha}{2}(x-x_0)^2} e^{\frac{i}{\hbar}p_0 x}, \tag{2.85}$$

where α is a constant to be determined later.

We first evaluate the integrals

$$\langle W|W\rangle = N^2 \int_{-\infty}^{\infty} e^{-\alpha(x-x_0)^2} dx = N^2 \sqrt{\frac{\pi}{\alpha}} = 1 \tag{2.86}$$

$$\begin{aligned}\langle W|\hat{x}|W\rangle &= N^2 \int_{-\infty}^{\infty} (x-x_0)e^{-\alpha(x-x_0)^2} dx \\ &\quad + x_0 N^2 \int_{-\infty}^{\infty} e^{-\alpha(x-x_0)^2} dx = x_0,\end{aligned} \tag{2.87}$$

since the first integral on the r.h.s. of Eq. (2.87) is zero. Similarly,

$$\begin{aligned}\langle W|\hat{x}^2|W\rangle &= N^2\int_{-\infty}^{\infty}(x-x_0)^2 e^{-\alpha(x-x_0)^2}dx \\ &\quad + 2x_0N^2\int_{-\infty}^{\infty}(x-x_0)e^{-\alpha(x-x_0)^2}dx \\ &\quad + x_0^2N^2\int_{-\infty}^{\infty}e^{-\alpha(x-x_0)^2}dx \\ &= \frac{N^2}{2\alpha}\sqrt{\frac{\pi}{\alpha}} + 0 + x_0^2N^2\sqrt{\frac{\pi}{\alpha}} = \frac{1}{2\alpha} + x_0^2. \end{aligned} \tag{2.88}$$

From Eqs. (2.87)–(2.90) we see that

$$\begin{aligned}\langle W|\hat{p}|W\rangle &= \frac{\hbar}{i}\left\langle W\left|\frac{d}{dx}\right|W\right\rangle \\ &= \frac{\hbar}{i}\alpha N^2\int_{-\infty}^{\infty}(x-x_0)e^{-\alpha(x-x_0)^2}dx \\ &\quad + p_0N^2\int_{-\infty}^{\infty}e^{-\alpha(x-x_0)^2}dx = p_0, \end{aligned} \tag{2.89}$$

and

$$\begin{aligned} W|\hat{p}^2|W\rangle &= -\hbar^2\left\langle W\left|\frac{d^2}{dx^2}\right|W\right\rangle = \hbar^2\alpha N^2\int_{-\infty}^{\infty}e^{-\alpha(x-x_0)^2}dx \\ &\quad + 2i\hbar p_0\alpha N^2\int_{-\infty}^{\infty}(x-x_0)e^{-\alpha(x-x_0)^2}dx \\ &\quad + p_0^2N^2\int_{-\infty}^{\infty}e^{-\alpha(x-x_0)^2}dx \\ &\quad - \alpha^2N^2\int_{-\infty}^{\infty}(x-x_0)^2e^{-\alpha(x-x_0)^2}dx \\ &= p_0^2 + \hbar^2\alpha - \frac{\hbar}{2}N^2\sqrt{\pi\alpha}. \end{aligned} \tag{2.90}$$

Our most important results from the above are:

$$N = \left(\frac{\alpha}{\pi}\right)^{\frac{1}{4}} \tag{2.91a}$$

$$\langle x\rangle = x_0 \tag{2.91b}$$

$$\langle x^2\rangle = x_0^2 + \frac{1}{2\alpha} \tag{2.91c}$$

$$\langle p\rangle = p_0 \tag{2.91d}$$

$$\langle p^2\rangle = p_0^2 + \frac{1}{2}\hbar^2\alpha. \tag{2.91e}$$

Then from Eqs. (2.91), we have for the uncertainties

$$|\Delta x| = \left[\langle\hat{x}^2\rangle - \langle\hat{x}\rangle^2\right] = \left(\frac{1}{2\alpha}\right)^{\frac{1}{2}} \tag{2.92a}$$

$$|\Delta p| = \left[\langle\hat{p}^2\rangle - \langle\hat{p}\rangle^2\right] = \left(\frac{1}{2}\hbar^2\alpha\right)^{\frac{1}{2}}. \tag{2.92b}$$

Thus Eqs. (2.92) show that

$$\alpha = \frac{1}{2(\Delta x)^2}, \quad |\Delta x||\Delta p| = \frac{\hbar}{2}, \tag{2.93}$$

so that the uncertainty product has the minimum value for all values of α. If we choose $\alpha = \mu\omega/\hbar$ and $x_0 = 0$, the wave function ψ_0 is identical to the ground-state harmonic-oscillator energy eigenfunction ψ_0. For all other values of α and x_0, the minimal wave-packet function ψ_0 is a superposition of the states $\psi_1(n = 0, 1, 2, \ldots)$, as discussed in Section 2.1 above. Using Eq. (2.83), we write

$$|\alpha, x_0, p_0\rangle = \sum_{n=0}^{\infty} |n\rangle\langle n|\alpha, x_0, p_0\rangle = \sum_{n=0}^{\infty} |n\rangle\langle n|W\rangle. \tag{2.94}$$

To evaluate the coefficients $\langle n|W\rangle$, we use Eq. (2.65):

$$\langle n|\alpha, x_0, p_0\rangle = \frac{1}{\sqrt{n!}}\langle 0|\hat{a}^n|\alpha, x_0, p_0\rangle. \tag{2.95}$$

Eq. (2.95) appears at first to be difficult to evaluate for arbitrary n; this is indeed the case for arbitrary values of α, but the task is simplified considerably if we "tune" α to the value $\mu\omega/\hbar$. If x_0 and p_0 are also both set equal to zero, the $\hat{a}^n$ operators will "annihilate" the *ket* $|\mu\omega/\hbar, 0, 0\rangle$, since it is the same as the ground-state *ket*; that is, $|0\rangle$. For non-zero p_0, the *ket* $|\mu\omega/\hbar, 0, 0\rangle$ is equivalent to

$$|\mu\omega/\hbar, 0, p_0\rangle = e^{\frac{i}{\hbar}p_0 x}|0\rangle. \tag{2.96}$$

Then

$$\hat{a}e^{\frac{i}{\hbar}p_0 x}|0\rangle = \frac{1}{\sqrt{2}}\left[\left(\frac{\hbar}{\mu\omega}\right)^{\frac{1}{2}}\frac{d}{dx} + \left(\frac{\mu\omega}{\hbar}\right)^{\frac{1}{2}}x\right]e^{\frac{i}{\hbar}p_0 x}|0\rangle$$

$$= e^{\frac{i}{\hbar}p_0 x}\left[\left(\frac{\hbar}{2\mu\omega}\right)^{\frac{1}{2}}\frac{i}{\hbar}p_0 + \hat{a}\right]|0\rangle \tag{2.97a}$$

$$\hat{a}^2 e^{\frac{i}{\hbar}p_0 x}|0\rangle = e^{\frac{i}{\hbar}p_0 x}\left[\left(\frac{\hbar}{2\mu\omega}\right)^{\frac{1}{2}}\frac{i}{\hbar}p_0 + \hat{a}\right]^2|0\rangle \tag{2.97b}$$

$$\hat{a}^3 e^{\frac{i}{\hbar}p_0 x}|0\rangle = e^{\frac{i}{\hbar}p_0 x}\left[\left(\frac{\hbar}{2\mu\omega}\right)^{\frac{1}{2}}\frac{i}{\hbar}p_0 + \hat{a}\right]^3|0\rangle \tag{2.97c}$$

$$\vdots$$

$$\hat{a}^n e^{\frac{i}{\hbar}p_0 x}|0\rangle = e^{\frac{i}{\hbar}p_0 x}\left[\left(\frac{\hbar}{2\mu\omega}\right)^{\frac{1}{2}}\frac{i}{\hbar}p_0 + \hat{a}\right]^n|0\rangle. \tag{2.97d}$$

From the binomial theorem,

$$e^{\frac{i}{\hbar}p_0^x}\left[\left(\frac{\hbar}{2\mu\omega}\right)^{\frac{1}{2}}\frac{i}{\hbar}p_0 + \hat{a}\right]^n|0\rangle$$

$$= e^{\frac{i}{\hbar}p_0^x}\sum_{n=0}^{n}\frac{n!}{l!(n-1)!}\left(\frac{ip_0}{\sqrt{2\mu\hbar\omega}}\right)^{(n-l)}\hat{a}^l|0\rangle. \tag{2.98}$$

But since Eq. (2.29) shows that $\hat{a}^l|0\rangle = 0$ for all $l \neq 0$, Eqs. (2.97d) and (2.98) result in

$$\hat{a}^n|\mu\omega/\hbar, 0, p_0\rangle = \left(\frac{ip_0}{\sqrt{2\mu\hbar\omega}}\right)^n|\mu\omega/\hbar, 0, p_0\rangle. \tag{2.99}$$

In this case, the matrix element we seek can be obtained from Eqs. (2.95) and (2.99); that is,

$$\langle n|\mu\omega/\hbar, 0, p_0\rangle = \frac{1}{\sqrt{n!}}\left(\frac{ip_0}{\sqrt{2\mu\hbar\omega}}\right)^n \langle 0|e^{\frac{1}{\hbar}p_0^x}|0\rangle. \tag{2.100}$$

The integral on the r.h.s. of Eq. (2.100) is evaluated by completing the square of the exponent in the integrand, as follows:

$$\begin{aligned}\langle 0|e^{\frac{i}{\hbar}p_0 x}|0\rangle &= \left(\frac{\mu\omega}{\pi\hbar}\right)^{\frac{1}{2}}\int_{-\infty}^{\infty} e^{-\frac{\mu\omega}{\hbar}x^2} e^{\frac{i}{\hbar}p_0 x}\,dx \\ &= \left(\frac{\mu\omega}{\pi\hbar}\right)^{\frac{1}{2}}\int_{-\infty}^{\infty} e^{-\frac{\mu\omega}{\hbar}\left(x-\frac{ip_0}{2\mu\omega}\right)^2}dx\, e^{-\frac{p_0^2}{4\mu\hbar\omega}} = e^{-\frac{p_0^2}{4\mu\hbar\omega}}.\end{aligned} \tag{2.101}$$

The matrix element in Eq. (2.100) is, therefore, given by

$$\langle n|\mu\omega/\hbar, 0, p_0\rangle = \frac{1}{\sqrt{n!}}\left(\frac{ip_0}{\sqrt{2\mu\hbar\omega}}\right)^n e^{-\frac{p_0^2}{4\mu\hbar\omega}}, \tag{2.102}$$

so that according to Eq. (2.94) the expansion of $|\mu\omega/\hbar, 0, p_0\rangle$ is

$$|\mu\omega/\hbar, 0, p_0\rangle = \sum_{n=0}^{\infty}|n\rangle\frac{1}{\sqrt{n!}}\left(\frac{ip_0}{\sqrt{2\mu\hbar\omega}}\right)^n e^{-\frac{{p_0}^2}{4\mu\hbar\omega}}, \tag{2.103}$$

for a minimal wave packet centered on the minimum of a harmonic potential, with momentum p_0 and with the width of a ground-state energy eigenfunction. It can be shown that the *ket* of Eq. (2.103) is normalized: where we have made use of the series expansion $e^x = \sum_{n=0}^{\infty} x^n/n!$.

For $x_0 \neq 0$ the position of the minimal wave packet is displaced from the potential minimum by the distance x_0. Although the wave function has the simple form of Eq. (2.85), it is convenient to express the displacement as an operation upon $|0\rangle$ with a translation operator as follows. First, we expand the function $f(x - x_0)$ about x_0 in a

Maclaurin series

$$\begin{aligned}
f(x-x_0) &= f(x) + \left[\frac{d}{dx_0} f(x-x_0)\right]_{x_0=0} \\
&\quad \times x_0 + \frac{1}{2!}\left[\frac{d^2}{dx_0^2} f(x-x_0)\right]_{x_0=0} x_0^2 + \cdots \\
&\quad + \frac{1}{n!}\left[\frac{d^n}{dx_0^n} f(x-x_0)\right]_{x_0=0} x_0^n + \cdots \\
&= f(x) - \frac{d}{dx} f(x) x_0 + \frac{1}{2!}\frac{d^2}{dx^2} f(x) x_0^2 + \cdots \\
&\quad + \frac{(-1)^n}{n!}\frac{d^n}{dx^n} f(x) x_0^n + \cdots \\
&= e^{-\frac{d}{dx}x_0} f(x) = e^{-\frac{i}{\hbar}\hat{p}x_0} f(x).
\end{aligned} \tag{2.104}$$

We see from Eq. (2.104) that the translation operator has the form $e^{-\frac{i}{\hbar}\hat{p}x_0}$; thus from Eqs. (2.85) and (2.104), we may write the displaced minimal wave packet *ket* as a combination of operators operating on the ground-state harmonic oscillator *ket* $|0\rangle$,

$$|\mu\omega/\hbar, x_0, p_0\rangle = e^{\frac{i}{\hbar}p_0 x} e^{-\frac{i}{\hbar}\hat{p}x_0}|0\rangle, \tag{2.105}$$

and from Eq. (2.95), coefficients for its expansion in the harmonic-oscillator energy eigenfunctions are given by

$$\langle n|\mu\omega/\hbar, x_0, p_0\rangle = \frac{1}{\sqrt{n!}}\langle 0|\hat{a}^n e^{\frac{i}{\hbar}p_0 x} e^{-\frac{i}{\hbar}\hat{p}x_0}|0\rangle. \tag{2.106}$$

From Eqs. (2.97) and (2.98) we have

$$\hat{a}^n e^{\frac{i}{\hbar}p_0 x} = e^{\frac{i}{\hbar}p_0 x}\sum_{l=0}^{n}\frac{n!}{l!(n-l)!}\left(\frac{ip_0}{\sqrt{2\mu\hbar\omega}}\right)^{(n-l)}\hat{a}^l, \tag{2.107}$$

so that Eq. (2.107) can be written

$$\langle n|\mu\omega/\hbar, x_0, p_0\rangle = \frac{1}{\sqrt{n!}}\sum_{l=0}^{n}\frac{n!}{l!(n-l)!}\left(\frac{i\hat{p}_0}{\sqrt{2\mu\hbar\omega}}\right)^{(n-l)}\langle 0|e^{\frac{i}{\hbar}p_0 x}\hat{a}^l e^{-\frac{i}{\hbar}\hat{p}x_0}|0\rangle. \tag{2.108}$$

To evaluate the bracket in Eq. (2.108), we need to find the *normal order* of the operator in the bracket; that is, we must express the operator as a sum of terms with powers of the annihilation operators $\hat{a}$ *on the right* of the creation operators $\hat{p} = \frac{\hbar}{i}\frac{d}{dx}$. First, we note that from the definition of the commutator

$$\hat{a}e^{-\frac{i}{\hbar}\hat{p}x_0} = e^{-\frac{i}{\hbar}\hat{p}x_0}\hat{a} + [\hat{a}, e^{-\frac{i}{\hbar}^p x_0}]_-. \tag{2.109}$$

From Eq. (2.7a) and the fact that $\hat{p}$ commutes with $e^{-\frac{i}{\hbar}\hat{p}x_0}$ the commutator can be written

$$\begin{aligned}[\hat{a}, e^{-\frac{i}{\hbar}\hat{p}x_0}]_- &= \frac{1}{\sqrt{2}}\left(\frac{\mu\omega}{\hbar}\right)^{\frac{1}{2}}[\hat{x}, e^{-\frac{i}{\hbar}\hat{p}x_0}]_- \\ &= \frac{1}{\sqrt{2}}\left(\frac{\mu\omega}{\hbar}\right)^{\frac{1}{2}}\sum_{l=0}^{\infty}\frac{1}{l!}\left(-\frac{i}{\hbar}x_0\right)^l[\hat{x}, \hat{p}^l]_-. \end{aligned} \tag{2.110}$$

The commutator $[\hat{x}, \hat{p}]_-$ can be found by induction:

$$\begin{aligned}[\hat{x}, \hat{p}]_- &= \frac{\hbar}{i} \\ [\hat{x}, \hat{p}^2]_- &= [\hat{x}, \hat{p}]_-\hat{p} + \hat{p}[\hat{x}, \hat{p}]_- = -2\frac{\hbar}{i}\hat{p} \\ [\hat{x}, \hat{p}^3]_- &= [\hat{x}, \hat{p}^2]_-\hat{p} + \hat{p}^2[\hat{x}, \hat{p}]_- = 3\frac{\hbar}{i}\hat{p}^2 \\ &\vdots \\ [\hat{x}, \hat{p}^l]_- &= -l\frac{\hbar}{i}\hat{p}^{(l-1)} \\ [\hat{x}, \hat{p}^{(l+1)}]_- &= [\hat{x}, \hat{p}^l]_-\hat{p} + \hat{p}^l[\hat{x}, \hat{p}]_- \\ &= -l\frac{\hbar}{i}\hat{p}^l - \frac{\hbar}{i}\hat{p}^l = -(l+1)\frac{\hbar}{i}\hat{p}^l. \end{aligned} \tag{2.111}$$

From Eqs. (2.110) and (2.111) we have

$$\begin{aligned}\hat{a}e^{-\frac{i}{\hbar}\hat{p}x_0} &= e^{-\frac{i}{\hbar}\hat{p}x_0}\hat{a} + \frac{1}{\sqrt{2}}\left(\frac{\mu\omega}{\hbar}\right)^{\frac{1}{2}}\sum_{l=1}^{\infty}\frac{1}{l-1!}\left(-\frac{i}{\hbar}x_0\right)^{(l-1)}\hat{p}^{(l-1)}x_0 \\ &= e^{-\frac{i}{\hbar}\hat{p}x_0}\left[\hat{a} + \frac{1}{\sqrt{2}}\left(\frac{\mu\omega}{\hbar}\right)\frac{1}{2}x_0\right], \end{aligned} \tag{2.112}$$

since $\sum_{l=1}^{\infty} \frac{1}{(l-1)}! \left(-\frac{i}{\hbar}x_0\right)^{(l-1)} \hat{p}^{(l-1)} = e^{-\frac{i}{\hbar}\hat{p}x_0}$, then

$$\begin{aligned}\hat{a}^l e^{-\frac{i}{\hbar}\hat{p}x_0} &= e^{-\frac{i}{\hbar}\hat{p}x_0}\left[\hat{a} + \frac{1}{\sqrt{2}}\left(\frac{\mu\omega}{\hbar}\right)^{\frac{1}{2}} x_0\right]^l \\ &= e^{-\frac{i}{\hbar}\hat{p}x_0}\sum_{k=0}^{l}\frac{l!}{k!(l-k)!}\left[\left(\frac{\mu\omega}{2\hbar}\right)^{\frac{1}{2}} x_0\right]^{(l-k)} \hat{a}^k,\end{aligned} \tag{2.113}$$

where we have used the binomial theorem to obtain Eq. (2.113). Insertion of Eq. (2.113) into Eq. (2.109) shows that only the $l = 0$ term fails to annihilate $|0\rangle$. Thus,

$$\begin{aligned}&\langle n|\mu\omega/\hbar, x_0, p_0\rangle \\ &= \frac{1}{\sqrt{n!}}\left[\left(\frac{\mu\omega}{2\hbar}\right)^{\frac{1}{2}}\left(x_0 + \frac{ip_0}{\mu\omega}\right)\right]\langle 0|e^{\frac{i}{\hbar}p_0 x} e^{-\frac{i}{\hbar}\hat{p}x_0}|0\rangle,\end{aligned} \tag{2.114}$$

where we have again used the binomial theorem to obtain the r.h.s. of Eq. (2.114). Since

$$e^{-\frac{i}{\hbar}\hat{p}x_0}\psi_0(x) = \psi_0(x - x_0), \tag{2.115}$$

the bracket on the r.h.s. in Eq. (2.114) becomes

$$\langle 0|e^{\frac{i}{\hbar}p_0 x} e^{-\frac{i}{\hbar}\hat{p}x_0}|0\rangle = e^{-\frac{\mu\omega}{4\hbar}\left(x_0^2 + \frac{p_0^2}{\mu^2\omega^2}\right)} e^{\frac{i}{2\hbar}p_0 x_0}, \tag{2.116}$$

where we have completed the square in the exponent so as to facilitate evaluation of the integral. Finally, Eqs. (2.114) and (2.116) give the result

$$\begin{aligned}\langle n|\mu\omega/\hbar, x_0, p_0\rangle &= \frac{1}{\sqrt{n!}}\left[\left(\frac{\mu\omega}{2\hbar}\right)^{\frac{1}{2}}\left(x_0 + \frac{ip_0}{\mu\omega}\right)\right]^n \\ &\quad \times e^{-\frac{\mu\omega}{4\hbar}\left(x_0^2 + \frac{p_0^2}{\mu^2\omega^2}\right)} e^{\frac{i}{2\hbar}p_0 x_0}.\end{aligned} \tag{2.117}$$

To check normalization, we write with the help of Eqs. (2.84) and (2.117)

$$\begin{aligned}
&\langle \mu\omega/\hbar, x_0, p_0 | \mu\omega/\hbar, x_0, p_0 \rangle \\
&= \sum_{n=0}^{\infty} \langle \mu\omega/\hbar, x_0, p_0 | n \rangle \langle n | \mu\omega/\hbar, x_0, p_0 \rangle \\
&= \sum_{n=0}^{\infty} \frac{1}{n!} \left[\frac{\mu\omega}{2\hbar} \left(x_0^2 + \frac{p_0^2}{\mu^2\omega^2} \right) \right]^n e^{-\frac{\mu\omega}{2\hbar}\left(x_0^2 + \frac{p_0^2}{\mu^2\omega^2}\right)} \\
&= e^{\frac{\mu\omega}{2\hbar}\left(x_0^2 + \frac{p_0^2}{\mu^2\omega^2}\right)} e^{-\frac{\mu\omega}{2\hbar}\left(x_0^2 + \frac{p_0^2}{\mu^2\omega^2}\right)} = 1. \qquad (2.118)
\end{aligned}$$

2.12. Time Evolution: The Schrödinger Picture

The minimal wave packet described in Section 2.1 is stationary only if p_0 and x_0 equal zero. In the more general case, the wave function evolves in time. To find the time dependence of ψ_W, we must solve the *time-dependent Schrödinger equation*,

$$\hat{H}\psi_W(x,t) = -\frac{i}{\hbar}\frac{\partial}{\partial t}\psi_W(x,t), \qquad (2.119)$$

subject to the boundary condition that at $t = 0$,

$$\psi_W(x,0) = \psi_W(\mu\omega/\hbar, x_0, p_0; x),$$

namely the Gaussian function of width $(\hbar/2\mu\omega)^{1/2}$, which is centered at x_0 and has initial momentum p_0. Formally, the solution to Eq. (2.119) with the required boundary condition is

$$\psi_W(x,t) = e^{-\frac{i}{\hbar}\hat{H}t}\psi_W(\mu\omega/\hbar, x_0, p_0; x). \qquad (2.120)$$

To obtain $\psi_W(x,t)$ in a more explicit form, we use the expansion of ψ_W in the eigenfunctions of $\hat{H}$ as given in Eqs. (2.94) and

(2.117),

$$\begin{aligned}
&\psi_W(\mu\omega/\hbar, x_0, p_0; x) \\
&= \sum_{n=0}^{\infty} \frac{1}{\sqrt{n!}} \left[\left(\frac{\mu\omega}{2\hbar} \right)^{\frac{1}{2}} \left(x_0 + \frac{ip_0}{\mu\omega} \right) \right]^n \\
&\quad \times e^{-\frac{\mu\omega}{4\hbar}\left(x_0 + \frac{p_0^2}{\mu^2\omega^2}\right)} e^{\frac{i}{2\hbar}p_0 x_0} \psi_n(x).
\end{aligned} \tag{2.121}$$

Using the harmonic oscillator eigenvalue equation

$$\hat{H}\psi_n(x) = \left(n + \frac{1}{2} \right) \hbar\omega\psi_n(x), \tag{2.122}$$

we can iterate with $\hat{H}$ to obtain for any power k,

$$\hat{H}^k\psi_n(x) = \left[\left(n + \frac{1}{2} \right) \hbar\omega \right]^k \psi_n(x); \tag{2.123}$$

thus, since $e^{-\frac{i}{\hbar}\hat{H}\,t}$ can be expanded as a power series in $-\frac{i}{\hbar}\hat{H}t$,

$$e^{-\frac{i}{\hbar}\hat{H}t}\psi_n(x) = e^{-i\left(n+\frac{1}{2}\right)\omega t}\psi_n(x). \tag{2.124}$$

The combination of Eqs. (2.120), (2.121), and (2.124) yields the result

$$\begin{aligned}
\psi_W(x,t) &= \sum_{n=0}^{\infty} \frac{1}{\sqrt{n!}} \left(\frac{\mu\omega}{2\hbar} \right)^{\frac{n}{2}} \left(x_0 + \frac{ip_0}{\mu\omega} \right)^n \\
&\quad \times e^{-\frac{\mu\omega}{4\hbar}} \left(x_0^2 + \frac{p_0^2}{\mu^2\omega^2} \right) e^{\frac{i}{2\hbar}p_0x_0}\, e^{-i\left(n+\frac{1}{2}\right)\omega t}\, \psi_n(x).
\end{aligned} \tag{2.125}$$

To see how the center of the wave packet moves with time, we calculate

$$\begin{aligned}
\langle W;t|\hat{x}|W;t\rangle &= \sum_{n=0}^{\infty}\sum_{m=0}^{\infty} \frac{1}{\sqrt{n!m!}} \left(\frac{\mu\omega}{2\hbar} \right)^{\frac{n+m}{2}} \left(x_0 - \frac{ip_0}{\mu\omega} \right)^n \left(x_0 + \frac{ip_0}{\mu\omega} \right)^m \\
&\quad \times e^{-\frac{\mu\omega}{2\hbar}} \left(x_0^2 + \frac{p_0^2}{\mu^2\omega^2} \right) e^{i(n-m)\omega t} \langle n|\hat{x}|m\rangle.
\end{aligned} \tag{2.126}$$

Substitution of the r.h.s. of Eq. (2.68) for $\langle n\,|\,x\,|m\rangle$ in Eq. (2.126) gives

$$\begin{aligned}
&\langle W;t|\hat{x}|W;t\rangle \\
&= \sum_{n=1}^{\infty} \frac{\sqrt{n}}{\sqrt{n!(n-1)!}} \left(\frac{\mu\omega}{2\hbar}\right)^{\frac{2n-1}{2}} \left(x_0 - \frac{ip_0}{\mu\omega}\right)^n \left(x_0 + \frac{ip_0}{\mu\omega}\right)^{(n-1)} \\
&\quad \times \left(\frac{\hbar}{2\mu\omega}\right)^{\frac{1}{2}} e^{-\frac{\mu\omega}{2\hbar}\left(x_0^2 + \frac{p_0^2}{\mu^2\omega^2}\right)} e^{i\omega t} \\
&\quad + \sum_{n=0}^{\infty} \frac{\sqrt{n+1}}{\sqrt{n(n+1)!}} \left(\frac{\mu\omega}{2\hbar}\right)^{\frac{2n+1}{2}} \left(x_0 - \frac{ip_0}{\mu\omega}\right)^n \left(x_0 + \frac{ip_0}{\mu\omega}\right)^{(n+1)} \\
&\quad \times \left(\frac{\hbar}{2\mu\omega}\right)^{\frac{1}{2}} e^{-\frac{\mu\omega}{2\hbar}\left(x_0^2 + \frac{p_0^0}{\mu^2\omega^2}\right)} e^{-i\omega t}. \qquad (2.127)
\end{aligned}$$

The first summation in Eq. (2.127) starts at $n = 1$ because Eq. (2.68) requires that $m = n - 1$ in the first term. We may combine the two summations by replacing n with $n + 1$ in the first sum and starting the sum with $n = 0$. This yields

$$\begin{aligned}
\langle W;t|\hat{x}|W;\rangle &= \left\{ \sum_{n=0}^{\infty} \frac{1}{n!} \left(\frac{\mu\omega}{2\hbar}\right)^n \left(x_0^2 + \frac{p_0^2}{\mu^2\omega^2}\right)^n e^{-\frac{\mu\omega}{2\hbar}\left(x_0^2 + \frac{p_0^2}{\mu^2\omega^2}\right)} \right\} \\
&\quad \times \frac{1}{2} \left[\left(x_0 - \frac{ip_0}{\mu\omega}\right) e^{i\omega t} + \left(x_0 + \frac{ip_0}{\mu\omega}\right) e^{-i\omega t} \right]. \qquad (2.128a)
\end{aligned}$$

Eq. (2.118a) shows that the factor in curly brackets is unity, so that we finally obtain

$$\begin{aligned}
\langle W;t|\hat{x}|W;t\rangle &= \frac{1}{2} x_0 \left(e^{i\omega t} + e^{-i\omega t}\right) + \frac{1}{2i} \frac{p_0}{\mu\omega} \left(e^{i\omega t} - e^{-i\omega t}\right) \\
&= x_0 \cos\omega t + \frac{p_0}{\mu\omega} \sin\omega t. \qquad (2.128b)
\end{aligned}$$

In a similar way, we find that

$$\langle W;t|\hat{p}|W;t\rangle = p_0 \cos\omega t - \mu\omega x_0 \sin\omega t, \qquad (2.129)$$

for which we used Eq. (2.69). Eqs. (2.128) and (2.129) are the *classical equations of motion* for a particle of mass μ in a potential $^1/_2\mu\omega^2 x^2$ with initial position x_0 and initial momentum p_0.

Thus, we see that classical mechanics properly describes the time evolution of the average values of position and momentum of a quantum oscillator that is initially displaced from its equilibrium position.

$$\begin{aligned}\langle W;t|\hat{x}^2|W;t\rangle &= \left(x_0^2 + \frac{\hbar}{2\mu\omega}\right)\cos^2\omega t \\ &\quad + \frac{2p_0x_0}{\mu\omega}\sin\omega t\cos\omega t + \left(\frac{p_0^2}{\mu^2\omega^2} + \frac{\hbar}{2\mu\omega}\right)\sin^2\omega t \\ &= \frac{\hbar}{2\mu\omega} + \left(x_0\cos\omega t + \frac{p_0}{\mu\omega}\sin\omega t\right)^2 \qquad (2.130a)\end{aligned}$$

$$\begin{aligned}\langle W;t|\hat{p}^2|W;t\rangle &= \left(p_0^2 + \frac{1}{2}\mu\hbar\omega\right)\cos^2\omega t + 2p_0x_0\mu\omega\sin\omega t\cos\omega t \\ &\quad + \mu^2\omega^2\left(x_0^2 + \frac{\hbar}{2\mu\omega}\right)\sin^2\omega t \\ &= \frac{1}{2}\mu\hbar\omega + (p_0\cos\omega t - \mu\omega x_0\sin\omega t)^2. \qquad (2.130b)\end{aligned}$$

To find the uncertainties Δx and Δp, we use Eqs. (2.72) and (2.73) to evaluate

$$|\Delta x| = \left(\frac{\hbar}{2\mu\omega}\right)^{\frac{1}{2}}, \quad |\Delta p| = \left(\frac{1}{2}\mu\hbar\omega\right)^{\frac{1}{2}}, \quad |\Delta x||\Delta p| = \frac{\hbar}{2}. \qquad (2.131)$$

From Eq. (2.131), we see that the uncertainties $|\Delta x|$, $|\Delta p|$, and the uncertainty product $|\Delta x||\Delta p|$ are *time independent* for a wave packet in a harmonic potential with α "tuned" to $\mu\omega/\hbar$.

The method we have used in this section, in which we calculate average values of operators from time-dependent *wave functions*, is called the *Schrödinger picture*. In the next section we examine the *Heisenberg picture*, in which the *operators* are considered to be time dependent.

2.13. Time Evolution: The Heisenberg Picture

The time-dependent average of an operator is $\hat{O}$ is

$$\langle O(t)\rangle = \langle W|e^{\frac{i}{\hbar}\hat{H}t}\hat{O}e^{-\frac{i}{\hbar}\hat{H}t}|W\rangle. \tag{2.132a}$$

In the Schrödinger picture, we have written $e^{-\frac{i}{\hbar}\hat{H}t}|W\rangle = |W;t\rangle$ so that Eq. (2.132a) becomes

$$O(t)\rangle = \langle W;t|\hat{O}|W;t\rangle; \tag{2.132b}$$

that is, the *wave functions* are time dependent, but the *operators* are time independent. In the Heisenberg picture we take the *operator* to be time dependent and write

$$\hat{O}(t) = e^{\frac{i}{\hbar}\hat{H}t}\hat{O}e^{-\frac{i}{\hbar}\hat{H}t}. \tag{2.133}$$

In the Fock operator treatment of the harmonic oscillator problem, operators are expressed in terms of $\hat{a}^\dagger$ and $\hat{a}$,

$$\hat{O} = \hat{O}(\hat{a}^\dagger, \hat{a}). \tag{2.134}$$

Assuming that $\hat{O}(\hat{a}^\dagger, \hat{a})$ can be written as a power series in $\hat{a}^\dagger$ and $\hat{a}$, we need to find

$$e^{\frac{i}{\hbar}\hat{H}t}(\hat{a}^\dagger)^l e^{-\frac{i}{\hbar}\hat{H}t}, \quad e^{\frac{i}{\hbar}\hat{H}t}\hat{a}^l e^{-\frac{i}{\hbar}\hat{H}t}. \tag{2.135}$$

But by inserting $\hat{1} = e^{\frac{i}{\hbar}\hat{H}t}e^{-\frac{i}{\hbar}\hat{H}t}$ between the operators, we have

$$\begin{aligned}
&e^{\frac{i}{\hbar}\hat{H}t}(\hat{a}^\dagger)^l e^{-\frac{i}{\hbar}\hat{H}t}\\
&= \left(e^{\frac{i}{\hbar}\hat{H}t}\hat{a}^\dagger e^{-\frac{i}{\hbar}\hat{H}t}\right)\left(e^{\frac{i}{\hbar}\hat{H}t}\hat{a}^\dagger e^{-\frac{i}{\hbar}\hat{H}t}\right)\ldots\left(e^{\frac{i}{\hbar}\hat{H}t}\hat{a}^\dagger e^{-\frac{i}{\hbar}\hat{H}t}\right)\\
&= [\hat{a}^\dagger(t)]^l,
\end{aligned} \tag{2.136}$$

and similarly for $\hat{a}$. Thus we need only find the explicit time dependence of $\hat{a}^\dagger$ and $\hat{a}$, which is

$$\hat{a}^\dagger(t) = e^{\frac{i}{\hbar}\hat{H}t}\hat{a}^\dagger e^{-\frac{i}{\hbar}\hat{H}t}$$

$$\hat{a}(t) = e^{\frac{i}{\hbar}\hat{H}t}\hat{a}e^{-\frac{i}{\hbar}\hat{H}t}.$$

In the case of $\hat{a}$, making use of Eq. (2.14) and a series expansion for $e^{-\frac{i}{\hbar}\hat{H}t}$ we have

$$\hat{a}(t) = e^{\frac{i}{\hbar}\hat{H}t}\,\hat{a}\sum_{l=0}^{\infty}\frac{1}{l!}\left(-\frac{i}{\hbar}t\right)^l(\hbar\omega)^l\left(\hat{a}^\dagger\hat{a}+\frac{1}{2}\right)^l. \tag{2.138}$$

We now find the normal order of $\hat{a}\hat{H}^l$ by induction:

$$\begin{aligned}
\hat{a}\hat{H} &= \hbar\omega\left(\hat{a}\hat{a}^\dagger\hat{a}+\frac{1}{2}\hat{a}\right) = \hbar\omega\left(\hat{a}^\dagger\hat{a}^2+[\hat{a},\hat{a}^\dagger]_-\hat{a}+\frac{1}{2}\hat{a}\right)\\
&= \hbar\omega\left[\left(\hat{a}^\dagger\hat{a}+\frac{1}{2}\right)+1\right]\hat{a} = (\hat{H}+\hbar\omega)\hat{a}
\end{aligned} \tag{2.139a}$$

$$\begin{aligned}
\hat{a}\hat{H}^2 &= (\hat{h}+\hbar\omega)\hat{a}\hat{H} = (\hat{H}+\hbar\omega)^2\hat{a}\\
&\ \ \vdots\\
\hat{a}\hat{H}^l &= (\hat{H}+\hbar\omega)^l\hat{a}\\
\hat{a}\hat{H}^{(l+1)} &= (\hat{h}+\hbar\omega)^l\hat{a}\hat{H} = (\hat{H}+\hbar\omega)^{(l+1)}\hat{a}.
\end{aligned} \tag{2.139b}$$

Thus Eq. (2.138) becomes simply

$$\begin{aligned}
\hat{a}(t) &= e^{\frac{i}{\hbar}\hat{H}t}\sum_{l=0}^{\infty}\frac{1}{l!}\left(-\frac{i}{\hbar}t\right)^l(\hat{H}+\hbar\omega)^l\hat{a}\\
&= e^{\frac{i}{\hbar}\hat{H}t}\,e^{-\frac{i}{\hbar}\hat{H}t}\,e^{-i\omega t}\hat{a} = e^{-i\omega t}\hat{a}.
\end{aligned} \tag{2.140a}$$

Since $\hat{a}^\dagger$ is the Hermitian conjugate of $\hat{a}$,

$$\hat{a}^\dagger(t) = e^{i\omega t}\hat{a}^\dagger. \tag{2.140b}$$

Then the operator $\hat{O}(t)$ has the form

$$\hat{O}(t) = \hat{O}(e^{\omega t}\hat{a}^\dagger, e^{-i\omega t}\hat{a}). \tag{2.141}$$

For example,

$$\hat{x}(t) = \left(\frac{\hbar}{2\mu\omega}\right)^{\frac{1}{2}} \left(e^{i\omega t}\hat{a}^\dagger + e^{-i\omega t}\hat{a}\right) \tag{2.142a}$$

$$\hat{x}^2(t) = \left(\frac{\hbar}{2\mu\omega}\right) \left[e^{2i\omega t}(\hat{a}^\dagger)^2 + 2\hat{a}^\dagger\hat{a} + e^{-2i\omega t}\hat{a}^2 + 1\right] \tag{2.142b}$$

$$\hat{p}(t) = i\left(\frac{\mu\hbar\omega}{2}\right)^{\frac{1}{2}} \left(e^{i\omega t}\hat{a}^\dagger - e^{-i\omega t}\hat{a}\right) \tag{2.142c}$$

$$\hat{p}^2(t) = -\left(\frac{\mu\hbar\omega}{2}\right) \left[e^{2i\omega t}(\hat{a}^\dagger)^2 - 2\hat{a}^\dagger\hat{a} + e^{-2i\omega t}\hat{a}^2 - 1\right], \tag{2.142d}$$

where we have used Eq. (2.10) to obtain Eqs. (2.142b) and (2.142d). It is instructional to write the r.h.s. of Eqs. (2.142) in terms of $\hat{x}$ and $\hat{p}$. Making use of Eqs. (2.7) and (2.8) and simplifying the result, we have

$$\hat{x}(t) = \hat{x}\cos\omega t + \frac{\hat{p}}{\mu\omega}\sin\omega t \tag{2.143a}$$

$$\hat{x}^2(t) = \hat{x}^2\cos^2\omega t + \frac{1}{\mu\omega}(\hat{x}\hat{p} + \hat{p}\hat{x})\sin\omega t\cos\omega t + \frac{\hat{p}^2}{\mu^2\omega^2}\sin^2\omega t \tag{2.143b}$$

$$\hat{p}(t) = \hat{p}\cos\omega t - \mu\omega\sin\omega t \tag{2.143c}$$

$$\hat{p}^2(t) = \hat{p}^2\cos^2\omega t - \mu\omega(\hat{x}\hat{p} + \hat{p}\hat{x})\sin\omega t\cos\omega t + \mu^2\omega^2\sin^2\omega t. \tag{2.143d}$$

Evaluation of the averages of the operators in Eqs. (2.143) is straightforward with use of Eqs. (2.91); we obtain, for example,

$$\begin{aligned}\langle W|\hat{x}\hat{p} + \hat{p}\hat{x}|W\rangle &= \frac{\hbar}{i}\left\langle W\left|x\frac{d}{dx} + \frac{d}{dx}x\right|W\right\rangle \\ &= \frac{\hbar}{i}\left\langle W\left|2x\frac{d}{dx} + 1\right|W\right\rangle.\end{aligned} \tag{2.144}$$

From Eq. (2.85), we have

$$\frac{d}{dx}\psi_W(x) = \left[-\alpha(x - x_0) + \frac{i}{\hbar}p_0\right]\psi_W(x). \tag{2.145}$$

We also find that

$$\langle W|\hat{x}\hat{p} + \hat{p}\hat{x}|W\rangle = -\frac{\hbar}{i}2\alpha\left(\langle\hat{x}^2\rangle - x_0^2\right) + 2p_0x_0 + \frac{\hbar}{i} = 2p_0x_0. \tag{2.146}$$

Substitution of Eq. (2.146) into Eq. (2.145) and use of Eq. (2.91c) gives

$$\langle\hat{x}(t)\rangle = x_0\cos\omega t + \frac{p_0}{\mu\omega}\sin\omega t \tag{2.147a}$$

$$\langle\hat{p}(t)\rangle = p_0\cos\omega t - \mu\omega x_0\sin\omega t \tag{2.147b}$$

$$\begin{aligned}\langle\hat{x}^2(t) = {} & \left(\frac{1}{2\alpha} + x_0^2\right)\cos^2\omega t + \frac{2}{\mu\omega}p_0x_0\sin\omega t\cos\omega t \\ & + \left(\frac{\hbar^2}{2\mu^2\omega^2}\alpha + \frac{p_0^2}{\mu^2\omega^2}\right)\sin^2\omega t\end{aligned} \tag{2.147c}$$

$$\begin{aligned}\langle\hat{p}^2(t)\rangle = {} & \left(\frac{\hbar^2}{2}\alpha + p_0^2\right)\cos^2\omega t - 2\mu\omega p_0x_0\sin\omega t\cos\omega t \\ & + 2\mu^2\omega^2\alpha + \mu^2\omega^2\left(\frac{1}{2\alpha}\right)\sin^2\omega t.\end{aligned} \tag{2.147d}$$

A great advantage of the Heisenberg picture over the Schrödinger picture is the simplicity of the form of $\hat{a}^\dagger(t)$ and $\hat{a}(t)$, permitting a simple evaluation of the time evolution of operator averages for a system with a given wave function at a specified time. This simplicity of the time evolution of the Fock operators is also a feature of the Dirac picture in time-dependent perturbation theory, which we treat in the next chapter.

2.14. Summary for Chapter 2

This section summarizes some of the most important results from Chapter 2 without including the derivations.

The (time-independent) Schrödinger equation in one dimension is

$$\hat{H}\psi(x) = E\psi(x), \tag{S2.1}$$

where $\hat{H}$ is the Hamiltonian operator, $\psi(x)$ is the wave function, and E is the energy of the system. For the harmonic oscillator with mass μ and force constant k, the Hamiltonian is

$$\hat{H} = -\frac{\hbar^2}{2\mu}\frac{d^2}{dx^2} + \frac{1}{2}kx^2, \tag{S2.2}$$

where $\hbar$ is (the reduced) Planck's constant. The force constant can be given in terms of the (angular) frequency ω as $k = \mu\omega^2$,

We replace x by the dimension-less variable

$$\xi = \left(\frac{\mu\omega}{\hbar}\right)^{\frac{1}{2}} x. \tag{S2.3a}$$

Then we have for the reduced Hamiltonian

$$\hat{h} = \frac{\hat{H}}{\hbar\omega} = \frac{1}{2}\left(-\frac{d^2}{d\xi^2} + \xi^2\right). \tag{S2.4}$$

With the notation $\varphi(\xi) = \psi(\sqrt{\hbar/\mu\omega}\xi)$ and $E/\hbar\omega$, we have the reduced Schrödinger equation

$$\hat{h}\varphi(\xi) = \varepsilon\varphi(\xi). \tag{S2.5}$$

The Fock annihilation and creation operators, respectively, are defined by

$$\hat{a} = \frac{1}{\sqrt{2}}\left(\frac{d}{d\xi} + \xi\right) \tag{S2.6}$$

$$\hat{a}^\dagger = \frac{1}{\sqrt{2}}\left(-\frac{d}{d\xi} + \xi\right). \tag{S2.7}$$

They have the commutation relation

$$[\hat{a}, \hat{a}^\dagger]_- = 1. \tag{S2.8}$$

In terms of the Fock operators,

$$\hat{p} = i\left(\frac{\mu\hbar\omega}{2}\right)^{\frac{1}{2}}(\hat{a}^\dagger - \hat{a}), \quad \hat{x} = \left(\frac{\hbar}{2\mu\omega}\right)^{\frac{1}{2}}(\hat{a}^\dagger + \hat{a}) \tag{S2.9}$$

and the reduced Hamiltonian becomes

$$\hat{h} = \frac{1}{2}\left(-\frac{d^2}{d\xi^2} + \xi^2\right) = \hat{a}^\dagger\hat{a} + \frac{1}{2} = \hat{n} + \frac{1}{2}, \tag{S2.10}$$

where $\hat{a}^\dagger\hat{a}$ is identified with the number operator $\hat{n}$.

For the harmonic oscillator, the energy eigenvalues are discrete and their eigenfunctions have the form

$$\varphi_n(\xi) = N_n H_n(\xi) e^{-\frac{1}{2}\xi^2}, \tag{S2.11}$$

with n taking non-negative integral values. The normalization constant is

$$N_n = \left(\frac{\mu\omega}{\pi\hbar}\right)^{\frac{1}{4}} \frac{1}{\sqrt{(2^n n!)}} \tag{S2.12}$$

so that the normalization requirement

$$\int_{-\infty}^{\infty} [\psi_n(x)]^2 dx = 1 \tag{S2.13}$$

is met. The $H_n(\xi)$ in Eq. (S2.11) are the Hermite polynomials; they may be generated from the recursion relation

$$H_{(n+1)}(\xi) = 2\xi H_n(\xi) - 2nH_{(n-1)}(\xi)$$

with $H_n(\xi) = 0$ for negative n and $H_0(\xi) = 1$. The eigenfunctions $\psi_n(x)$ are orthogonal in the sense that

$$\begin{aligned}\int_{-\infty}^{\infty} \psi_n(x)\psi_m(x)dx &= N_n Nm \left(\frac{\hbar}{\mu\omega}\right)^{\frac{1}{2}} \int_{-\infty}^{\infty} H_n(\xi)H_m(\xi)e^{-\xi^2}d\xi \\ &= N_n^2\left(\frac{\hbar}{\mu\omega}\right)\frac{1}{2}2^n n!\sqrt{\pi}\delta_{nm}. \end{aligned} \tag{S2.14}$$

In the Dirac bracket notation,

$$\int_{-\infty}^{\infty} \psi_n(x)\psi_m(x)dx = \langle n|m\rangle, \tag{S2.15}$$

and the separated *bra* and *ket* are generated from the ground state by

$$|n\rangle = \frac{1}{\sqrt{n!}}(\hat{a}^{\dagger})^n|0\rangle, \quad \langle n| = \frac{1}{\sqrt{n!}}\langle 0|\hat{a}^n. \tag{S2.16}$$

The matrix elements of x and p are

$$x_{nm} = \langle n|\hat{x}|m\rangle = \left(\frac{\hbar}{2\mu\omega}\right)^{\frac{1}{2}} \left[\sqrt{n}\delta_{m,n-1} + \sqrt{n+1}\delta_{m,n+1}\right] \tag{S2.17a}$$

$$p_{nm} = \langle n|\hat{p}|m\rangle = i\left(\frac{\mu\hbar\omega}{2}\right)^{\frac{1}{2}} \left[\sqrt{n}\delta_{m,n-1} - \sqrt{n+1}\delta_{m,n+1}\right]. \tag{S2.17b}$$

The energy and uncertainty product for state n are

$$E = \left(n + \frac{1}{2}\right)\hbar\omega \tag{S2.18}$$

$$p_{nm} = \langle n|\hat{p}|m\rangle = i\left(\frac{\mu\hbar\omega}{2}\right)^{\frac{1}{2}} \langle n|(\hat{a}^{\dagger} - \hat{a})|m\rangle$$

$$= i\left(\frac{\mu\hbar\omega}{2}\right)^{\frac{1}{2}} \left[\sqrt{n}\delta_{m,n-1} - \sqrt{n+1}\delta_{m,n+1}\right]. \tag{S2.19}$$

The displacement operator is

$$f(x - x_0) = e^{-\frac{i}{\hbar}\hat{p}x_0} f(x). \tag{S2.20}$$

The identity operator in *ket, bra* form is

$$\sum_{n=0}^{\infty} |n\rangle\langle n| = \hat{1}. \tag{S2.21}$$

In the Schrödinger picture the wave functions are time-dependent; thus the average value of a variable represented by the operator $\hat{O}$ is

$$\langle O(t)\rangle = \langle\psi(x,t)|\hat{O}|\psi(x,t)\rangle. \tag{S2.22}$$

In the Heisenberg picture the operators are time-dependent:

$$\langle O(t)\rangle\rangle = \langle\psi(x)|\hat{O}(t)|\psi(x)\rangle. \tag{S2.23}$$

A minimum Gaussian wave packet at x_0 with momentum p_0 and width $\Delta x = 1/\sqrt{2\alpha}$ has the form at $t = 0$

$$\psi_W(\alpha, x_0, p_0; x) = (\alpha/\pi)^{\frac{1}{4}} e^{-\frac{\alpha}{2}(x-x_0)^2}\, e^{\frac{i}{\hbar}p_0 x}. \tag{S2.24}$$

In a harmonic potential, at t, the packet has the form

$$\psi_W(x,t) = \sum_{n=0}^{\infty} \frac{1}{\sqrt{n!}} \left(\frac{\mu\omega}{2\hbar}\right)^{\frac{n}{2}} \left(x_0 + \frac{ip_0}{\mu\omega}\right)^n \times e^{-\frac{\mu\omega}{4\hbar}\left(x_0^2 + \frac{p_0^2}{\mu^2\omega^2}\right)} e^{\frac{i}{2\hbar}p_0 x_0} e^{-i(n+\frac{1}{2})\omega t} \psi_n(x,0) \tag{S2.25}$$

and the average values of x and p move along the classical path

$$\langle W; t|\hat{x}|W; t\rangle = x_0 \cos\omega t + \frac{p_0}{\mu\omega}\sin\omega t \tag{S2.26a}$$

$$\langle W; t|\hat{p}|W; t\rangle = p_0 \cos\omega t - \mu\omega x_0 \sin\omega t. \tag{S2.26b}$$

But the uncertainties depend upon the reciprocal width parameter as in Eqs. (2.148). When α is "tuned" to the uncertainty product is minimized to give

$$|\Delta x| = \left(\frac{\hbar}{2\mu\omega}\right)^{\frac{1}{2}}, \quad |\Delta p| = \left(\frac{1}{2}\mu\hbar\omega\right)^{\frac{1}{2}}, \quad |\Delta x||\Delta p| = \frac{\hbar}{2}. \tag{S2.27}$$

Chapter 3

Time-Dependent Perturbation Expansions

3.1. Introduction

The purpose of Chapter 3 is to present time-dependent perturbation theory in the Dirac picture and to illustrate its use by applying it to the problem of a harmonic oscillator whose potential-energy function has been displaced a fixed distance. This problem is a physically simple one with an easily obtained exact solution; nevertheless, it affords an opportunity to show how one may use diagrammatic analysis to great advantage.

In the course of our development of the general theory, we present and prove Wick's theorem, which is the basis for graphical representation of the perturbation expansions. As an example of the power of Wick's theorem, we derive Glauber's representation of minimal wave packets from the operator form we introduced in Chapter 2. Some of the properties of the Glauber or *coherent* states are derived here in anticipation of our use of a generalization of them in Chapter 8 for our treatment of coherent photon ensembles. Moreover, the Glauber states are the starting point for a "random phase" treatment in Chapter 6 of *thermal* ensembles of photons and molecules.

Graphs based directly upon the application of Wick's theorem to the displaced-oscillator problem are developed, with their interpretation and evaluation given in some detail. The form of the graphs in this chapter is limited to problems in which both the perturbation and the operator whose average value is to be calculated are linear functions of the Fock operators. The grounding in the methods of graphical construction afforded the reader here in connection with a simple problem will, however, carry over to more general applications.

The connected-graph rule is introduced in a natural way in Chapter 3 as a means of understanding why the first-order perturbation treatment of the displaced oscillator is exact. We "derive" the rule by a detailed examination of the way the time integrations of two disconnected pieces may be disentangled when a sum is taken of a complete class of graphs of a given order, followed by the summation of the disconnected parts to infinite order, to obtain a normalization factor for the connected graphs. For the displaced oscillator, the factor is unity.

The reader may find some of the details tedious, especially the summations leading to Wick's theorem and the Glauber states. To some extent this is due to the present author's attempt to use mathematical methods that are relatively unsophisticated to make the powerful tools of quantum field theory more widely accessible. Most readers may very well skip the more challenging of the explicit developments. On the other hand, some tedium is unavoidable. It would be disingenuous to claim that quantum theory is easy! But the remarkable simplicity of some of the results — and the insights these results provide — should repay readers for the investment of their time and trouble.

3.2. The Dirac or Interaction Picture

As we saw in Section 2.13, the average value of an operator at time t when the wave function is known at time $t = 0$ is given by

$$\begin{aligned}\langle O(t)\rangle &= \langle\psi(t=0)|e^{\frac{i}{\hbar}\hat{H}t}\hat{O}e^{-\frac{i}{\hbar}\hat{H}t}|\psi(t=0)\rangle \\ &= \langle\psi(t=0)|\hat{O}(t)|\psi(t=0)\rangle.\end{aligned} \tag{3.1}$$

When the initial wave function $\psi(t = 0)$ can be expanded in terms of the eigenfunctions of $\hat{H}$, as in Section 2.12, or when the time-dependent operator $\hat{O}(t)$ can be worked out explicitly, as in Section 2.13, the exact value of $\langle O(t)\rangle$ can sometimes be obtained. These are the Schrödinger and Heisenberg pictures, respectively. As is more often the case, the initial wave function is not easily expanded in the eigenvalues of $\hat{H}$, but is an eigenfunction of some simpler Hamiltonian $\hat{H}_0$, or can be expanded in its eigenfunctions; or the time-dependent operator $\hat{O}(t)$ cannot be easily obtained from the time-evolution operator $e^{-\frac{i}{\hbar}\hat{H}t}$, but has a simple form if $e^{-\frac{i}{\hbar}\hat{H}_0 t}$ is used instead. In these cases, perturbation expansions can be attempted. The perturbation treatment we use is similar to one proposed by Dirac. Let us write the (full) time-evolution operator in the form

$$e^{-\frac{i}{\hbar}\hat{H}t} = e^{-\frac{i}{\hbar}\hat{H}_0 t}\hat{U}(t), \tag{3.2}$$

so that Eq. (3.1) becomes

$$\begin{aligned} \langle O(t)\rangle &= \langle\psi(t=0)|\hat{U}^\dagger(t)[e^{-\frac{i}{\hbar}\hat{H}_0 t}\hat{O}_e e^{-\frac{i}{\hbar}\hat{H}t}]\hat{U}(t)|\psi(t=0)\rangle \\ &= \langle\psi(t=0)|\hat{U}^\dagger(t)\hat{O}_\mathrm{I}(t)\hat{U}(t)|\psi(t=0)\rangle, \end{aligned} \tag{3.3}$$

where we have defined $\hat{O}_\mathrm{I}(t)$ as the operator in square brackets in the first line of Eq. (3.3); *i.e.*,

$$\hat{O}_\mathrm{I}(t) = e^{\frac{i}{\hbar}\hat{H}_0 t}\hat{O}e^{-\frac{i}{\hbar}\hat{H}_0 t}. \tag{3.4}$$

Calculations of the time evolution of average values of operators based upon Eqs. (3.2)–(3.4) are said to be in the *Dirac or interaction* picture. It is very important to note that the operators $\hat{O}_\mathrm{I}(t)$ in the interaction picture treated in this chapter and the operators $\hat{O}_\mathrm{H}(t)$ in the Heisenberg picture treated in Chapter 2 are *not the same* in general; in the Heisenberg picture, the Hamiltonian in the time evolution operators from which we obtain the time-dependent operators is the *full* Hamiltonian $\hat{H}$, while in the Dirac picture, the Hamiltonian is the "non-interacting" Hamiltonian $\hat{H}_0$. In the applications to follow, however, the Hamiltonian $\hat{H}_0$ happens to be the same as the Hamiltonian $\hat{H}$ of Chapter 2, to which we add a "perturbation".

We solve for $\hat{U}(t)$ by multiplying Eq. (3.2) from the left by $e^{\frac{i}{\hbar}\hat{H}_0 t}$ to obtain

$$\hat{U}(t) = e^{\frac{i}{\hbar}\hat{H}_0 t}e^{-\frac{i}{\hbar}\hat{H}t}. \tag{3.5}$$

The time derivative of $\hat{U}(t)$ is thus

$$\begin{aligned}\frac{d}{dt}\hat{U}(t) &= \frac{i}{\hbar}e^{\frac{i}{\hbar}\hat{H}_0 t}(\hat{H}_0 - \hat{H})e^{-\frac{i}{\hbar}\hat{H}t} \\ &= -\frac{i}{\hbar}e^{\frac{i}{\hbar}\hat{H}_0 t}(\hat{H} - \hat{H}_0)e^{-\frac{i}{\hbar}\hat{H}t}. \end{aligned} \tag{3.6a}$$

(It is important to note that $e^{\frac{i}{\hbar}\hat{H}_0 t}$ and $e^{\frac{i}{\hbar}\hat{H}t}$ do not necessarily commute.)

Insertion of $\hat{\mathrm{I}} = e^{-\frac{i}{\hbar}\hat{H}_0 t}e^{\frac{i}{\hbar}\hat{H}_0 t}$ into Eq. (3.6a) and use of Eq. (3.5) give

$$\begin{aligned}\frac{d}{dt}\hat{U}t &= \frac{i}{\hbar}e^{\frac{i}{\hbar}\hat{H}_0 t}(\hat{H} - \hat{H}_0)e^{-\frac{i}{\hbar}\hat{H}_0 t}e^{\frac{i}{\hbar}\hat{H}_0 t}e^{-\frac{i}{\hbar}\hat{H}t} \\ &= -\frac{i}{\hbar}e^{\frac{i}{\hbar}\hat{H}_0 t}(\hat{H} - \hat{H}_0)e^{-\frac{i}{\hbar}\hat{H}_0 t}\hat{U}(t) \\ &= -\frac{i}{\hbar}\lambda\hat{H}_1(t)\hat{U}(t), \end{aligned} \tag{3.6b}$$

where we have written $\lambda\hat{H}_1(t)$ for $e^{-\frac{i}{\hbar}\hat{H}_0 t}(\hat{H} - \hat{H}_0)e^{\frac{i}{\hbar}\hat{H}_0 t}$.

The operator $\hat{H}_{\mathrm{I}}(t)$ is called the *Interaction Hamiltonian*, since in many applications $\hat{H}_0$ is the Hamiltonian of a system of non-interacting particles, while $\hat{H}$ is the Hamiltonian for the particles with their mutual interaction potentials "turned on"; λ is a dimensionless scaling or *perturbation parameter* which gives the strength of the interaction. A formal solution to the differential equation in Eq. (3.6) under the boundary condition that $\hat{U}(0) = 1$, as is required by Eq. (3.5), is

$$\hat{U}(t) = 1 - \frac{i}{\hbar}\lambda\int_0^t \hat{H}_1(t')\hat{U}(t')dt'. \tag{3.7}$$

The importance of Eq. (3.7) is that it may be used to expand $\hat{U}(t)$ in powers of λ. We begin by substituting the r.h.s. of Eq. (3.7) for $\hat{U}(t')$ in the integral to obtain

$$\hat{U}(t) = 1 - \frac{i}{\hbar}\lambda\int_0^t dt''\hat{H}_1(t'')\left[1 - \frac{i}{\hbar}\lambda\int_0^{t''} dt'\hat{H}_{\mathrm{I}}(t')\hat{U}(t')\right]. \tag{3.8}$$

$$\hat{U}(t) = 1 + \sum_{j=1}^{\infty}\lambda^j\hat{U}^{(j)}(t). \tag{3.9}$$

It is clear from Eq. (3.8) that the term that is first order in λ, *i.e.*, linear in the interaction strength, is

$$\lambda\hat{U}^{(1)}(t) = -\frac{1}{\hbar}\lambda\int_0^t dt_1\hat{H}_{\mathrm{I}}(t_1), \tag{3.10}$$

where we have replaced the time integration variable t'' by t_1. The next term is found by once more inserting the r.h.s. of Eq. (3.7) into Eq. (3.9),

$$\hat{U}(t) = 1 - \frac{i}{\hbar}\lambda\int_0^t dt_3\hat{H}_{\mathrm{I}}(t_3) \times \left\{1 - \frac{i}{\hbar}\lambda\int_0^{t_3} dt_2\hat{H}_{\mathrm{I}}(t_2)\right.$$
$$\left.\times\left[1 - \frac{i}{\hbar}\lambda\int_0^{t_2} dt_1\hat{H}_{\mathrm{I}}(t_1)\hat{U}(t_1)\right]\right\}, \tag{3.11}$$

where we have changed the subscripts on the time integration variables so that the later time has the higher numerical subscript. The coefficient of λ^2 in the series is seen to be

$$\lambda^2\hat{U}^{(2)}(t) = \left(-\frac{i}{\hbar}\right)^2\lambda^2\int_0^t dt_2\int_0^{t_2} dt_1\hat{H}_{\mathrm{I}}(t_2)\hat{H}_{\mathrm{I}}(t_1). \tag{3.12}$$

This process may be continued indefinitely to obtain the Dyson series

$$\begin{aligned}\hat{U}(t) &= 1 + \sum_{j=1}^{\infty}\left(-\frac{i}{\hbar}\right)^j\lambda^j\int_0^t dt_j\int_0^{t_j} dt_{j-1}\cdots\\ &\quad\times\int_0^{t_2} dt_1\hat{H}_{\mathrm{I}}(t_j)\hat{H}_{\mathrm{I}}(t_{j-1})\cdots\hat{H}_{\mathrm{I}}(t_1)\\ &= 1 + \sum_{j=1}^{\infty}\lambda^j\hat{U}^{(j)}(t).\end{aligned} \tag{3.13}$$

Note that for each order j, the variables are $t_1\cdots t_j$ with $t_j > t_{j-1}\cdots t_2 > t_1$. The next step is to find the explicit form of $\hat{H}_{\mathrm{I}}(t)$. In our immediate applications of Eq. (3.13), $\hat{H}_0$ is the Hamiltonian for the harmonic oscillator, or the sum of such Hamiltonians, and the difference, $\hat{H}_0 - \hat{H}$, can be expressed in terms of the operators of Chapter 2. In this case, $H_{\mathrm{I}}(t)$ turns out to be a string of sums of $\hat{a}_{\mathrm{I}}(t)$ and $\hat{a}^\dagger(t)$.

In the next section, we see how such strings can be put into *normal order*, thus greatly simplifying the evaluation of the resulting brackets.

3.3. Normal Ordering of Strings of Fock Operators: Wick's Theorem

We saw in Section 2.6 that evaluation of a bracket in which the operator a string of Fock operators is very simple if the string is first put into *normal order*; that is, a sum of terms in which all of the $\hat{a}^\dagger$'s appear to the left of all the $\hat{a}$'s. Eq. (2.48) gives the normal order of the special case

$$\hat{a}^n(\hat{a}^\dagger)^m = \sum_{k=0}^{\{n,m\}_<} \frac{n!m!}{k!(m-k)!(n-k)!}(\hat{a}^\dagger)^{(m-k)}\hat{a}^{(n-k)}, \tag{3.14}$$

where we use $\{n, m\}_<$ to indicate the lesser of n, m.

Here, we complete the induction proof of Eq. (2.48) [*i.e.*, Eq. (3.14)] outlined in Chapter 2. We showed there that Eq. (2.48) holds for powers $n, m = 0, 1, 2$. Assuming that Eq. (2.48) holds for all values of the powers up to the values n, m, we have

$$\hat{a}\hat{a}^n(\hat{a}^\dagger)^m = \sum_{k=0}^{\{n,m\}_<} \frac{n!m!}{k!(m-k)!(n-k)!}[\hat{a}(\hat{a}^\dagger)^{(m-k)}]\hat{a}^{(n-k)}. \tag{3.15}$$

Now Eq. (2.48) must hold for the factor in square brackets, since the powers of the operators are in the range of assumed validity of Eq. (2.48). By the commutation relation,

$$\hat{a}(\hat{a}^\dagger)^{(m-k)} = (\hat{a}^\dagger)^{(m-k)}\hat{a} + (m-k)(\hat{a}^\dagger)^{(m-k-1)}. \tag{3.16}$$

Substituting Eq. (3.16) for the square brackets in Eq. (3.15) gives

$$\hat{a}\hat{a}^n(\hat{a}^\dagger)^m = \sum_{k=0}^{\{n,m\}_<} \frac{n!m!}{k!(m-k)!(n-k)!}$$
$$[(\hat{a}^\dagger)^{(m-k)}\hat{a} + (m-k)(\hat{a}^\dagger)^{(m-k-1)}]\hat{a}^{(n-k)}$$

$$= \sum_{k=0}^{\{n,m\}_<} \frac{n!m!}{k!(m-k)!(n-k)!} (\hat{a}^\dagger)^{(m-k)} \hat{a}^{(n+1-k))}$$

$$+ \sum_{k=0}^{\{n,m-1\}_<} \frac{n!m!}{k!(m-k-1)!(n-k)!}$$

$$\times (\hat{a}^\dagger)^{(m-k-1)} \hat{a}^{(n-k)}. \tag{3.17}$$

Now change the summation variable in the last sum from k to $l = k+1$ and in the first sum change k to l:

$$\hat{a}\hat{a}^n(\hat{a}^\dagger)^m = \sum_{l=0}^{\{n,m\}_<} \frac{n!m!}{l!(m-l)!(n-l)!} (\hat{a}^\dagger)^{(m-l)} \hat{a}^{(n+1-1)}$$

$$+ \sum_{l=1}^{\{n+1,m\}_<} \frac{n!m!}{(l-l)!(m-l)!(n+1-l)!}$$

$$\times (\hat{a}^\dagger)^{(m-l)} \hat{a}^{(n+1-l)}. \tag{3.18}$$

The two sums in Eq. (3.18) can be combined for $1 < l \leq \{n, m\}$ as follows:

$$\sum_{l=1}^{\{n,m\}_<} \frac{n!m!}{l!(m-l)!(n-l)!} (\hat{a}^\dagger)^{(m-l)} \hat{a}^{(n+1-1)}$$

$$+ \sum_{l=1}^{\{n,m\}_<} \frac{n!m!}{(1-1)!(m-l)!(n+1-l)} (\hat{a}^\dagger)^{m-1} \hat{a}^{(n+1-l)}$$

$$= \sum_{l=1}^{\{n,m\}_<} \frac{(n+1)!m!}{l!(m-l)!(n+1-l)!} (\hat{a}^\dagger)^{(m-1)} \hat{a}^{(n+1-l)}. \tag{3.19}$$

The $l = 0$ term in the first sum on the r.h.s. of Eq. (3.18) that we have omitted in Eq. (3.19) is obtained if l is set equal to zero in the sum in the last line of Eq. (3.19). Inclusion of this term thus gives

$$\sum_{l=0}^{\{n,m\}_<} \frac{n!m!}{l!(m-l)!(n-l)!}(\hat{a}^\dagger)^{(m-l)}\hat{a}^{(n+1-l)} + \sum_{l=1}^{\{n,m\}_<} \frac{n!m!}{(l-1)!(m-l)!(n+1-l)!}(\hat{a}^\dagger)^{(m-l)}\hat{a}^{(n+1-l)} = \sum_{l=0}^{\{n,m\}_<} \frac{(n+1)!m!}{l!(m-l)!(n+1-l)!}(\hat{a}^\dagger)^{(m-1)}\hat{a}^{(n+1-l)}. \tag{3.20}$$

The only remaining matter is that of the upper limit in the second sum in Eq. (3.18). In the case that $m > n+1$, the term with $l = n+1$ is seen to be identical in the last sum of both Eq. (3.18) and Eq. (3.20); thus we finally obtain

$$\hat{a}\hat{a}^n(\hat{a}^\dagger)^m = \sum_{l=0}^{\{n+1,m\}_<} \frac{(n+1)!m!}{l!(m-l)!(n+1-l)!}(\hat{a}^\dagger)^{(m-l)}\hat{a}^{(n+1-l)}. \tag{3.21}$$

Comparison of Eq. (3.21) with Eq. (3.14) shows that if Eq. (3.14) holds for some n, m, then it holds also for $n+1, m$. Next we use Eq. (3.14) to write

$$\hat{a}^n(\hat{a}^\dagger)^m\hat{a}^\dagger = \sum_{k=0}^{\{n,m\}_<} \frac{n!m!}{k!(m-k)!(n-k)!}(\hat{a}^\dagger)^{(m-k)}[\hat{a}^{(n-k)}\hat{a}^\dagger]. \tag{3.22}$$

Again, Eq. (3.16) must hold for the factor in square brackets:

$$\hat{a}^{(n-k)}\hat{a}^\dagger = \hat{a}^\dagger\hat{a}^{(n-k)} + (n-k)\hat{a}^{(n-k-1)}. \tag{3.23}$$

Substituting Eq. (3.23) for the square brackets in Eq. (3.22) gives

$$\hat{a}^n(\hat{a}^\dagger)^m\hat{a}^\dagger = \sum_{k=0}^{\{n,m\}_<} \frac{n!m!}{k!(m-k)!(n-k)!}(\hat{a}^\dagger)^{(m-k)} \times [\hat{a}^\dagger\hat{a}^{(n-k)} + (n-k)\hat{a}^{(n-k-1)}]$$

$$
\begin{aligned}
&= \sum_{k=0}^{\{n,m\}_<} \frac{n!m!}{k!(m-k)!(n-k)!}(\hat{a}^\dagger)^{(m+1-k)}\hat{a}^{(n-k)} \\
&\quad + \sum_{k=0}^{\{n,-1,m\}_<} \frac{nm}{k!(m-k)!(n-1-k))!}(\hat{a}^\dagger)^{(m-k)}\hat{a}^{(n-k-1)}.
\end{aligned}
\tag{3.24}
$$

We change the summation variable in the last sum from k to $l = k+1$ and in the first sum change k to l, combine the two sums, and proceed as for Eqs. (3.19)–(3.21) to obtain

$$
\begin{aligned}
\hat{a}^n(\hat{a}^\dagger)^m\hat{a}^\dagger &= \sum_{l=0}^{\{n,m\}_<} \frac{n!m!}{l!(m-l)!(n-l)!}(\hat{a}^\dagger)^{(m-l)}\hat{a}^{(n+1-1)} \\
&\quad + \sum_{l=1}^{\{n,m+1\}_<} \frac{n!m!}{(l-1)!(m+1-l)!(n-1)}(\hat{a}^\dagger)^{(m+1-l)}\hat{a}^{(n-l)} \\
&\quad + \sum_{l=0}^{\{n,m+1\}_<} \frac{n!(m+1)!}{l!(m+1-l)!(n-1)!}(\hat{a}^\dagger)^{(m+1-l)}\hat{a}^{(n-l)}.
\end{aligned}
\tag{3.25}
$$

Eq. (3.25) shows that if Eq. (3.14) holds for some n, m, it also holds for $n, m+1$. This completes the induction proof of Eq. (3.14), which will now form the basis for Wick's theorem.

The kth term on the r.h.s. of Eq. (3.14) is the number of ways k pairs $\hat{a}, \hat{a}^\dagger$ can be formed in the string $\hat{a}^n(\hat{a}^\dagger)^m$, multiplied by the normal-ordered string of the remaining operators *not* contained in the k pairs. This follows from the facts that $n!/[k!(n-k)!]$ is the number of ways that the k operators $\hat{a}$ can be chosen, $m!/[k!(-k)!]$, and $k!$ is the number of ways k pairs can be formed from these chosen operators. This means that for any finite string $\hat{a}^n(\hat{a}^\dagger)^m$, we may obtain the coefficients of each term by counting the number of ways that k pairs can be formed. A diagram is

$$
\begin{aligned}
\hat{a}\hat{a}^\dagger\hat{a}\hat{a}^\dagger\hat{a}^\dagger &= \mathbb{N}\{\hat{a}\hat{a}^\dagger\hat{a}\hat{a}^\dagger\hat{a}^\dagger\} \\
&\quad + \mathbb{N}\{\overbrace{\hat{a}\hat{a}^\dagger}\hat{a}\hat{a}^\dagger\hat{a}^\dagger\} + \mathbb{N}\{\overbrace{\hat{a}\hat{a}^\dagger\hat{a}\hat{a}^\dagger}\hat{a}^\dagger\} + \mathbb{N}\{\overbrace{\hat{a}\hat{a}^\dagger\hat{a}\hat{a}^\dagger\hat{a}^\dagger}\}
\end{aligned}
$$

$$+\mathbb{N}\{\hat{a}\hat{a}^\dagger\hat{a}\hat{a}^\dagger\hat{a}^\dagger\} + \mathbb{N}\{\hat{a}\hat{a}^\dagger\hat{a}\hat{a}^\dagger\hat{a}^\dagger\}$$

$$+\mathbb{N}\{\hat{a}\hat{a}^\dagger\hat{a}\hat{a}^\dagger\hat{a}^\dagger\} + \mathbb{N}\{\hat{a}\hat{a}^\dagger\hat{a}\hat{a}^\dagger\hat{a}^\dagger\} + \mathbb{N}\{a\hat{a}^\dagger\hat{a}\hat{a}^\dagger\hat{a}^\dagger\}$$

$$+\mathbb{N}\{a\hat{a}^\dagger\hat{a}\hat{a}^\dagger\hat{a}^\dagger\}$$

$$= \hat{a}^\dagger\hat{a}^\dagger\hat{a}^\dagger\hat{a}\hat{a} + 5\hat{a}^\dagger\hat{a}^\dagger\hat{a} + 4\hat{a}^\dagger. \tag{3.26}$$

A line connecting an operator $\hat{a}$ on the left with an operator $\hat{a}^\dagger$ on the right is called a *contraction.* We use the symbol $\mathbb{N}\{\hat{O}\}$ for the *normal-ordered product* of the string $\hat{O}$, which means "eliminate the contracted pairs and normal-order the remaining operators in the string $\hat{O}$". This definition of $\mathbb{N}\{\hat{O}\}$ is used to obtain the last line of Eq. (3.26).

To prove Wick's theorem,[1] we begin by assuming the theorem holds for some string of operators,

$$\{\hat{o}_1\hat{o}_2\cdots\hat{o}_s\} = \sum_{l=0}^{l_{\max}} C_l^{jk}(\hat{a}^\dagger)^{(j-1)}\hat{a}^{(k-1)}, \tag{3.27}$$

where l is the number of contracted pairs of operators and the integers C_l^{jk} are the number of possible pairing schemes for l contractions. The upper limit of the summation index, $l_{\max}$, is the maximum possible number of contractions for the given string of operators. For example, in Eq. (3.27) $j = 3$, $k = 2$, $l_{\max} = 2$, $C_0^{32} = 1$, $C_1^{32} = 5$ and $C_2^{32} = 4$. Now we multiply Eq. (3.28) on the left or on the right by an additional operator. If we operate on the left by $\hat{a}^\dagger$ there are no new contractions possible, since all contractions require an $\hat{a}$ on the left, and the r.h.s. is still in normal order. If we operate on the right by $\hat{a}$, again no new contractions and the r.h.s. is in normal order. In these cases, Wick's theorem is clearly obeyed by the new string of operators. If we operate on the left by $\hat{a}$, we have from Eqs. (3.14) and (3.28)

$$\hat{a}\{\hat{o}_1\hat{o}_2\cdots\hat{o}_s\} = \sum_{l=0}^{l_{\max}} C_l^{jk}[\hat{a}(\hat{a}^\dagger)^{(j-1)}]\hat{a}^{(k-l)}. \tag{3.28}$$

[1] Wick, G. C., *Phys. Rev.* **80**, 268 (1950).

The factor in square brackets is a special case of Eq. (3.14), and thus can be replaced with

$$\hat{a}(\hat{a}^\dagger)^{(j-l)} = (\hat{a}^\dagger)^{(j-1)}\hat{a} + (j-1)(\hat{a}^\dagger)^{(j-l-1)}. \tag{3.29}$$

Insertion of Eq. (3.28) into Eq. (3.29) gives

$$\begin{aligned}
&\hat{a}\{\hat{o}_1\hat{o}_2\cdots\hat{o}_s\} \\
&= \sum_{l=0}^{l_{\max}} C_l^{jk}[(\hat{a}^\dagger)^{(j-l)}\hat{a} + (j-l)(\hat{a}^\dagger)^{(j-l-1)}]\hat{a}^{(k-l)} \\
&= \sum_{l=0}^{l_{\max}} C_l^{jk}[(\hat{a}^\dagger)^{(j-l)}\hat{a}^{(k+1-l)} + (j-l)(\hat{a}^\dagger)^{(j-l-1)}\hat{a}^{(k-l)}].
\end{aligned} \tag{3.30}$$

Since the first term in brackets in the last line of Eq. (3.30) contains an extra factor $\hat{a}$, it is clear that the sum of these terms represents all of the contractions *not* involving the newly introduced operator $\hat{a}$. The second term in Eq. (3.30) contains the same number of $\hat{a}$ operators as are present in the corresponding term of Eq. (3.29); therefore, the sum of these terms represents all the contractions that *involve* the new operator. Thus, the string of operators on the l.h.s. of Eq. (3.30) obeys Wick's theorem.

To see how Eq. (3.30) works in a simple case, let us start with the string $\{\hat{a}^\dagger\hat{a}\hat{a}^\dagger\hat{a}^\dagger\}$. Using Wick's theorem, we may write it as

$$\begin{aligned}
\hat{a}^\dagger\hat{a}\hat{a}^\dagger\hat{a}^\dagger &= \sum_{l=0}^{1} C_l^{31}(\hat{a}^\dagger)^{(3-l)}\hat{a}^{(1-l)} \\
&= C_0^{31}(\hat{a}^\dagger)^3\hat{a} + C_1^{31}(\hat{a}^\dagger)^2 \\
&= 1\times(\hat{a}^\dagger)^3\hat{a} + 2\times(\hat{a}^\dagger)^2.
\end{aligned}$$

Then according to Eq. (3.30), the string $\hat{a}\{\hat{a}^\dagger\hat{a}\hat{a}^\dagger\hat{a}^\dagger\}$ is given by

$$\begin{aligned}
\hat{a}\hat{a}^\dagger\hat{a}\hat{a}^\dagger\hat{a}^\dagger &= \sum_{l=0}^{1} C_l^{31}[(\hat{a}^\dagger)^{(3-l)}\hat{a}^{(2-l)} + (3-l)(\hat{a}^\dagger)^{(2-l)}\hat{a}^{(1-l)}] \\
&= 1\times[(\hat{a}^\dagger)^3\hat{a}^2 + 3(\hat{a}^\dagger)^2\hat{a}] + 2\times[(\hat{a}^\dagger)^2\hat{a} + 2\hat{a}^\dagger] \\
&= \hat{a}^\dagger\hat{a}^\dagger\hat{a}^\dagger\hat{a}\hat{a} + 5\hat{a}^\dagger\hat{a}^\dagger\hat{a} + 4\hat{a}^\dagger,
\end{aligned} \tag{3.31}$$

which is the same result as is given by Eq. (3.26).

Similarly, if we operate on the right of Eq. (3.28) by $\hat{a}^\dagger$, we have

$$
\begin{aligned}
\{\hat{o}_1\hat{o}_2\cdots\hat{o}_s\}\hat{a}^\dagger &= \sum_{i=0}^{l_{\max}} C_l^{jk}(\hat{a}^\dagger)^{(j-1)}[\hat{a}^{(k-l)}\hat{a}^\dagger] \\
&= \sum_{i=0}^{l_{\max}} C_l^{jk}(\hat{a}^\dagger)^{(j-1)}[\hat{a}^\dagger\hat{a}^{(k-l)} + (k-l)\hat{a}^{(k-1-l)}] \\
&= \sum_{i=0}^{l_{\max}} C_l^{jk}[(\hat{a}^\dagger)^{(j+1-l)}\hat{a}^{(k-l)} \\
&\quad + (k-l)(\hat{a}^\dagger)^{(j-l)}\hat{a}^{(k-1-l)}]. \qquad (3.32)
\end{aligned}
$$

Again we see that the first term in the last line of Eq. (3.32) contains all the contractions *not* involving the newly introduced operator while the second term represents all the contractions that *involve* the new operator, so that Wick's theorem is also obeyed by the string of operators on the l.h.s. of Eq. (3.32). It should be clear that we may build up any desired string by continuing to insert operators on the left and on the right, and that in each instance Wick's theorem is obeyed; this completes the proof.

The power of Wick's theorem, which is the basis for diagrammatic perturbation theory, will become evident throughout the remainder of this work.

3.4. Application of Normal Ordering by Wick's Theorem: Glauber States

As an example of the utility of normal ordering by means of Wick's theorem, we return to the normalized minimal wave packet given in Eq. (2.106). The normal-order form of the operator product that produces a wave packet centered at x_0 and with momentum p_0 is simple and facilitates calculations that use the Fock operators.

We begin by finding the normal-order form of the operator

$$e^{\frac{i}{\hbar}p_0\hat{x}} = e^{i\gamma(\hat{a}^\dagger+\hat{a})} = \sum_{n=0}^{\infty} \frac{(i\gamma)^n}{n!}(\hat{a}^\dagger + \hat{a})^n, \tag{3.33}$$

where we have used Eq. (2.9) to write the operator $\hat{x}$ in terms of the Fock operators and have written γ for $p_0/(2\mu\hbar\omega)^{1/2}$. To find the normal-order form of $(\hat{a}^\dagger + \hat{a})^n$, we use Wick's theorem to write it as the sum of normal-ordered products:

$$(\hat{a}^\dagger + \hat{a})^n = \sum_{k=0}^{[n/2]} C_k \mathbb{N}\{(\hat{a}^\dagger + \hat{a})^{(n-2k)}\}, \tag{3.34}$$

where k is the number of contractions of the type $\overparen{(\hat{a}^\dagger + \hat{a}) \cdots (\hat{a}^\dagger + \hat{a})}$, the maximum number being $[n/2]$, defined as the largest integer less than or equal to $n/2$; C_k is the number of ways the $2k$ operator factors to be paired with one another may be selected from the n operator products $(\hat{a}^\dagger + \hat{a})^n$, times the number of ways the $2k$ operator factors can be paired in k contractions; that is,

$$C_k = \left[\frac{n!}{(2k)!(n-2k)!}\right] \times \left[\frac{(2k)!}{2^k k!}\right] = \frac{n!}{2^k k!(n-2k)!}. \tag{3.35}$$

The first factor in square brackets in Eq. (3.35) is the number of ways that $2k$ objects can be selected from n objects. The second factor, the number of ways that k pairs can be formed from $2k$ objects, is obtained as follows: begin by choosing any one of the $2k$ objects (there are $2k$ ways to do this) and then choose another object to be paired with it (there are $2k-1$ ways to choose the second object); now choose a third object ($2k-2$ ways) and a fourth object ($2k-3$ ways) to be paired with it, and so on until all $2k$ objects have been paired. There are $2k(2k-1)(2k-2)(2k-3)\cdots 2 \cdot 1 = (2k)!$ ways of pairing the objects in this way. However, the *order* in which the pairs are chosen is irrelevant; since there are $k!$ ways to arrange k pairs in some definite order, we have over counted by a factor of $k!$. Furthermore, the order in which the two components of each pair are chosen is also irrelevant; since there are 2^k ways to order the components of each of the k pairs, we have over counted by a factor of 2^k. Thus, the number of distinct pairing schemes is $(2k)!/(2^k k!)$.

Another way to arrive at this number is as follows: take one of the $2k$ objects at random; it must be paired with one of the others, but there are $2k-1$ ways to choose its "mate". Now take one of the remaining objects; it may be paired with any one of the $2k-3$ objects that have not yet been chosen for pairing. Continue this process until only two objects are left; they can be paired in only one way. Thus, the number of pairing schemes is $(2k-1)(2k-3)\cdots 1$. This number is equal to $(2k)! \div 2k(2k-2)\cdots 2 = (2k)!/(2^k k!)$, our previous result.

The normal-ordered product, $\mathbb{N}\{(\hat{a}^\dagger + \hat{a})^{(n-2k)}\}$ is obtained by use of the binomial theorem and placing the $\hat{a}^\dagger$'s to the left of the $\hat{a}$'s:

$$\mathbb{N}\{(\hat{a}^\dagger + \hat{a})^{(n-2k)}\} = \sum_{s=0}^{n-2k} \frac{(n-2k)!}{s!(n-2k-s)!} (\hat{a}^\dagger)^{(n-2k-s)} \hat{a}^s. \tag{3.36}$$

From Eqs. (3.33)–(3.36), we have

$$e^{\frac{i}{\hbar} p_0 \hat{x}} = \sum_{n=0}^{\infty} \sum_{k=0}^{[n/2]} \sum_{s=0}^{n-2k} (i\gamma)^n \frac{(\hat{a}^\dagger)^{n-2k-s} \hat{a}^s}{2^k k!(n-2k-s)!s!}. \tag{3.37}$$

We treat the operator $e^{-\frac{i}{\hbar}\hat{p}x_0}$ similarly to obtain

$$\begin{aligned} e^{-\frac{i}{\hbar}\hat{p}x_0} = e^{\beta(\hat{a}^\dagger - \hat{a})} &= \sum_{m=0}^{\infty} \frac{\beta^m}{m!} (\hat{a}^\dagger - \hat{a})^m \\ &= \sum_{m=0}^{\infty} \sum_{1=0}^{[m/2]} \sum_{r=0}^{m-2l} \beta^m \frac{(-1)^l (\hat{a}^\dagger)^{(m-2l-r)} \hat{a}^r}{2^l l!(m-2l-r)!r!}. \end{aligned} \tag{3.38}$$

In Eq. (3.38) we have made use of Eq. (2.8) for $\hat{p}$ and have written β for $(\mu\omega/2\hbar)^{1/2} x_0$. The factor $(-1)^l$ is due to the sign of $\hat{a}$ in the contracting factors $\overline{(\hat{a}^\dagger - \hat{a}) \cdots (\hat{a}^\dagger - \hat{a})}$. Since the operator in Eq. (3.38) operates on the ground-state *ket* $|0\rangle$ [see Eq. (2.106)], all terms with $r > 0$ annihilate $|0\rangle$; thus,

$$e^{-\frac{i}{\hbar}\hat{p}x_0}|0\rangle = \sum_{m=0}^{\infty}\sum_{l=0}^{[m/2]} \beta^m \frac{(-1)^l(\hat{a}^\dagger)^{(m-2l)}}{2^l l!(m-2l)!}|0\rangle. \tag{3.39}$$

From Eqs. (3.37) and (3.39), we have for the minimal wave packet

$$\begin{aligned} e^{\frac{i}{\hbar}p_0\hat{x}}e^{-\frac{i}{\hbar}\hat{p}x_0}|0\rangle &= \sum_{n=0}^{\infty}\sum_{k=0}^{[n/2]}\sum_{s=0}^{n-2k}\sum_{m=0}^{\infty}\sum_{l=0}^{[m/2]} \beta^m (i\gamma)^n \\ &\quad \times \frac{(-1)^l(\hat{a}^\dagger)^{(n-2k-s)}\hat{a}^s(\hat{a}^\dagger)^{(m-2l)}}{2^k k l! s!(n-2k-s)!(m-2l)!}|0\rangle. \end{aligned} \tag{3.40}$$

It remains to find the normal-ordered form for $\hat{a}^s(\hat{a}^\dagger)^{(m-l)}|0\rangle$. If $s > m-2l$ not all the $\hat{a}$'s can be contracted; this will lead to annihilation of $|0\rangle$ and thus a zero result. For $s \le m-2l$, we have for the only non-zero term, namely the one in which all s of the $\hat{a}$ operators are contracted with s of the $\hat{a}^\dagger$'s,

$$\hat{a}^s(\hat{a}^\dagger)^{(m-2l)}|0\rangle = \left[\frac{(m-2l)!}{s!(m-2l-s)!}\right] s!(\hat{a}^\dagger)^{(m-2l-s)}|0\rangle, \tag{3.41a}$$

where the factor in square brackets is the number of ways of selecting s operators $\hat{a}^\dagger$'s from the string $(\hat{a}^\dagger)^{(m-2l)}$ to be contracted with the s operators $\hat{a}$, and $s!$ is the number of ways the s pairs may be chosen from two sets of s objects each. Alternatively, if we used Eq. (2.48) to obtain

$$\begin{aligned} \hat{a}^s(\hat{a}^\dagger)^{(m-2l)}|0\rangle &= \sum_{k=0}^{s_<}\left[\frac{s!(m-2l)!}{k!(m-2l-s)!(s-k)!}\right] \\ &\quad \times (\hat{a}^\dagger)^{(m-2l-k)}\hat{a}^{(s-k)}|0\rangle, \end{aligned} \tag{3.41b}$$

where $s_<$ is the lesser of s and $m-2l$. But since $\hat{a}^{(s-k)}|0\rangle = 0$ unless $s-k=0$, only the term with $k=s$ survives. Thus Eq. (3.41b) becomes identical with Eq. (3.41a), as Wick's theorem requires. The result from Eqs. (3.40) and (3.41a) is

$$e^{\frac{i}{\hbar}p_0\hat{x}}e^{-\frac{i}{\hbar}\hat{p}x_0}|0\rangle = \sum_{n=0}^{\infty}\sum_{k=0}^{[n/2]}\sum_{s=0}^{s_{\max}}\sum_{m=0}^{\infty}\sum_{l=0}^{[m/2]}\beta^m(i\gamma)^n \times \frac{(-1)^l(\hat{a}^\dagger)^{(n+m-2k-2l-2s)}}{2^{(k+l)}k!l!s!(n-2k-s)!(m-2l-s)!}|0\rangle, \tag{3.42a}$$

where $s_{\max}$ is the lesser of $n-2k$ or $m-2l$. To simplify Eq. (3.42a), we change the summation indices as follows:

$$\begin{aligned} n' &= n+m-2k-2l-2s, & m &= n'-m'+2(l'-k')+s' \\ m' &= n-2k-s, & n &= m'+2k'+s' \\ l' &= k+l, & l &= l'-k' \\ k' &= k \\ s' &= s \end{aligned}$$

To determine the ranges of the *primed* variables, we note that the ranges of the *un-primed* variables in Eq. (3.42a) are restricted only by the constraint that the factorial arguments must be non-negative. These substitutions give

$$e^{\frac{i}{\hbar}p_0\hat{x}}e^{-\frac{i}{\hbar}\hat{p}x_0}|0\rangle = \sum_{n'=0}^{\infty}\sum_{m'=0}^{n'}\sum_{l'=0}^{\infty}\sum_{k'=0}^{l'}\sum_{s'=0}^{\infty}\frac{\beta^{[n-m'+2(l'-k')+s']}(i\gamma)^{(m'+2k'+s')}(-1)^{(l'-k')}}{2^l(k')!(l'-k')!(s')!(m')!(n'-m')!}(\hat{a}^{\dagger>n'}|0\rangle, \tag{3.42b}$$

where the finite upper limits of m' and k' are required by the restriction of the factorial arguments $l'-k'$ and $n'-m$ to non-negative values. Regrouping the factors in the summand and using the fact that $i^{2k} = -i^k$, we obtain

$$
\begin{aligned}
&e^{\frac{i}{\hbar}p_0\hat{x}}e^{-\frac{i}{\hbar}\hat{p}\hat{x}_0}|0\rangle \\
&= \left[\sum_{s=0}^{\infty}\frac{(i\beta\gamma)^{s'}}{(s')!}\right] \times \left[\sum_{l'=0}^{\infty}\sum_{k'=0}^{l'}\frac{(-1)^{l'}}{2^{l'}(k')!(l'-k')!}\beta^{2(l'-k')}(i\gamma)^{2k'}\right] \\
&\quad \times \sum_{n'=0}^{\infty}\left[\sum_{m'=0}^{n'}\frac{\beta^{n'-m'}(i\gamma)^{m'}}{(m')!(n'-m')!}\right](\hat{a}^\dagger)^{n'}|0\rangle. \qquad (3.43)
\end{aligned}
$$

From the series expansion of an exponential and the binomial theorem, we can make the identifications

$$
\sum_{s'=0}^{\infty}\frac{(i\beta\gamma)^{s'}}{(s')!} = e^{i\beta\gamma} \qquad (3.44)
$$

$$
\begin{aligned}
&\sum_{l'=0}^{\infty}\sum_{k'=0}^{l'}\frac{(-1)^{l'}}{2^{l'}(k')!(l'-k')!}\beta^{2(l'-k')}(i\gamma)^{2k'} \\
&\quad = \sum_{l'=0}^{\infty}\frac{(-1)^{l'}}{2^{l'}(l')!}\sum_{k'=0}^{l'}\frac{(l')!}{(k')!(l'-k')!}\beta^{2(l'-k')}(i\gamma)^{2k'} \\
&\quad = \sum_{l'=0}^{\infty}\frac{(-1)^{l'}}{2^{l'}(l')!}(\beta^2+\gamma^2)^{l'} = e^{-\frac{1}{2}(\beta^2+\gamma^2)} \qquad (3.45)
\end{aligned}
$$

$$
\begin{aligned}
&\sum_{n'=0}^{\infty}\sum_{m'=0}^{n'}\frac{\beta^{(n'-m')}(i\gamma)^{m'}}{(m')!(n'-m')!}(\hat{a}^\dagger)^{n'} \\
&\quad = \sum_{n'=0}^{\infty}\frac{1}{(n')!}\sum_{m'=0}^{n'}\frac{(n')!\beta^{(n!-m')}(i\gamma)^{m'}}{(m')!(n'-m')!}(\hat{a}^\dagger)^{n'} \\
&\quad = \sum_{n'=0}^{\infty}\frac{1}{(n')!}(\beta+i\gamma)^{n'}(\hat{a}^\dagger)^{n'} = e^{(\beta+i\gamma)\hat{a}^\dagger}. \qquad (3.46)
\end{aligned}
$$

From Eqs. (3.43)–(3.46) we finally obtain

$$
e^{\frac{i}{\hbar}p_0\hat{x}}e^{-\frac{i}{\hbar}\hat{p}x_0}|0\rangle = e^{i\beta\gamma}e^{-\frac{1}{2}(\beta^2+\gamma^2)}e^{(\beta+i\gamma\hat{a}^\dagger}|0\rangle. \qquad (3.47)
$$

The first exponential factor on the r.h.s. of Eq. (3.47) is a phase factor that in many cases may be ignored; the second factor is a normalization constant, which we shall see below. The third factor is an operator operating on $|0\rangle$ to give the minimal wave packet. Writing α for the complex number $\beta + i\gamma$ we define the *ket* $|\alpha\rangle$ by

$$e^{\alpha\hat{a}^\dagger}|0\rangle = |\alpha\rangle; \tag{3.48}$$

it is an *eigenket* of the (non-Hermitian!) operator $\hat{a}$ with eigenvalue α. To see this, we make use once more of Wick's theorem [or Eq. (3.14)]:

$$\begin{aligned}
\hat{a}e^{\alpha\hat{a}^\dagger}|0\rangle &= \hat{a}\sum_{n=0}^{\infty}\frac{\alpha^n}{n!}(\hat{a}^\dagger)^n|0\rangle = \sum_{n=0}^{\infty}\frac{\alpha^n}{n!}n(\hat{a}^\dagger)^{n-1}|0\rangle \\
&= \alpha\sum_{n=1}^{\infty}\frac{\alpha^{n-1}}{(n-1)!}(\hat{a}^\dagger)^{n-1}|0\rangle \\
&= \alpha\sum_{m=0}^{\infty}\frac{\alpha^m}{m!}(\hat{a}^\dagger)^m|0\rangle = \alpha e^{\alpha\hat{a}^\dagger}|0\rangle,
\end{aligned} \tag{3.49}$$

where we have used the fact that the $n = 0$ term vanishes in the last expression on the first line and have replaced the summation variable n with $m = n - 1$ in the last line. The states defined by Eq. (3.47), which have been described in great detail by Glauber[2], are the basis for treating coherent ensembles of bosons. We will use a generalization of these states in Chapter 6 where we treat molecular ensembles and in Chapter 8 in our discussion of photon ensembles in a laser beam. The normalization of a Glauber, or coherent, state is easily demonstrated, since

$$\begin{aligned}
\langle\alpha|\alpha\rangle &= \langle 0|e^{\alpha^*\hat{a}}e^{\alpha\hat{a}^\dagger}|0\rangle \\
&= \sum_{n=0}^{\infty}\sum_{m=0}^{\infty}\frac{(\alpha^*)}{n!}\frac{\alpha^m}{m!}\langle 0|\hat{a}^n(\hat{a}^\dagger)^m|0\rangle.
\end{aligned} \tag{3.50}$$

[2]Glauber, R. J., *Phys. Rev.* **131**, 2766 (1963).

Unless $n = m$ and the operators on the l.h.s. of Eq. (3.50) are fully contracted, the ground-state *ket* or *bra* will be annihilated. Since there are $n!$ ways to effect the n contractions,

$$\langle\alpha|\alpha\rangle = \sum_{n=0}^{\infty} \frac{(\alpha^*\alpha)^n}{(n!)^2} \langle 0|n!|0\rangle$$

$$= \sum_{n=0}^{\infty} \frac{(\alpha^*\alpha)^n}{n!} = e^{\alpha^*\alpha} = e^{(\beta^2+\gamma^2)}. \tag{3.51a}$$

From Eq. (3.51a), the normalization factor is seen to be $e^{-\frac{1}{2}(\beta^2+\gamma^2)}$, as we have indicated previously.

Unlike eigenfunctions of Hermitian operators, Glauber states with distinct eigenvalues are *not orthogonal*, since wave packets with different (x_0, p_0) overlap one another. To find the overlap, we use the same method by which we obtained Eq. (3.51a) to show that

$$\langle\alpha|\alpha'\rangle = e^{\alpha^*\alpha'} \neq 0. \tag{3.51b}$$

Note that Eqs. (3.51b) can be derived easily from the eigenvalue relation of Eq. (3.49), since by iteration we obtain

$$\begin{aligned}
&\langle 0|e^{\alpha^*\hat{a}} e^{\alpha'\hat{a}^\dagger}|0\\
&= \left\langle 0 \left| \sum_{n=0}^{\infty} \frac{(\alpha^*)^n}{n!} \hat{a}^n e^{\alpha'\hat{a}^\dagger} \right| 0 \right\rangle \\
&= \left\langle 0 \left| \sum_{n=0}^{\infty} \frac{(\alpha^*)^n}{n!} (\alpha')^n e^{\alpha'\hat{a}^\dagger} \right| 0 \right\rangle \\
&= \langle 0|e^{\alpha^*\alpha'} e^{\alpha'\hat{a}^\dagger}|0\rangle = e^{\alpha^*\alpha'} \langle 0|e^{\alpha'\hat{a}^\dagger}|0\rangle = e^{\alpha^*\alpha'},
\end{aligned} \tag{3.52}$$

where the final result follows from expanding the exponential operator in the bracket as a power series in $\hat{a}^\dagger$ and keeping only the first term (unity), since the higher terms annihilate $\langle 0|$.

It should now be clear from these simple examples that the Glauber form of the minimal wave packet, given in Eq. (3.48), greatly facilitates both calculations and analysis. We now return to time-dependent perturbation expansions.

3.5. Perturbation Treatment of the Time Evolution of a Wave Packet

As a simple example of the application of time-dependent perturbation expansions, we consider a harmonic oscillator whose potential energy function is displaced by the distance x_0. In this case,

$$\hat{H} = \frac{\hat{p}^2}{2\mu} + \frac{1}{2}k(x - x_0)^2 \tag{3.53a}$$

$$\hat{H}_0 = \frac{\hat{p}^2}{2\mu} + \frac{1}{2}kx^2. \tag{3.53b}$$

Therefore, according to Eqs. (2.14) and (2.9),

$$\hat{H}_0 = \hbar\omega\left(\hat{a}^\dagger\hat{a} + \frac{1}{2}\right) \tag{3.54a}$$

$$\begin{aligned}\hat{H} - \hat{H}_0 &= \frac{1}{2}k[(x - x_0)^2 - x^2] = \frac{1}{2}k(x_0^2 - 2x_0 x)\\ &= kx_0\left[\left(\frac{\hbar}{2\mu\omega}\right)^{\frac{1}{2}}(\hat{a}^\dagger + \hat{a}) - \frac{1}{2}x_0\right].\end{aligned} \tag{3.54b}$$

If we include the constant $1/2kx_0^2$ term with $\hat{H}_0$, the calculation is much simpler; thus we have instead of Eqs. (3.54),

$$\begin{aligned}\hat{H}_0 &= \hbar\omega\left(\hat{a}^\dagger\hat{a} + \frac{1}{2}\right) + \frac{1}{2}kx_0^2\\ &= \hbar\omega\left[\hat{a}^\dagger\hat{a} + \frac{1}{2}(1 + \xi_0^2)\right]\end{aligned} \tag{3.55a}$$

$$\begin{aligned}\hat{H} - \hat{H}_0 &= -kx_0\left(\frac{\hbar}{2\mu\omega}\right)^{\frac{1}{2}}(\hat{a}^\dagger + \hat{a})\\ &= -\hbar\omega\frac{\xi_0}{\sqrt{2}}(\hat{a}^\dagger + \hat{a}),\end{aligned} \tag{3.55b}$$

where we have used Eqs. (2.3) and (2.4) to obtain the second lines of Eqs. (3.55a) and (3.55b).

We may find the time evolution of the average value of x by applying Eqs. (3.2) and (3.13). On the other hand, this problem is identical to the one treated in Sections 2.12 and 2.13 for a displaced "tuned" wave packet with zero initial momentum. We repeat the Heisenberg-picture calculation here for this special case, so that we may compare our perturbation calculation with the exact result. The full Hamiltonian is given by the addition of Eqs. (3.55a) and (3.55b):

$$\hat{H} = \hbar\omega \left[\hat{a}^\dagger \hat{a} - \frac{\xi_0}{\sqrt{2}}(\hat{a}^\dagger + \hat{a}) + \frac{1}{2}(1 + \xi_0^2) \right]. \tag{3.56}$$

To calculate the time-dependent operator $\hat{x}_{\mathrm{H}}(t)$ in the Heisenberg picture, we use Eq. (2.9) to write

$$\hat{x}_{\mathrm{H}}(t) = \left(\frac{\hbar}{2\mu\omega} \right)^{\frac{1}{2}} [\hat{a}_{\mathrm{H}}^\dagger(t) + \hat{a}_{\mathrm{H}}(t)] \tag{3.57}$$

and can proceed as in Eqs. (2.138)–(2.141) to obtain $\hat{a}_{\mathrm{H}}(t)$ and $\hat{a}_{\mathrm{H}}^\dagger(t)$.

Alternatively, we can find by $\hat{a}_{\mathrm{H}}(t)$ and $\hat{a}_{\mathrm{H}}^\dagger(t)$ by integration of a differential equation:

$$\begin{aligned} \frac{d}{dt}\hat{a}_{\mathrm{H}}(t) &= \frac{d}{dt} e^{\frac{i}{\hbar}\hat{H}t} \hat{a} e^{-\frac{i}{\hbar}\hat{H}t} \\ &= \frac{i}{\hbar} e^{\frac{i}{\hbar}\hat{H}t} [\hat{H}\hat{a} - \hat{a}\hat{H}] e^{-\frac{i}{\hbar}\hat{H}t}. \end{aligned} \tag{3.58}$$

From Eqs. (3.56) and (2.10), we have

$$\hat{H}\hat{a} - \hat{a}\hat{H} = -[\hat{a}, \hat{H}]_- = -\hbar\omega \left(\hat{a} - \frac{\xi_0}{\sqrt{2}} \right). \tag{3.59}$$

Then substitution of Eq. (3.59) into Eq. (3.58) gives

$$\frac{d}{dt}\hat{a}_{\mathrm{H}}(t) = -i\omega \left[\hat{a}_{\mathrm{H}}(t) - \frac{\xi_0}{\sqrt{2}} \right]. \tag{3.60}$$

We solve Eq. (3.60) subject to the boundary condition that $\hat{a}_{\mathrm{H}}(0) = \hat{a}$ to obtain

$$\hat{a}_{\mathrm{H}}(t) = e^{-i\omega t}\hat{a} + \frac{\xi_0}{\sqrt{2}}(1 - e^{-i\omega t}). \tag{3.61a}$$

Taking the Hermitian conjugate of Eq. (3.61a) gives

$$\hat{a}_{\mathrm{H}}^{\dagger}(t) = e^{i\omega t}\hat{a}^{\dagger} + \frac{\xi_0}{\sqrt{2}}(1 - e^{i\omega t}). \tag{3.61b}$$

Substituting Eqs. (3.61) into Eq. (3.57), we have

$$\hat{x}_{\mathrm{H}}(t) = \left(\frac{\hbar}{2\mu\omega}\right)^{\frac{1}{2}} \Big\{ e^{i\omega t}\hat{a}^{\dagger} + e^{-i\omega t}\hat{a} + \sqrt{2}\xi_0 \left[1 - \frac{1}{2}(e^{i\omega t} + e^{-i\omega t})\right]\Big\}. \tag{3.62a}$$

With use of the identity $e^{i\omega t} = \cos\omega t + i\sin\omega t$, Eq. (3.62a) can be written

$$\begin{aligned}\hat{x}_{\mathrm{H}}(t) &= \left(\frac{\hbar}{2\mu\omega}\right)^{\frac{1}{2}} \{(\hat{a}^{\dagger} + \hat{a})\cos(\omega t) \\ &\quad + i(\hat{a}^{\dagger} - \hat{a})\sin(\omega t) + \sqrt{2}\xi_0(1 - \cos\omega t)\} \\ &= \hat{x}\cos\omega t - \frac{\hat{p}}{\mu\omega}\sin(\omega t) + x_0(1 - \cos\omega t),\end{aligned} \tag{3.62b}$$

where the second line of Eq. (3.62b) is obtained with the help of Eqs. (2.8) and (2.9). If the initial wave function is the ground state of the *un-displaced* potential energy, $1/2kx^2$, we have

$$\begin{aligned}\langle x(t)\rangle &= \langle 0|\hat{x}_{\mathrm{H}}(t)|0\rangle \\ &= \langle 0|\hat{x}|0\rangle\cos\omega t - \frac{1}{2\mu\omega}\langle 0|\hat{p}|0\rangle\sin\omega t \\ &\quad + \langle 0|0\rangle x_0(1 - \cos\omega t).\end{aligned} \tag{3.63a}$$

But since the operator averages in the second line of Eq. (3.63a) are zero in the symmetric potential and with zero initial momentum, only the last term survives, giving the expected result

$$\langle x(t)\rangle = x_0(1 - \cos\omega t). \tag{3.63b}$$

Now we return to the perturbation calculation. According to Eq. (3.3), we need to find

$$\langle x(t)\rangle = \langle 0|\hat{U}^{\dagger}(t)\hat{x}_{\mathrm{I}}(t)\hat{U}(t)|0\rangle, \tag{3.64}$$

with $\hat{U}(t)$ given by Eq. (3.13) and with $\hat{H}(t)$ and $\hat{x}(t)$ calculated as follows:

$$\begin{aligned}\frac{d}{dt}\hat{a}_{\mathrm{I}}(t) &= \frac{d}{dt}e^{\frac{i}{\hbar}\hat{H}_0 t}\hat{a}e^{-\frac{i}{\hbar}\hat{H}_0 t} \\ &= -\frac{i}{\hbar}e^{\frac{i}{\hbar}\hat{H}_0 t}[\hat{a},\hat{H}_0]_- e^{-\frac{i}{\hbar}\hat{H}_0 t} \\ &= -i\omega\hat{a}_{\mathrm{I}}(t), \end{aligned} \tag{3.65}$$

where we have used Eqs. (3.54a) and (2.140a) to obtain the last line. Eq. (3.65) integrates to give

$$\hat{a}_{\mathrm{I}}(t) = e^{-i\omega t}\hat{a}; \tag{3.66a}$$

the Hermitian conjugate is

$$\hat{a}_{\mathrm{I}}^\dagger(t) = e^{i\omega t}\hat{a}^\dagger. \tag{3.66b}$$

We now substitute Eqs. (3.66) into Eq. (3.55b) to obtain

$$\lambda\hat{H}_{\mathrm{I}}(t) = \hbar\omega\frac{\xi_0}{\sqrt{2}}(\hat{a}^\dagger e^{i\omega t} + \hat{a}e^{-i\omega t}) \tag{3.67a}$$

$$\hat{H}_{\mathrm{I}}(t) = -\frac{\hbar\omega}{\sqrt{2}}(\hat{a}^\dagger e^{i\omega t} + \hat{a}e^{-i\omega t}); \quad \lambda = \xi_0. \tag{3.67b}$$

The identification of the dimension-less parameter λ with ξ_0 in Eq. (3.67b) is a natural choice, since it is a measure of the displacement of the wave packet; *i.e.*, the perturbation. In many of the equations to follow, we will omit the factor λ. This should cause no confusion, since unless λ appears, the factor ξ_0 is included in $\hat{H}_{\mathrm{I}}$. From Eqs. (2.9), (3.4) and (3.66), we have

$$\begin{aligned}\hat{x}_{\mathrm{I}}(t) &= \left(\frac{\hbar}{2\mu\omega}\right)^{\frac{1}{2}}[\hat{a}_{\mathrm{I}}^\dagger(t) + \hat{a}_{\mathrm{I}}(t)] \\ &= \left(\frac{\hbar}{2\mu\omega}\right)^{\frac{1}{2}}(\hat{a}^\dagger e^{i\omega t} + \hat{a}e^{-i\omega t}). \end{aligned} \tag{3.68}$$

We now use Eq. (3.13) to expand Eq. (3.64) in powers of ξ_0; note that each of the operators $\hat{U}^\dagger$ and $\hat{U}$ are to be expanded, so that to each order in ξ_0 except zero, we will have operators $\hat{H}(t_j)$ on each side of $x_{\rm I}(t)$. The zero-order result is zero, since $\langle 0|\hat{x}|0\rangle = 0$. The contributions to the term linear in ξ_0 are

$$\begin{aligned}
&\frac{i}{\hbar}\int_0^t dt_1 \langle 0|\hat{H}_{\rm I}(t_1)\hat{x}_{\rm I}(t)|0\rangle \\
&\quad = \frac{i}{\hbar}\left(\frac{\hbar}{2\mu\omega}\right)^{\frac{1}{2}}(-\hbar\omega)\frac{\xi_0}{\sqrt{2}}\int_0^t dt_1 \langle 0|(\hat{a}^\dagger e^{i\omega t_1} + \hat{a}e^{-i\omega t_1}) \\
&\quad \times (\hat{a}^\dagger e^{i\omega t} + \hat{a}e^{-i\omega t})|0\rangle \\
&\quad = -\frac{1}{2}i\omega x_0 \int_0^t dt_1 \langle 0|[(\hat{a}^\dagger)^2 e^{i\omega(t_1+t)} + \hat{a}\hat{a}^\dagger e^{-i\omega(t_1-t)} \\
&\quad + \hat{a}^\dagger \hat{a} e^{i\omega(t_1-t)} + \hat{a}^2 e^{-i\omega(t_1+t)}]|0\rangle
\end{aligned} \tag{3.69a}$$

and

$$\begin{aligned}
&-\frac{i}{\hbar}\int_0^t dt_1 \langle 0|\hat{x}_{\rm I}(t)\hat{H}_{\rm I}(t_1)|0\rangle \\
&\quad = -\frac{i}{\hbar}\left(\frac{\hbar}{2\mu\omega}\right)^{\frac{1}{2}}(-\hbar\omega)\frac{\xi_0}{\sqrt{2}}\int_0^t dt_1 \langle 0|(\hat{a}^\dagger e^{i\omega t} + \hat{a}e^{-i\omega t}) \\
&\quad \times (\hat{a}^\dagger e^{i\omega t_1} + \hat{a}e^{-i\omega t_1})|0\rangle \\
&\quad = \frac{1}{2}i\omega x_0 \int_0^t dt_1 \langle 0|[(\hat{a}^\dagger)^2 e^{i\omega(t_1+t)} + \hat{a}\hat{a}^\dagger e^{i\omega(t_1-t)} \\
&\quad + \hat{a}^\dagger \hat{a} e^{-i\omega(t_1-t)} + \hat{a}^2 e^{-i\omega(t_1+t)}]|0\rangle.
\end{aligned} \tag{3.69b}$$

The only non-zero term in Eqs. (3.69) is $\langle 0|\hat{a}\hat{a}^\dagger|0\rangle$, while the ground-state *ket* or *bra* is annihilated by the others.

Since $\hat{a}|0\rangle = 0$ and $\langle 0|\hat{a}^\dagger = (\hat{a}|0\rangle)^\dagger = 0$, we may use the commutation relation to write

$$-\frac{1}{2}i\omega x_0 \int_0^t dt_1 e^{-i\omega(t_1-t)} = \frac{1}{2}x_0(1 - e^{i\omega t}). \tag{3.70a}$$

It follows also that

$$-\frac{i}{\hbar}\int_0^t dt_1 \langle 0|\hat{x}_\mathrm{I}(t)\hat{H}_\mathrm{I}(t_1)|0\rangle$$

$$= -\frac{1}{2}i\omega x_0 \int_0^t dt_1 e^{-i\omega(t_1-t)} = \frac{1}{2}x_0(1-e^{-i\omega t}). \quad (3.70\text{b})$$

Adding the two contributions gives the first-order result

$$\langle x(t)\rangle^{(1)} = \frac{1}{2}x_0(2-e^{i\omega t}-e^{-i\omega t}) = x_0(1-\cos\omega t). \quad (3.71)$$

Although Eq. (3.71) is the *first-order* result, we know from Eq. (3.63b) that it is also the *exact* result! If we calculate the higher orders, we thus expect them all to vanish. Before going to the higher orders, we may greatly simplify the writing, the bookkeeping, and the analysis by a simple graphical representation of the integrals. We begin by noting that the selection of the one non-zero term in Eqs. (3.70) can be accomplished very simply by use of Wick's theorem; that is,

$$\frac{i}{\hbar}\int_0^t dt_1 \langle 0|\hat{H}_\mathrm{I}(t_1)\hat{x}_\mathrm{I}(t)|0\rangle$$

$$= \frac{i}{\hbar}\left(\frac{\hbar}{2\mu\omega}\right)^{\frac{1}{2}}(-\hbar\omega)\frac{\xi_0}{\sqrt{2}}$$

$$\times \int_0^t dt_1 \langle 0|(\hat{a}^\dagger e^{i\omega t_1} + \overbracket{\hat{a}e^{-i\omega t_1})(\hat{a}^\dagger}e^{i\omega t} + \hat{a}e^{-i\omega t})|0\rangle$$

$$= -\frac{1}{2}i\omega x_0 \int_0^t dt_1 e^{-i\omega(t_1-t)}\langle 0|0\rangle = \frac{1}{2}x_0(1-e^{i\omega t}) \quad (3.72\text{a})$$

and

$$-\frac{i}{\hbar}\int_0^t dt_1 \langle 0|\hat{x}_\mathrm{I}(t)\hat{H}_\mathrm{I}(t_1)|0\rangle$$

$$= -\frac{i}{\hbar}\left(\frac{\hbar}{2\mu\omega}\right)^{\frac{1}{2}}(-\hbar\omega)\frac{\xi_0}{\sqrt{2}}$$

$$\times \int_0^t dt_1 \langle 0|(\hat{a}^\dagger e^{i\omega t} + \overline{\hat{a} e^{-i\omega t})(\hat{a}^\dagger e^{i\omega t_1}} + \hat{a} e^{-i\omega t_1})|0\rangle$$

$$= \frac{1}{2} i\omega x_0 \int_0^t dt_1 e^{-i\omega(t_1 - t)} \langle 0|0\rangle = \frac{1}{2} x_0 (1 - e^{-i\omega t}). \quad (3.72b)$$

Notice that there can be only *one* contraction between the operators "into" or "out of" a factor consisting of a sum of a single $\hat{a}^\dagger$ and a single $\hat{a}$, since a contraction must have $\hat{a}$ on the left. We now proceed to draw Eqs. (3.72) as graphs. We denote the operator $\hat{x}_I(t)$ by an open circle and the operator $\hat{H}_I(t)$ by a closed circle; contractions are represented by a line between them. Thus, we have for the first-order contribution

$$\langle x(t)\rangle^{(1)} = \underset{t \quad t_1}{\circ\!\!-\!\!\bullet} + \underset{t_1' \quad t}{\bullet\!\!-\!\!\circ} \quad (3.73)$$

Each $\circ$ provides a factor $\left(\frac{\hbar}{2\mu\omega}\right)^{\frac{1}{2}}$; each $\bullet$ provides a factor $i\omega\frac{\xi_0}{\sqrt{2}}$ and each $\bullet$ on the left of the $\circ$ provides an additional factor of -1. The time-integration variables for the $\bullet$'s on the right of the $\circ$ are t_j's, while those for the $\bullet$'s to the left of the $\circ$ are $t_{j'}$'s; the time variables are numbered from the *outside* towards the $\circ$, which is assigned the time t. In Eq. (3.73) we have placed the time variables under the dots for illustrative purposes.

Each contraction contributes a factor $e^{-i\omega t_j}$ from the contraction origin and a factor $e^{i\omega t_k}$ from the contraction terminus. In the example of Eq. (3.73), the contraction for the first term contributes $e^{i\omega(t_1 - t)}$ and that for the second term is $e^{i\omega(t - t_1')}$. The nested time integrations are done as in Eq. (3.13), that is,

$$\int_0^t dt_n \int_0^{t_n} dt_{n-1} \cdots \int_0^{t_2} dt_1, \quad \int_0^t dt_n' \int_0^{t_n'} dt_{n-1}' \cdots \int_0^{t_2'} dt_1'.$$

Following these rules, we interpret the first term of Eq. (3.73) as

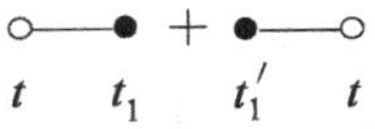

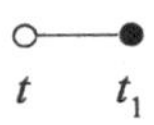

$$\underset{t \quad t_1}{\circ\!\!-\!\!\bullet} = \left(\frac{\hbar}{2\mu\omega}\right)^{\frac{1}{2}} \left(i\omega\frac{\xi_0}{\sqrt{2}}\right) \int_0^t dt_1 e^{i\omega(t_1 - t)}$$

$$= i\omega\frac{x_0}{2} \left(\frac{1}{i\omega}\right)(1 - e^{-i\omega t}) = \frac{1}{2} x_0 (1 - e^{-i\omega t}) \quad (3.74a)$$

and for the second term

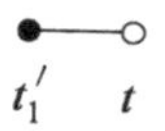

$$= \left(\frac{\hbar}{2\mu\omega}\right)^{\frac{1}{2}} \left(-i\omega\frac{\xi_0}{\sqrt{2}}\right) \int_0^t dt_1' e^{-\omega(t_1'-t)}$$

$$= i\omega\frac{x_0}{2}\frac{1}{-i\omega}[1 - e^{i\omega t}] = \frac{1}{2}x_0(1 - e^{i\omega t}). \qquad (3.74b)$$

It is useful to note that the complex conjugate of a graph is its mirror image. Now we look at the second-order term in the expansion of $\langle x(t)\rangle$ in powers of ξ_0:

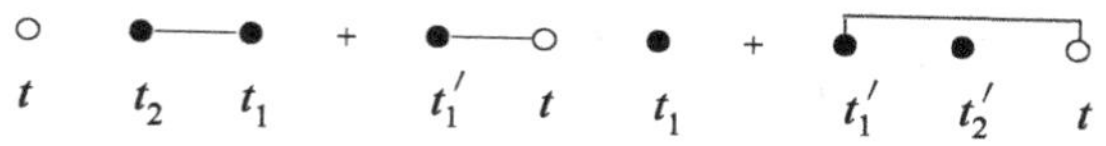

There is a problem with these second-order diagrams, however: they cannot be *fully contracted*, which means that an annihilation or a creation operator will remain which will annihilate the *ket* or the *bra*, respectively, yielding a zero value for each second-order term.

There are twelve third-order terms:

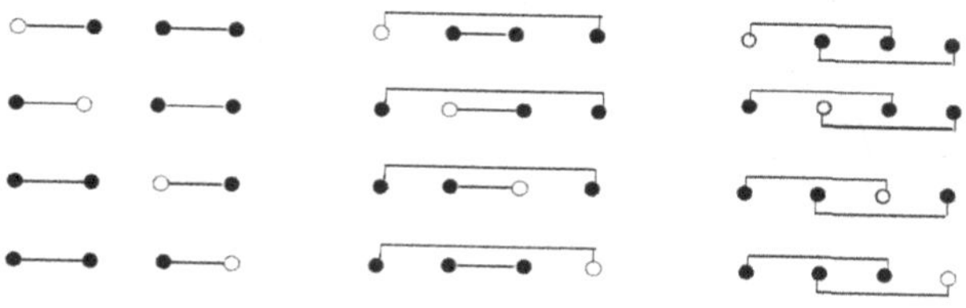

These graphs shows that there are only four independent ones:

This is because the last two rows are mirror images of the first two rows. Thus the operator products are Hermitian conjugates and therefore the graphs are complex conjugates. In addition, the rules show that the contractions in the last two graphs in each row yield the same integrands. All twelve graphs must be included in the calculation, but they may be derived from the four independent ones, as follows:

$$\langle x(t) \rangle = (\circ\!\!-\!\!\bullet \quad \bullet\!\!-\!\!\bullet) + (\circ\!\!-\!\!\bullet \quad \bullet\!\!-\!\!\bullet)^*$$

$$+ 2 \times [(\!\!\!\!\!\!\begin{smallmatrix}\circ\end{smallmatrix}\!\!\!\!\!\bullet\!\!-\!\!\bullet \quad \bullet) + (\!\!\!\!\!\!\begin{smallmatrix}\circ\end{smallmatrix}\!\!\!\!\!\bullet\!\!-\!\!\bullet \quad \bullet)^*]$$

$$+ (\bullet\!\!-\!\!\circ \quad \bullet\!\!-\!\!\bullet) + (\bullet\!\!-\!\!\circ \quad \bullet\!\!-\!\!\bullet)^*$$

$$+ 2 \times [(\!\!\!\bullet \quad \circ\!\!-\!\!\bullet \quad \bullet) + (\!\!\!\bullet \quad \circ\!\!-\!\!\bullet \quad \bullet)^*]$$

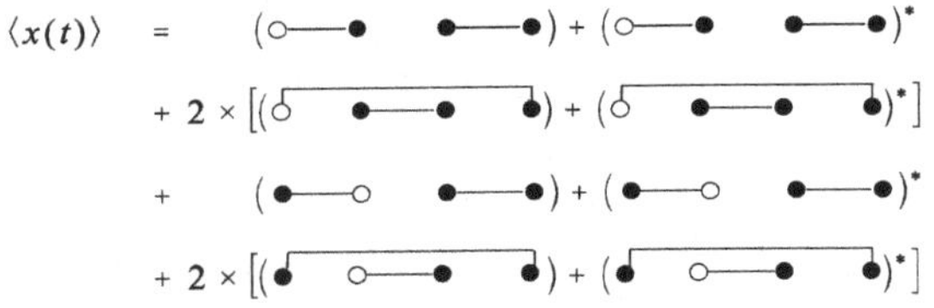

We now use the rules to evaluate the independent third-order graphs.

$$\underset{t}{\circ}\!\!-\!\!\underset{t_1}{\bullet} \quad \underset{t_2}{\bullet}\!\!-\!\!\underset{t_3}{\bullet}$$

$$= \left(\frac{\hbar}{2\mu\omega}\right)^{\frac{1}{2}} \left(i\omega\frac{\xi_0}{\sqrt{2}}\right)^3 e^{i\omega t} \int_0^t dt_3 e^{i\omega t_3} \int_0^{t_3} dt_2 e^{i\omega t_2} \int_0^{t_2} dt_1 e^{i\omega t_1}$$

$$= (i\omega)^3 x_0 \left(\frac{\xi_0}{2}\right)^2 e^{-i\omega t} \int_0^t dt_3 e^{i\omega t_3} \int_0^{t_3} dt_2 e^{-\omega t_2} \frac{1}{i\omega}(e^{i\omega t_2} - 1)$$

$$= (i\omega)^2 x_0 \left(\frac{\xi_0}{2}\right)^2 e^{i\omega t} \int_0^t dt_3 e^{i\omega t_3} \int_0^{t_3} dt_2 (1 - e^{-i\omega t_2})$$

$$= (i\omega)^2 x_0 \left(\frac{\xi_0}{2}\right)^2 e^{-i\omega t} \int_0^t dt_3 e^{i\omega t_3} \left[t_3 - \left(-\frac{1}{i\omega}\right)(e^{-i\omega t_3} - 1)\right]$$

$$= (i\omega)^2 x_0 \left(\frac{\xi_0}{2}\right)^2 e^{-i\omega t} \int_0^t dt_3 \left[e^{i\omega t_3} + \left(\frac{1}{i\omega}\right)(1 - e^{i\omega t_3})\right]. \tag{3.75}$$

The sum of this graph and its complex conjugate is

$$\circ\!\!-\!\!\bullet \quad \bullet\!\!-\!\!\bullet \quad + \quad \bullet\!\!-\!\!\bullet \quad \bullet\!\!-\!\!\circ$$

$$= \circ\!\!-\!\!\bullet \quad \bullet\!\!-\!\!\bullet \quad + \quad (\circ\!\!-\!\!\bullet \quad \bullet\!\!-\!\!\bullet)^*$$

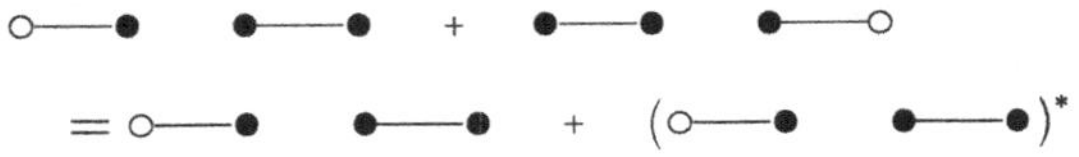

$$= x_0 \xi_0^2 \left(\frac{1}{2}\omega t \sin\omega t + \cos\omega t - 1\right). \tag{3.76}$$

A calculation similar to that of Eq. (3.76) gives

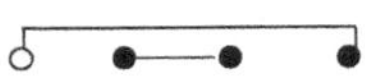

$$\!\!\!\begin{smallmatrix}\circ\end{smallmatrix}\!\!\!\bullet\!\!-\!\!\bullet \quad \bullet = ix_0 \left(\frac{\xi_0}{2}\right)^2 (\sin\omega t - \omega t). \tag{3.77}$$

Since the graph in Eq. (3.77) is pure imaginary, we have

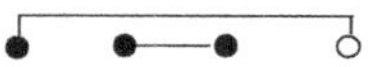
$$= -ix_0 \left(\frac{\xi_0}{2}\right)^2 (\sin \omega t - \omega t), \qquad (3.78)$$

so that the sum of the graphs in Eqs. (3.77) and (3.78) is zero. Similarly,

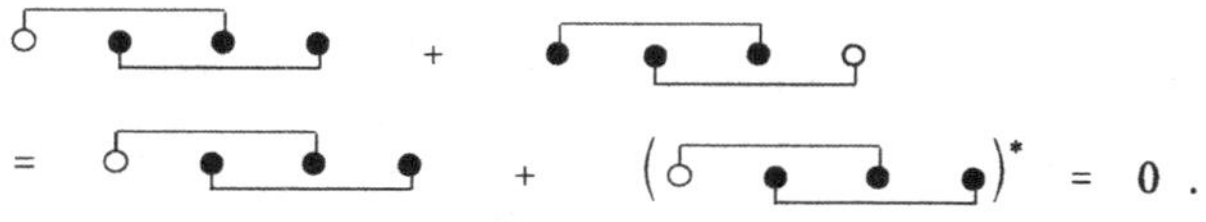
$$= 0\,. \qquad (3.79)$$

Since the complex conjugate of a graph is its mirror image, we have

$$= -x_0 \left(\frac{\xi_0}{2}\right)^2 [(1 - e^{i\omega t})(1 + i\omega t e^{i\omega t})]. \qquad (3.80a)$$

$$- x_0 \left(\frac{\xi_0}{2}\right)^2 [2e^{-i\omega t} - 1 - e^{-2i\omega t} + i\omega t(e^{-2i\omega t} e^{-i\omega t})].$$
(3.80b)

$$= x_0 \left(\frac{x_0}{2}\right)^2 (1 - e^{-\omega t})(1 - i\omega t - e^{iital\omega t})$$

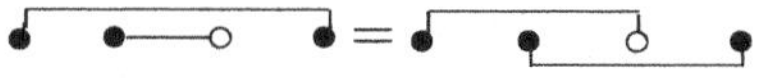
$$= x_0 \left(\frac{x_0}{2}\right)^2 [2e^{i\omega t} - 1 - e^{2i\omega t} + i\omega t(e^{2i\omega t} - e^{i\omega t})].$$
(3.81a,b)

The sum of all twelve third-order graphs is thus equal to zero, as we expected:

$\langle x(t) \rangle^{(3)} =$
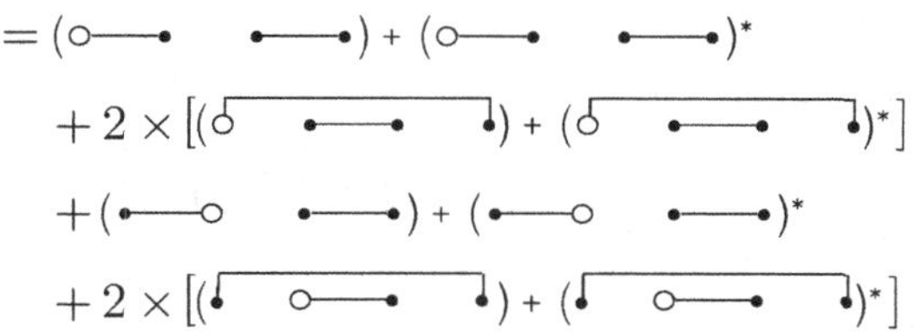

$$= x_0\xi_0^2\left[\frac{1}{2}\omega t\sin(\omega t)+\cos(\omega t)-1\right]+2\times(0)$$

$$+x_0\xi_0^2\left\{\cos(\omega t)-\frac{1}{2}[1+\cos(2\omega t)+\omega t\sin(\omega t)]\right\}$$

$$+2\times x_0\xi_0^2\left[\frac{1}{4}\cos(2\omega t)-\cos(\omega t)\frac{3}{4}\right]=0. \tag{3.82}$$

The calculations for the sum of all graphs to third order in the perturbation are seen to be quite tedious. Suppose we did *not* have the exact solution in hand and therefore were not aware that the first-order calculation is exact for this problem. It would appear that we must proceed as we have been doing — that is, showing that the calculation for each higher order is zero. If the reader attempts to do this, it will become evident that the opera is a very long one, even at fifth order, until at last the diva sings her long whole note, namely zero! As it turns out, there is an alternative that is aided by the graphical representations of the time integrals. We can in fact take the calculation to *infinite* order and show that the result is to multiply the first-order result by unity. The exercise will teach us an essential fact about perturbation expansions.

3.5.1. *Summation to Infinite Order; Connected Graphs*

As we saw in Section 3.4, the third-order graphs for the problem of a displaced harmonic oscillator sum to zero. Indeed, *all* graphs of higher order other than the first are expected to sum to zero, since the first-order solution is identical to the exact solution. In this section, we use graphical analysis to understand the basis of throwing out all of the higher-order graphs in this problem. We start by considering the third-order graph

$$=\left(+\frac{i}{\hbar}\right)\lambda^2\int_0^t dt_2'\int_0^{t_2'}dt_1'\mathbb{C}\left\{\hat{H}_{\rm I}(t_1')\hat{H}_{\rm I}(t_2')\left(-\frac{i}{\hbar}\right)\lambda\mathbb{C}\right.$$

$$\left.\times\left\{\hat{x}(t)\int_0^t dt_1\hat{H}_{\rm I}(t_1)\right\}\right\}. \tag{3.83}$$

Here we use general notation for the perturbation expansions and indicate the sum of all full contractions of a string of operators $\{\hat{o}_1\hat{o}_2\cdots\}$ by $\mathbb{C}\{\hat{o}_1\hat{o}_2\cdots\}$. In the case of Eq. (3.83), the "sums of full contractions" consist of only one contraction each. As a second example, consider the case in which the two-point graph on the left of the $\circ$ is replaced with the sum of all fully contracted four-point graphs:

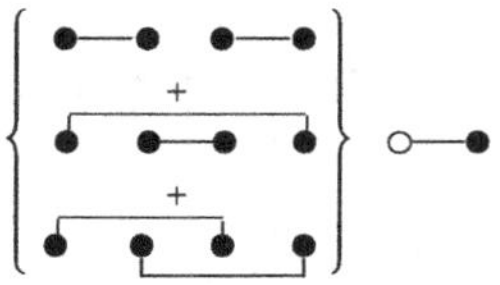

$$= \left(+\frac{i}{\hbar}\right)^4 \lambda^4 \int_0^t dt'_4 \int_0^{t'_4} dt'_3 \int_0^{t'_3} dt'_2 \int_0^{t_2} dt'_1$$
$$\times \mathbb{C}\{\hat{H}_{\mathrm{I}}(t'_1)\hat{H}_{\mathrm{I}}(t'_2)\hat{H}_{\mathrm{I}}(t'_3)\hat{H}_{\mathrm{I}}(t'_4)\}$$
$$\times \left(-\frac{i}{\hbar}\right) \lambda \mathbb{C}\left[\hat{x}(t)\int_0^t dt_1 \hat{H}_{\mathrm{I}}(t_1)\right]. \tag{3.84}$$

We remind the reader that in the first line of Eq. (3.84), $\mathbb{C}$ indicates the sum of the three fourth-order graphs on the left of the $\circ$, since the symbol means the sum of all full contractions of the operators.

Now replace the graphs to the left of the $\circ$ with the sum of all fully contracted graphs to all orders — including the zero-order graph, unity — which we represent with the symbol $\blacksquare^\dagger$, where

$$\blacksquare^\dagger = 1 + \sum_{n=2}^{\infty}\left(+\frac{i}{\hbar}\right)^n \lambda^n \int_0^{t_1} dt'_n \int_0^{t_n} dt'_{n-1}\cdots$$
$$\times \int_0^{t_3} dt'_2 \int_0^{t_2} dt'_1 \mathbb{C}\{\hat{H}_{\mathrm{I}}(t'_1)\hat{H}_{\mathrm{I}}(t'_2)\cdots\hat{H}_{\mathrm{I}}(t'_{n-1})\hat{H}_{\mathrm{I}}(t'_n)\}. \tag{3.85a}$$

Note that only even orders will appear on the r.h.s. of Eq. (3.85a) because all terms are fully contracted. Comparison of Eq. (3.85a) with Eq. (3.13) show that the sum of these fully contracted graphs is just the fully contracted operator $\hat{U}^\dagger(t)$; that is,

$$\blacksquare^\dagger = \mathbb{C}\{\hat{U}^\dagger(t)\} = \langle 0|\hat{U}^\dagger(t)|0\rangle. \tag{3.85b}$$

From Eqs. (3.85) we have for the generalization of Eq. (3.84)

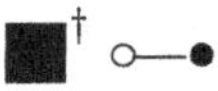

$$\blacksquare^{\dagger} \circ\!\!-\!\!\bullet = \mathbb{C}\{\hat{U}^{\dagger}(t)\} \left(-\frac{i}{\hbar}\right) \lambda \mathbb{C} \left[\hat{x}(t) \int_0^t dt_1 \hat{H}_{\mathrm{I}}(t_1)\right]. \qquad (3.86)$$

Now look at the third-order graph:

$$\circ\!\!-\!\!\bullet \quad \bullet\!\!-\!\!\bullet$$

$$= \left(-\frac{i}{\hbar}\right) \lambda \mathbb{C} \left\{\hat{x}(t) \int_0^t dt_3 \hat{H}_{\mathrm{I}}(t_3)\right\} \left(-\frac{i}{\hbar}\right)^2 \lambda^2$$

$$\times \int_0^{t_3} dt_2 \int_0^{t_2} dt_1 \mathbb{C}\{\hat{H}_{\mathrm{I}}(t_2)\hat{H}_{\mathrm{I}}(t_1)\}. \qquad (3.87)$$

By the same procedure with which we obtained Eq. (3.86), if we replace the disconnected pair of contracted dots on the right of the graph with the sum over all fully contracted graphs to all (even) orders, we obtain

$$\circ\!\!-\!\!\bullet \quad \blacksquare^{\dagger} = \left(-\frac{i}{\hbar}\right) \lambda \mathbb{C} \left\{\hat{x}(t) \int_0^t dt_1 \hat{H}_{\mathrm{I}}(t_1)\right\}$$

$$\times \left[1 + \sum_{n=2}^{\infty} \left(-\frac{i}{\hbar}\right)^n \lambda^n \int_0^{t_1} dt_n \int_0^{t_1} dt_{n-1} \cdots \int_0^{t_3} dt_2\right.$$

$$\left.\times \int_0^{t_2} dt_1'' \mathbb{C}\{\hat{H}_{\mathrm{I}}(t_n)\hat{H}_{\mathrm{I}}(t_{n-1}) \cdots \hat{H}_{\mathrm{I}}(t_2)\hat{H}_{\mathrm{I}}(t_1'')\}\right]$$

$$= \left(-\frac{i}{\hbar}\right) \lambda \mathbb{C} \left\{\hat{x}(t) \int_0^t dt_1 \hat{H}_{\mathrm{I}}(t_1)\right\} \mathbb{C}\{\hat{U}(t_1)\}, \qquad (3.88)$$

where we have renamed the time integration variable that runs from 0 to t so that it remains the same for all orders; this requires that we also rename the variable that runs from 0 to t_2. Although Eq. (3.88) looks very much like Eq. (3.86), there is a significant difference, namely that $\mathbb{C}\{\hat{U}(t_1)\}$ is linked to the factor on its left by integration in Eq. (3.88), while in Eq. (3.86) the two factors are separable. We emphasize this distinction by placing a prime on the box representing the sum to infinite order. To see how we may disentangle the factors in Eq. (3.88), consider the two third-order graphs:

$$= \left(-\frac{i}{\hbar}\right)^3 \lambda^3 \int_0^t dt_3 \int_0^{t_3} dt_2 \int_0^{t_2} dt_1 \times \mathbb{C}\{\hat{x}(t)\hat{H}_{\rm I}(t_1)\}\mathbb{C}\{\hat{H}_{\rm I}(t_3)\hat{H}_{\rm I}(t_2)\}$$

$$= \left(-\frac{i}{\hbar}\right)^3 \lambda^3 \int_0^t dt_3 \int_0^{t_3} dt_2 \int_0^{t_2} dt_1 \times \mathbb{C}\{\hat{x}(t)\hat{H}_{\rm I}(t_2)\}\mathbb{C}\{\hat{H}_{\rm I}(t_3)\hat{H}_{\rm I}(t_1)\}. \quad (3.89)$$

Note that we are treating these two graphs separately, even though they are equal in value for the displaced oscillator. Also note that we have grouped the fully contracted operator strings together; we are allowed to do this, since after full contractions the result is a number, so that the *ordering* of the factors is irrelevant. These two graphs are similar except for the juxtaposition of t_1 and t_2. These variables may not be simply interchanged, since their limits are different. However, the fact that $t_1 \leq t_2$ may be expressed either by

$$\int_0^{t_3} dt_2 \int_0^{t_2} dt_1 \quad \text{or by} \quad \int_0^{t_3} dt_1 \int_{t_1}^{t_3} dt_2.$$

We may therefore write the second graph in Eqs. (3.89) as

$$= \left(-\frac{i}{\hbar}\right)^3 \lambda^3 \int_0^t dt_3 \int_0^{t_3} dt_1 \int_{t_1}^{t_3} dt_2 \mathbb{C}\{\hat{x}(t)\hat{H}_{\rm I}(t_2)\}\mathbb{C}\{\hat{H}_{\rm I}(t_3)\hat{H}_{\rm I}(t_1)\}.$$

Now interchange t_1 and t_2 in the first graph in Eqs.(3.89):

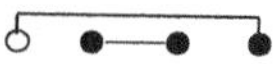

$$= \left(-\frac{i}{\hbar}\right)^3 \lambda^3 \int_0^t dt_3 \int_0^{t_3} dt_1 \int_0^{t_1} dt_2 \mathbb{C}\{\hat{x}(t)\hat{H}_{\rm I}(t_2)\}\mathbb{C}\{\hat{H}_{\rm I}(t_3)\hat{H}_{\rm I}(t_1)\}.$$

Adding these two graphs gives

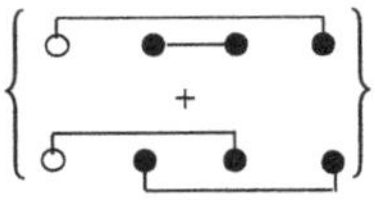

$$= \left(-\frac{i}{\hbar}\right)^3 \lambda^3 \int_0^t dt_3 \int_0^{t_3} dt_1 \int_0^{t_3} dt_2 \mathbb{C}\{\hat{x}(t)\hat{H}_{\mathrm{I}}(t_2)\}\mathbb{C}\{\hat{H}_{\mathrm{I}}(t_3)\hat{H}_{\mathrm{I}}(t_1)\}. \tag{3.90a}$$

Now replace by $\int_0^t dt_3 \int_0^{t_3} dt_2$ by $\int_0^t dt_2 \int_t^{t_2} dt_3$:

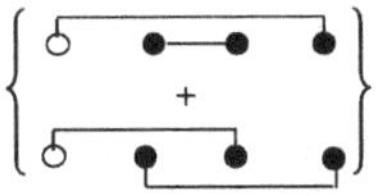

$$= \left(-\frac{i}{\hbar}\right)^3 \lambda^3 \int_0^t dt_3 \int_{t_3}^{t} dt_2 \int_0^{t_2} dt_1 \mathbb{C}\{\hat{x}(t)\hat{H}_{\mathrm{I}}(t_3)\}\mathbb{C}\{\hat{H}_{\mathrm{I}}(t_2)\hat{H}_{\mathrm{I}}(t_1)\}. \tag{3.90b}$$

Interchanging t_2 and t_3 gives

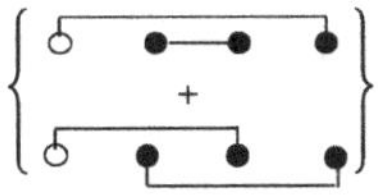

$$= \left(-\frac{i}{\hbar}\right)^3 \lambda^3 \int_0^t dt_3 \int_{t_3}^{t} dt_2 \int_0^{t_2} dt_1 \mathbb{C}\{\hat{x}(t)\hat{H}_{\mathrm{I}}(t_3)\}\mathbb{C}\{\hat{H}_{\mathrm{I}}(t_2)\hat{H}_{\mathrm{I}}(t_1)\}. \tag{3.90c}$$

Now add Eq. (3.87) to Eq. (3.90c) to obtain

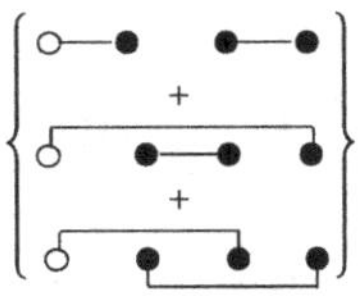

$$\begin{aligned} &= \left(-\frac{i}{\hbar}\right)^3 \lambda^3 \int_0^t dt_3 \int_0^t dt_2 \int_0^{t_2} dt_1 \mathbb{C}\{\hat{x}(t)\hat{H}_{\mathrm{I}}(t_3)\}\mathbb{C}\{\hat{H}_{\mathrm{I}}(t_2)\hat{H}_{\mathrm{I}}(t_1)\}. \\ &= -\frac{i}{\hbar}\lambda \int_0^t dt_3\{\hat{x}(t)\hat{H}_{\mathrm{I}}(t_3)\} \left(-\frac{i}{\hbar}\right)^2 \lambda^2 \int_0^t dt_2 \int_0^{t_2} dt_1 \\ &\quad \times \mathbb{C}\{\hat{H}_{\mathrm{I}}(t_2)\hat{H}_{\mathrm{I}}(t_1)\}. \end{aligned} \tag{3.91}$$

Eq. (3.91) shows that by including all the third-order graphs with the ∘ on the left we have disentangled the integration and rendered the result as the product of two independent factors! Now we may replace the second-order fully contracted factor on the right with the sum of all fully contracted graphs of all orders, including the zero-order term, unity, to obtain the separable form of Eq. (3.88):

$$= \left(-\frac{i}{\hbar}\right)\lambda\mathbb{C}\left\{\hat{x}(t)\int_0^t dt_1\hat{H}_{\mathrm{I}}(t_1)\right\}$$
$$\times\left[1+\sum_{n=2}^{\infty}\left(-\frac{i}{\hbar}\right)^n\lambda^n\int_0^t dt_n\int_0^{t_n}dt_{n-1}\cdots\int_0^{t_3}dt_2\right.$$
$$\left.\times\int_0^{t_2}dt_1''\mathbb{C}\{\hat{H}_{\mathrm{I}}(t_n)\hat{H}_{\mathrm{I}}(t_{n-1})\cdots\hat{H}_{\mathrm{I}}(t_2)\hat{H}_{\mathrm{I}}(t_1'')\}\right]$$
$$= \left(-\frac{1}{\hbar}\right)\lambda\mathbb{C}\left\{\hat{x}(t)\int_0^t dt_1\hat{H}_{\mathrm{I}}(t_1)\right\}\mathbb{C}\{\hat{U}(t)\}. \qquad (3.92)$$

It may seem to the reader that we have oversimplified things by assuming that the infinite-order expansion can proceed in this way. What about graphs of the type , for example? The answer is that we disentangle such a graph by adding in all graphs in which the ∘ is connected to different •'s, after making the appropriate changes in the time variables. In the most general case, we have

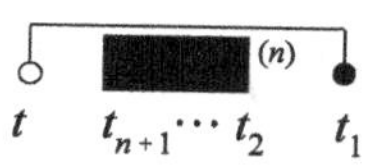

$$= -\frac{i}{\hbar}\int_0^t dt_{n+1}\int_0^{t_{n+1}}dt_n\cdots\int_0^{t_3}dt_2\int_0^{t_2}dt_1\mathbb{C}\{\hat{x}(t)\hat{H}_{\mathrm{I}}(t_1)\}$$
$$\times\left(-\frac{i}{\hbar}\right)^n\mathbb{C}\{\hat{H}_{\mathrm{I}}(t_{n+1})\cdots\hat{H}_{\mathrm{I}}(t_2)\}, \qquad (3.93a)$$

where the black box represents the sum of all possible full contractions in an nth order expansion of $\hat{U}$. Since the names of the time variables are of importance in this demonstration, they are explicitly identified under the diagram. Now we make the replacement

$$\int_0^{t_3} dt_2 \int_0^{t_2} dt_1 \to \int_0^{t_3} dt_1 \int_{t_1}^{t_3} dt_2$$

and then interchange t_1 and t_2 to obtain

$G_1 =$ 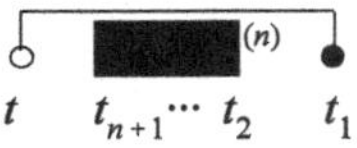

$$= -\frac{i}{\hbar}\int_0^{t} dt_{n+1}\int_0^{t_{n+1}} dt_n \cdots \int_0^{t_3} dt_2 \int_{t_2}^{t_3} dt_1 \mathbb{C}\{\hat{x}(t)\hat{H}_{\mathrm{I}}(t_2)\}$$

$$\times \left(-\frac{i}{\hbar}\right)^n \mathbb{C}\{\hat{H}_{\mathrm{I}}(t_{n+1})\cdots\hat{H}_{\mathrm{I}}(t_3)\hat{H}_{\mathrm{I}}(t_1)\}. \qquad (3.93\text{b})$$

Now add

$G_2 =$ 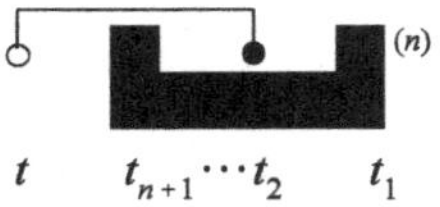

$$= -\frac{i}{\hbar}\int_0^{t} dt_{n+1}\int_0^{t_{n+1}} dt_n \cdots \int_0^{t_3} dt_2 \int_0^{t_2} dt_1 \mathbb{C}\{\hat{x}(t)\hat{H}_{\mathrm{I}}(t_2)\}$$

$$\times \left(-\frac{i}{\hbar}\right)^n \mathbb{C}\{\hat{H}_{\mathrm{I}}(t_{n+1})\cdots\hat{H}_{\mathrm{I}}(t_3)\hat{H}_{\mathrm{I}}(t_1)\} \qquad (3.94)$$

to obtain

$$G_1 + G_2 = -\frac{i}{\hbar}\int_0^{t} dt_{n+1}\int_0^{t_{n+1}} dt_n \cdots \int_0^{t_4} dt_3 \int_0^{t_3} dt_2$$

$$\times \int_0^{t_3} dt_1 \mathbb{C}\{\hat{x}(t)\hat{H}_{\mathrm{I}}(t_2)\}\left(-\frac{i}{\hbar}\right)^n$$

$$\times \mathbb{C}\{\hat{H}_{\mathrm{I}}(t_{n+1})\cdots\hat{H}_{\mathrm{I}}(t_4)\hat{H}_{\mathrm{I}}(t_3)\hat{H}_{\mathrm{I}}(t_1)\}. \qquad (3.95\text{a})$$

Now make the replacements

$$\int_0^{t_4} dt_3 \int_0^{t_3} dt_2 \int_0^{t_3} dt_1 \to \int_0^{t_4} dt_2 \int_{t_2}^{t_4} dt_3 \int_0^{t_3} dt_1$$

and then interchange t_2 and t_3 to obtain

$$G_1 + G_2 = -\frac{i}{\hbar}\int_0^t dt_{n+1}\int_0^{t_{n+1}} dt_n \cdots \int_0^{t_4} dt_3 \times \int_{t_3}^{t_4} dt_2 \int_0^{t_2} dt_1 \mathbb{C}\{\hat{x}(t)\hat{H}_{\rm I}(t_3)\}\left(-\frac{i}{\hbar}\right)^n \times \mathbb{C}\{\hat{H}_{\rm I}(t_{n+1})\cdots\hat{H}_{\rm I}(t_4)\hat{H}_{\rm I}(t_2)\hat{H}_{\rm I}(t_1)\}, \quad (3.95b)$$

to which we add

$G_3 =$ 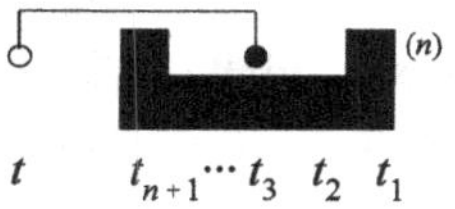

$$= -\frac{i}{\hbar}\int_0^t dt_{n+1}\int_0^{t_{n+1}} dt_n \cdots \int_0^{t_4} dt_3 \int_0^{t_3} dt_2 \times \int_0^{t_2} dt_1 \mathbb{C}\{\hat{x}(t)\hat{H}_{\rm I}(t_3)\} \times \left(-\frac{i}{\hbar}\right)^n \mathbb{C}\{\hat{H}_{\rm I}(t_{n+1})\cdots\hat{H}_{\rm I}(t_4)\hat{H}_{\rm I}(t_2)\hat{H}_{\rm I}(t_1)\} \quad (3.96)$$

to get

$$G_1 + G_2 + G_3 = -\frac{i}{\hbar}\int_0^t dt_{n+1}\int_0^{t_{n+1}} dt_n \cdots \times \int_0^{t_4} dt_3 \int_0^{t_3} dt_2 \int_0^{t_2} dt_1 \mathbb{C}\{\hat{x}(t)\hat{H}_{\rm I}(t_3)\} \times \left(-\frac{i}{\hbar}\right)^n \mathbb{C}\{\hat{H}_{\rm I}(t_{n+1})\cdots\hat{H}_{\rm I}(t_4)\hat{H}_{\rm I}(t_2)\hat{H}_{\rm I}(t_1)\}. \quad (3.97a)$$

Then we make the replacement

$$\int_0^{t_5} dt_3 \int_{t_3}^{t_5} dt_4 \int_0^{t_4} dt_2 \to \int_0^{t_5} dt_4 \int_0^{t_4} dt_3 \int_0^{t_4} dt_2$$

and afterwards interchange t_3 and t_4 to obtain

$$G_1 + G_2 + G_3 = -\frac{i}{\hbar}\int_0^t dt_{n+1}\int_0^{t_{n+1}} dt_n \cdots \\ \times \int_0^{t_5} dt_4 \int_{t_4}^{t_5} dt_3 \int_0^{t_3} dt_2 \int_0^{t_2} dt_1 \mathbb{C}\{\hat{x}(t)\hat{H}_{\mathrm{I}}(t_4)\} \\ \times \left(-\frac{i}{\hbar}\right)^n \mathbb{C}\{\hat{H}_{\mathrm{I}}(t_{n+1})\cdots\hat{H}_{\mathrm{I}}(t_5)\hat{H}_{\mathrm{I}}(t_3)\hat{H}_{\mathrm{I}}(t_2)\hat{H}_{\mathrm{I}}(t_1)\}. \tag{3.97b}$$

It should be clear that with each addition of a graph in which the operator $\hat{x}(t)$ is contracted successively with $\hat{H}_{\mathrm{I}}(t_4)$, $\hat{H}_{\mathrm{I}}(t_5)$, *etc.*, followed by the appropriate t-variable limit replacements and label interchanges, we move the repeated upper limit to higher and higher t-variable labels, until we finally reach

$$\sum_{j=1}^{n} G_j = -\frac{i}{\hbar}\int_0^t dt_{n+1}\int_{t_{n+1}}^{t} dt_n \cdots \int_0^{t_3} dt_2 \int_0^{t_2} dt_1 \mathbb{C}\{\hat{x}(t)\hat{H}_{\mathrm{I}}(t_{n+1})\} \\ \times \left(-\frac{i}{\hbar}\right)^n \mathbb{C}\{\hat{H}_{\mathrm{I}}(t_n)\cdots\hat{H}_{\mathrm{I}}(t_1)\}, \tag{3.98}$$

to which we add

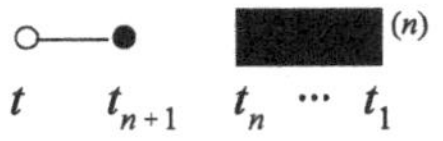

$$= -\frac{i}{\hbar}\int_0^t dt_{n+1}\int_0^{t_{n+1}} dt_n \cdots \int_0^{t_3} dt_2 \int_0^{t_2} dt_1 \mathbb{C}\{\hat{x}(t)\hat{H}_{\mathrm{I}}(t_{n+1})\} \\ \times \left(-\frac{i}{\hbar}\right)^n \mathbb{C}\{\hat{H}_{\mathrm{I}}(t_n)\cdots\hat{H}_{\mathrm{I}}(t_1)\}, \tag{3.99}$$

to get

$$\sum_{j=1}^{n} G_n +$$

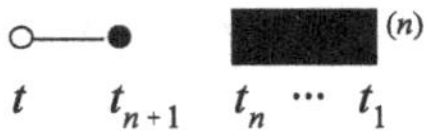

$$= -\frac{i}{\hbar}\int_0^t dt_{n+1}\mathbb{C}\{\hat{x}(t)\hat{H}_{\mathrm{I}}(t_{n+1})\} \\ \times \left(-\frac{i}{\hbar}\right)^n \int_0^t dt_n \cdots \int_0^{t_2} dt_1 \mathbb{C}\{\hat{H}_{\mathrm{I}}(t_n)\cdots\hat{H}_{\mathrm{I}}(t_1)\}. \tag{3.100}$$

Now the graphs o—• and ■■■■■$^{(n)}$ have been disentangled and both are functions of the single time variable t. If we sum Eq. (3.100) over all n, including the $n = 0$ term, (o—•), we obtain Eq. (3.92). This result is the same as replacing •—• in Eq. (3.87) by the sum of all nth order graphs. Thus, the procedure we used to obtain Eq. (3.92) is entirely justified.

Combining Eqs. (3.86) and (3.92), we can write

■† o—• ■

$$= \mathbb{C}\{\hat{U}^\dagger(t)\} \left(-\frac{i}{\hbar}\right) \lambda \mathbb{C}\left\{\hat{x}(t) \int_0^t dt_1 \hat{H}_\mathrm{I}(t_1)\right\} \mathbb{C}\{\hat{U}(t)\},$$

$$(3.101)$$

where all contractions are contained *within* each block; in other words, there are *no contractions of operators in the left-hand block with any in the right-hand block*. We have thus shown that the sum of all the disconnected graphs with •'s either to the right or to the left of the first-order graph is a *constant* times the first-order graph. This constant is still not particularly easy to evaluate, since it is given by

$$\blacksquare^\dagger\ \blacksquare = \mathbb{C}\{\hat{U}^\dagger(t)\}\mathbb{C}\{\hat{U}(t)\} = \langle 0|\hat{U}^\dagger|0\rangle\langle 0|\hat{U}|0\rangle. \qquad (3.102)$$

If we can remove the *ket-bra* $|0\rangle\langle 0|$, the constant would be *unity*, since by Eq. (3.5) we have

$$\langle 0|\hat{U}^\dagger\hat{U}|0\rangle = \langle 0|[e^{\frac{i}{\hbar}\hat{H}t}e^{\frac{i}{\hbar}\hat{H}_0 t}][e^{\frac{i}{\hbar}\hat{H}_0 t}e^{-\frac{i}{\hbar}\hat{H}_0 t}]|0\rangle$$

$$= 1. \qquad (3.103)$$

What is missing are those graphs in which parts of the $\hat{U}^\dagger$ and $\hat{U}$ operators are contracted *across* the first-order graph. The third-order graphs of this type are

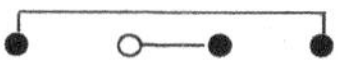

• o—• •

$$= 1 + \frac{i}{\hbar}\int_0^t dt_1' \left(-\frac{i}{\hbar}\right)\int_0^t dt_2 \left(-\frac{i}{\hbar}\right)\int_0^{t_2} dt_1$$

$$\times \mathbb{C}\{\hat{x}(t)\hat{H}_\mathrm{I}(t_2)\}\mathbb{C}\{\hat{H}_\mathrm{I}(t_1')\hat{H}_\mathrm{I}(t_1)\} \qquad (3.104\mathrm{a})$$

and

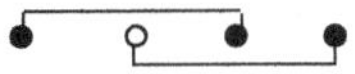

$$= +\frac{i}{\hbar}\int_0^t dt_1' \left(-\frac{i}{\hbar}\right)\int_0^t dt_2 \left(-\frac{i}{\hbar}\right)\int_0^{t_2} dt_1$$
$$\times\, \mathbb{C}\{\hat{x}(t)\hat{H}_{\mathrm{I}}(t_1)\}\mathbb{C}\{\hat{H}_{\mathrm{I}}(t_1')\hat{H}_{\mathrm{I}}(t_2)\}. \qquad (3.104b)$$

We can disentangle these integrations in a manner similar to the methods we have used for the other third-order graphs. First we replace

$$\int_0^t dt_2 \int_0^{t_2} dt_1 \quad \text{by} \quad \int_0^t dt_1 \int_{t_1}^t dt_2$$

in Eq. (3.104b) to obtain

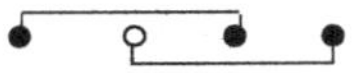

$$= +\frac{i}{\hbar}\int_0^t dt_1' \left(-\frac{i}{\hbar}\right)\int_0^t dt_1 \left(-\frac{i}{\hbar}\right)\int_{t_1}^t dt_2 \mathbb{C}\{\hat{x}(t)\hat{H}_{\mathrm{I}}(t_1)\}$$
$$\times\, \mathbb{C}\{\hat{H}_{\mathrm{I}}(t_1')\hat{H}_{\mathrm{I}}(t_2)\}. \qquad (3.105a)$$

Then interchange t_1 and t_2 in Eq. (3.105a):

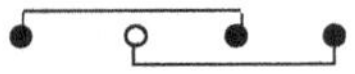

$$= +\frac{i}{\hbar}\int_0^t dt_1' \left(-\frac{i}{\hbar}\right)\int_0^t dt_2 \left(-\frac{i}{\hbar}\right)\int_{t_2}^t dt_1 \mathbb{C}\{\hat{x}(t)\hat{H}_{\mathrm{I}}(t_2)\}$$
$$\times\, \mathbb{C}\{\hat{H}_{\mathrm{I}}(t_1')\hat{H}_{\mathrm{I}}(t_1)\}. \qquad (3.105b)$$

Now add Eqs. (3.105a) and (3.105b):

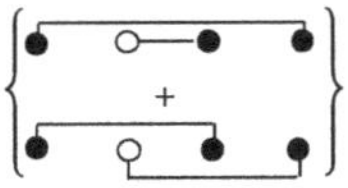

$$= +\frac{i}{\hbar}\int_0^t dt_1' \left(-\frac{i}{\hbar}\right)\int_0^t dt_2 \left(-\frac{i}{\hbar}\right)\int_0^t dt_1 \mathbb{C}\{\hat{x}(t)\hat{H}_{\mathrm{I}}(t_2)\}$$
$$\times\, \mathbb{C}\{\hat{H}_{\mathrm{I}}(t_1')\hat{H}_{\mathrm{I}}(t_1)\}. \qquad (3.106)$$

Now that the integrations have been disentangled, replace

$$\frac{i}{\hbar}\int_0^t dt_1'\,\hat{H}_I(t_1') \quad \text{and} \quad -\frac{i}{\hbar}\int_0^t dt_1\,\hat{H}_I(t_I)$$

by sums over all orders of the expansions of $\hat{U}^\dagger(t)$ and $\hat{U}(t)$, respectively, to obtain

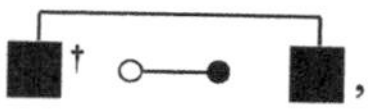

where the contraction bar between the blocks indicates that the set of full contractions includes *at least one* that is *between* the $\hat{U}^\dagger(t)$ and $\hat{U}(t)$ parts, since those graphs that are contracted entirely within $\hat{U}^\dagger(t)$ or $\hat{U}(t)$ have been included in Eq. (3.101). When this result is added to Eq. (3.101), we finally have

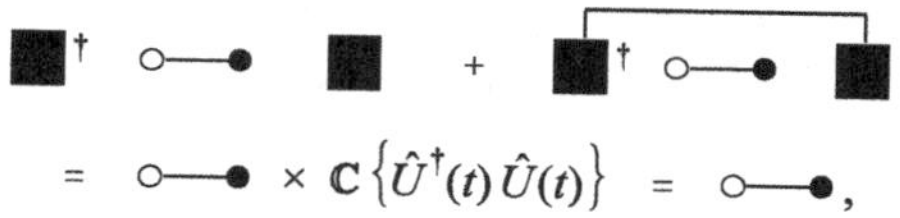

$$= \;\circ\!\!-\!\!\bullet\; \times\; \mathbb{C}\big\{\hat{U}^\dagger(t)\,\hat{U}(t)\big\} \;=\; \circ\!\!-\!\!\bullet, \qquad (3.107a)$$

which now includes *all fully contracted graphs of all orders* in which the $\circ$ is contracted with a $\bullet$ to its right. The complex conjugate of Eq. (3.107a) is the sum of all fully contracted graphs of all orders in which the $\circ$ is contracted with a $\bullet$ to its *left*:

$$= \;\bullet\!\!-\!\!\circ\; \times\; \mathbb{C}\big\{\hat{U}^\dagger(t)\,\hat{U}(t)\big\} \;=\; \bullet\!\!-\!\!\circ. \qquad (3.107b)$$

The sum of Eqs. (3.107a) and (3.107b) is thus the complete solution of the problem to infinite order of the perturbation expansion and is identical with both the Heisenberg result given in Eq. (3.62b) and with the first-order result given in Eq. (3.71), or in graphical form, Eq. (3.73).

The meaning of Eqs. (3.107) and their sum is important. It explains what happens when we attempt a calculation of $\hat{x}(t)$ to each order in a perturbation expansion: the first-order result is the exact result and all orders higher than the first give zero! To see how this follows from the results of Eqs. (3.107) we note that Eqs. (3.107)

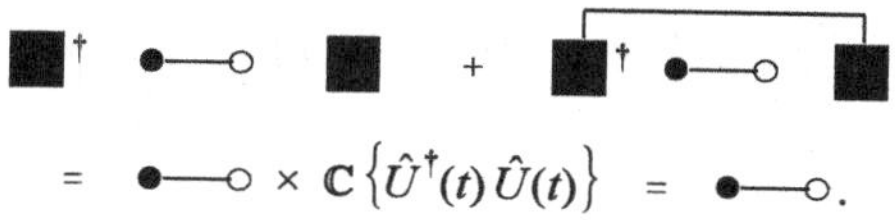

and the discussion following them allow us to write Eq. (3.63), which is the complete perturbation solution, in the form

$$\begin{aligned}\langle \hat{x}(t)\rangle &= \langle 0|\hat{U}^{\dagger}(t)(\circ\!-\!\bullet + \bullet\!-\!\circ)\hat{U}(t)|0\rangle \\ &= \langle 0|\hat{U}^{\dagger}(t)\hat{U}(t)|0\rangle(\circ\!-\!\bullet + \bullet\!-\!\circ) = (\circ\!-\!\bullet + \bullet\!-\!\circ), \end{aligned} \tag{3.108a}$$

where we have used the fact that the sum of the first-order graphs in parentheses is a number, since it is fully contracted, and have used Eq. (3.103) to obtain the last line. Now let us express $U^{\dagger}$ and $\hat{U}$ as perturbation sums as in Eq. (3.8), displaying the perturbation parameter explicitly, including the factor λ for the first-order graphs in parentheses:

$$\begin{aligned}\langle \hat{x}(t)\rangle = \langle 0| &\left[1 + \sum_{n=1}^{\infty} \lambda^{n}\hat{U}^{\dagger(n)}(t)\right] \lambda \\ &\times(\circ\!-\!\bullet + \bullet\!-\!\circ)\left[1 + \sum_{m=1}^{\infty} \lambda^{m}\hat{U}^{(m)}(t)\right]|0\rangle. \end{aligned} \tag{3.108b}$$

Separating out the first-order and higher-order terms in Eq. (3.108b) gives

$$\begin{aligned}\langle \hat{x}(t)\rangle &= \langle 0|\lambda(\circ\!-\!\bullet + \bullet\!-\!\circ)|0\rangle \\ &\quad + \sum_{n=1}^{\infty}\sum_{m=1}^{\infty} \lambda^{n+m+1}\langle 0|\hat{U}^{\dagger(n)}(t)(\circ\!-\!\bullet + \bullet\!-\!\circ)\hat{U}^{(m)}t)|0\rangle. \end{aligned} \tag{3.108c}$$

From Eq. (3.108a), we see that the expression on the second line of Eq. (3.108c), containing all the terms higher than those of first order, is equal to zero; that is

$$\begin{aligned}&\sum_{n=1}^{\infty}\sum_{m=1}^{\infty} \lambda^{(n+m+1)}\langle 0|\hat{U}^{\dagger(n)}(t) \\ &\qquad \times(\circ\!-\!\bullet + \bullet\!-\!\circ)\hat{U}^{(m)}(t)|0\rangle = 0. \end{aligned} \tag{3.109}$$

Not only is the sum of these higher-order terms zero, but the individual terms of a given order $j = n + m + 1$ are zero. This follows from the fact that Eq. (3.109) holds for all values of λ; for otherwise Eq. (3.109) would be an equation in λ which we could solve to give the *special* values of λ for which Eq. (3.109) holds. Thus,

$$\lambda_j \sum_{n=0}^{j-1} \sum_{m=0}^{j-1-n} \langle 0|\hat{U}^{\dagger(n)}(t)(\circ\!-\!\bullet + \bullet\!-\!\circ)\hat{U}^{(m)}(t)|0\rangle = 0. \qquad (3.110)$$

Eq. (3.110) shows why the third-order graphs summed to zero in Eq. (3.82) and why fifth-, seventh-, and higher odd-order graphs all sum to zero. (The *even*-order graphs are zero because they cannot be fully contracted.)

The calculations to infinite order in which we made use of graphical analysis — and in particular showed that disconnected graphs total to a normalization factor of unity — are an example of a general characteristic of diagrammatic perturbation theory; namely, that *disconnected graphs may be ignored.* This is the reason that the first-order result for the problem of the displaced oscillator is equal to the exact, infinite-order result: the first-order graphs are the only connected graphs that can be constructed! This will be the case for *any* problem in which both the perturbation and the operator whose average value is being calculated are linear functions of the Fock operators. The physical implication for the present problem is the relatively trivial one that *the frequency of motion of a wave packet in a harmonic potential is unchanged by a linear displacement of the potential.*

3.6. Summary for Chapter 3

In this chapter, we introduced the Dirac, or interaction, picture of time-dependent quantum mechanics, in which the operator $\hat{O}$ whose average value is to be determined is written as

$$\hat{O}_{\mathrm{I}}(t) = e^{\frac{i}{\hbar}\hat{H}_0 t}\hat{O}e^{-\frac{i}{\hbar}\hat{H}_0 t}, \qquad (S3.1)$$

where $\hat{H}_0$ is the non-perturbed Hamiltonian whose eigenfunctions and eigenvalues are known. The problem then becomes that of using perturbation expansions to find the operator

$$\hat{U}(t) = e^{\frac{i}{\hbar}\hat{H}_0 t} e^{-\frac{i}{\hbar}\hat{H}t} \tag{S3.2}$$

so that

$$\langle O(t)\rangle = \langle \psi(t=0)|\hat{U}^\dagger(t)\hat{O}_\mathrm{I}(t)\hat{U}(t)|\psi(t=0)\rangle. \tag{S3.3}$$

To accomplish this, we used the Dyson series

$$\begin{aligned}\hat{U}(t) &= 1 + \sum_{j=1}^{\infty}\left(-\frac{i}{\hbar}\right)^j \lambda^j \int_0^t dt_j \int_0^{t_j} dt_{(j-1)} \cdots \\ &\quad \times \int_0^{t_2} dt_1 \hat{H}_\mathrm{I}(t_j)\hat{H}_\mathrm{I}(t_{j-1})\cdots \hat{H}_\mathrm{I}(t_1) \\ &= 1 + \sum_{j=1}^{\infty} \lambda^j \hat{U}^{(j)}(t),\end{aligned} \tag{S3.4}$$

where λ is the perturbation parameter. To evaluate the operators $\hat{U}^j(t)$, we implemented Wick's theorem to evaluate averages of strings of operators in the Fock representation. A method of graphing the contractions between operators was developed in which $\circ$ represents the Fock operator $\hat{a}$ and $\bullet$ represents the pertubation Hamiltonian $\hat{H}_\mathrm{I} = \hat{H} - \hat{H}_0$. A line between the dots represents a contraction between an $\hat{a}$ and an $\hat{a}^\dagger$ in one of the $\hat{H}_\mathrm{I}$'s. An example in the case of third-order terms in the perturbation parameter λ is

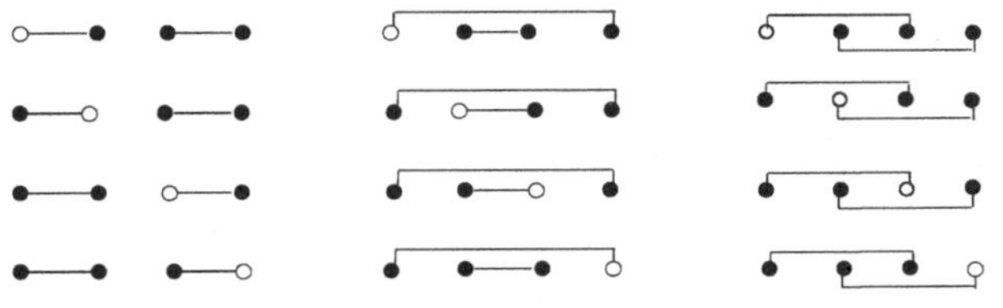

For details, the reader is referred to Section 3.2. Wick's theorem was then used to identify the Glauber states

$$e^{\alpha\hat{a}^\dagger}|0\rangle = |\alpha\rangle, \tag{S3.5}$$

which were shown to be minimal wave packets in a harmonic potential with frequency ω, mass μ, initial position x_0 and initial momentum p_0. The constant α is

$$\alpha = (\mu\omega/2\hbar)^{\frac{1}{2}} x_0 + ip_0/(2\mu\hbar\omega)^{\frac{1}{2}}. \tag{S3.6}$$

We showed these states to be eigenstates of the annihilation operator, $\hat{a}$: that is,

$$\hat{a}e^{\alpha\hat{a}^\dagger}|0\rangle = \alpha e^{\alpha\hat{a}^\dagger}|0\rangle. \tag{S3.7}$$

Glauber states are used to represent coherent ensembles of photons and, as we shall see in Chapter 6, molecules. In a harmonic potential, the wave packet was shown to oscillate with the classical frequency ω:

$$\langle x(t)\rangle = x_0(1 - \cos\omega t) \tag{S3.8}$$

and the oscillation to be independent of the position of the potential minimum. This was seen to be the first-order perturbation result as well as the infinite-order result. Graphical analysis was used to show that the summation of all terms of order two and higher vanish. In the general case, the important result, arrived at by explicit — though admittedly tedious — calculation, is that disconnected graphs contribute a constant factor to the sum of connected graphs, re-normalizing the wave functions. In the case that $\hat{H}_\mathrm{I}$ is linear in the Fock operators, the re-normalization factor is 1 and can therefore be ignored. See Section 3.5 for details.

Chapter 4

Spinless Particles

4.1. Introduction

In this chapter, we generalize the harmonic-oscillator model that has been developed in the previous chapters to treat spinless particles as quanta in quantized fields. To introduce the theory of quantized fields, we use as a guide the quantization of the harmonic oscillator. Starting with the Lagrangian function, we find the equations of motion by the Principle of Least Action and obtain the momentum conjugate to the coördinate. We identify the Hamiltonian and then quantize it by the application of the momentum-coördinate commutation relation.

Proceeding from this treatment of the harmonic oscillator, we extend Lagrangian mechanics to classical fields by replacing the coördinate by a field $\varphi(x, t)$; that is, a single-valued function of three spatial coordinates and time. The Lagrangian thus becomes a functional of this field and its space-time derivatives. The minimization of the action in a given region of space-time leads to the Klein-Gordon wave equation. The field momentum $\mathscr{P}(x)$ is defined and the conserved Hamiltonian is seen to be the spatial integral of the energy density. The fields are quantized by applying a generalization of the commutation rules for the fields $\mathscr{P}(x)$ and $\varphi(x)$. The Fock operators are seen to be creation and annihilation operators for particles of mass m and momentum $\boldsymbol{p}$.

4.2. Mechanics of the Harmonic Oscillator

In this section we "derive" the point at which we started Chapter 2, namely the quantum-mechanical Hamiltonian for the harmonic oscillator.

4.2.1. *Lagrangian Mechanics*

The Lagrangian L of a 1-dimensional mechanical system is a function of the coordinate x and its time derivative $\dot{x}$, and perhaps the time:

$$L = L(t, x, \dot{x}). \tag{4.1}$$

The action S is the time integral of the Lagrangian from an initial to a final point in space:

$$S = \int_{t_1}^{t_2} dt L(t, x, \dot{x}). \tag{4.2}$$

There are an infinite number of possible paths between the *fixed* points $x\,(t_1)$ and $x\,(t_2)$. The classical path followed by the system during the time interval t_1, t_2 is the one which minimizes the action; that is, we set the variation in S equal to zero:

$$\delta S = \int_{t_1}^{t_2} dt \left[\frac{\partial L}{\partial x}\delta x + \frac{\partial L}{\partial \dot{x}}\delta \dot{x}\right] = 0. \tag{4.3}$$

Now we can write the second term in the brackets above as

$$\frac{\partial L}{\partial \dot{x}}\delta \dot{x} = \frac{d}{dt}\left(\frac{\partial L}{\partial \dot{x}}\delta x\right) - \frac{d}{dt}\left(\frac{\partial L}{\partial \dot{x}}\right)\delta x \tag{4.4}$$

such that Eq. (4.4) becomes

$$\delta S = \int_{t_1}^{t_2} dt \left[\frac{\partial L}{\partial x} - \frac{d}{dt}\left(\frac{\partial L}{\partial \dot{x}}\right)\right]\delta x + \left[\frac{\partial L}{\partial \dot{x}}\delta x\right]_{t_1}^{t_2} = 0. \tag{4.5}$$

Since the variation in x vanishes at the end points of the integration, the last term in Eq. (4.5) is zero. On the other hand, the variation δx may be freely chosen *between* the end points so that we have the

Lagrangian equation of motion

$$\frac{\partial L}{\partial x} - \frac{d}{dt}\left(\frac{\partial L}{\partial \dot{x}}\right) = 0. \tag{4.6}$$

To prove that Eq. (4.6) follows from Eq. (4.5), we need note merely that if

$$\frac{\partial L}{\partial x} - \frac{d}{dt}\left(\frac{\partial L}{\partial \dot{x}}\right) \neq 0$$

we could choose

$$\delta x = g(t)\left[\frac{\partial L}{\partial x} - \frac{d}{dt}\left(\frac{\partial L}{\partial \dot{x}}\right)\right]^{-1},$$

where $g(t)$ is an integrable function that vanishes at t_1 and t_2 so that

$$\delta S = \int_{t_1}^{t_2} g(t)dt \neq 0,$$

in contradiction to our premise.

We need to find an explicit form for $L(t, x, \dot{x})$ so that Eq. (4.6) is consistent with Newton's equation of motion in one dimension for a particle with mass m, namely

$$m\ddot{x} = f(x) = -\frac{d}{dx}V(x), \tag{4.7}$$

where $f(x)$ is the force on the particle at point x, and $V(x)$ is the potential energy. As the reader can verify, Eqs. (4.6) and (4.7) are compatible if

$$L(t, x, \dot{x}) = \frac{1}{2}m\dot{x}^2 - V(x). \tag{4.8}$$

For the harmonic oscillator, with potential-energy function

$$V(x) = \frac{1}{2}kx^2, \tag{4.9}$$

Eq. (4.7) gives the second-order differential equation

$$\ddot{x} = -\frac{k}{m}x, \tag{4.10}$$

for which the general solution is

$$x(t) = x(0)\cos\omega t + \frac{1}{\omega}\dot{x}(0)\sin\omega t, \tag{4.11}$$

where the frequency ω is given by

$$\omega = \sqrt{\frac{k}{m}}. \tag{4.12}$$

4.2.2. *Constants of Motion: Momentum and the Hamiltonian*

A constant of motion is a variable that remains constant in time. The general technique for finding constants of motion from invariances of the Lagrangian is due to Emmy Noether.[1] The simplest to find in the present case of a particle moving in one dimension is the momentum. From Eq. (4.6), we see that if L is independent of x, we have

$$\frac{d}{dt}\frac{\partial L}{\partial \dot{x}} = 0, \tag{4.13}$$

that is, $\partial L/\partial \dot{x}$ is a constant of motion which we define as the momentum p:

$$p = \frac{\partial L}{\partial \dot{x}}. \tag{4.14}$$

Now let the Lagrangian be displaced in time by an interval Δt; by this we mean that as $L \to L + \Delta L$

$$t \to t + \Delta t, \quad \dot{x}(t) \to \dot{x}(t + \Delta t), \quad x(t) \to x(t + \Delta t).$$

If Δt is small, we have

$$\Delta L = \frac{dL}{dt}\Delta t = \left(\frac{\partial L}{\partial t} + \frac{\partial L}{\partial \dot{x}}\frac{d\dot{x}}{dt} + \frac{\partial L}{\partial x}\frac{dx}{dt}\right)\Delta t.$$

[1]Noether, E., *Nachrichten von der Gesellschaft der Wissenschaften zu Göttingen, Mathematisch-Physikalische Klasse* **1918**, 235 (1918).

Now suppose that L does not depend explicitly on time. This means that $\partial L/\partial t = 0$ and that

$$\frac{dL}{dt} = \frac{\partial L}{\partial \dot{x}}\frac{d\dot{x}}{dt} + \frac{\partial L}{\partial x}\frac{dx}{dt}. \tag{4.15}$$

We have seen in the previous Section 4.2.1 that the first term on the right-hand side of Eq. (4.15) can be written

$$\frac{\partial L}{\partial \dot{x}}\frac{d\dot{x}}{dt} = \frac{d}{dt}\left(\frac{\partial L}{\partial \dot{x}}\frac{dx}{dt}\right) - \left[\frac{d}{dt}\left(\frac{\partial L}{\partial \dot{x}}\right)\right]\frac{dx}{dt}. \tag{4.16}$$

Combination of Eq. (4.16) with Eq. (4.6) and a simple rearrangement allows us to write Eq. (4.15) as

$$\frac{d}{dt}\left(\frac{\partial L}{\partial \dot{x}}\frac{dx}{dt} - L\right) = 0. \tag{4.17}$$

But $dx/dt = \dot{x}$ and from Eqs. (4.14) and (4.8), $\partial L/\partial \dot{x} = p = m\dot{x}$. The constant of motion in Eq. (4.17) is therefore

$$\frac{\partial L}{\partial \dot{x}}\frac{dx}{dt} - L = m\dot{x}^2 - L = \frac{p^2}{2m} + V, \tag{4.18}$$

which we define as the Hamiltonian function, H:

$$H = \frac{p^2}{2m} + V. \tag{4.19}$$

In the case of the harmonic oscillator, where the potential energy is given by $V = \frac{1}{2}kx^2$, the Hamiltonian is

$$H = \frac{p^2}{2m} + \frac{1}{2}kx^2. \tag{4.20}$$

4.2.3. *Quantization of the Harmonic Oscillator*

The Hamiltonian H for the harmonic oscillator becomes a quantum-mechanical operator $\hat{H}$ when we replace the variables p and x by the operators $\hat{p}$ and $\hat{x}$ which obey the commutation rule

$$[\hat{p}, \hat{x}]_- = \frac{\hbar}{i}. \tag{4.21}$$

Second quantization is then accomplished by defining the Fock creation and annihilation operators, $\hat{a}^\dagger$ and $\hat{a}$, respectively, by

$$\hat{a}^\dagger = \left(\frac{1}{2\hbar m\omega}\right)^{\frac{1}{2}} (\hat{p} + im\omega\hat{x})e^{i\theta} \tag{4.22a}$$

$$\hat{a} = \left(\frac{1}{2\hbar m\omega}\right)^{\frac{1}{2}} (\hat{p} - im\omega\hat{x})e^{-i\theta}. \tag{4.22b}$$

The phase factor $e^{i\theta}$ is irrelevant for single particles, but will become important in the treatment of ensembles in Chapter 6. We therefore include it in the present discussion. Inversion of Eq. (4.22) gives

$$\hat{x} = \frac{i}{\sqrt{2}}\left(\frac{\hbar}{m\omega}\right)^{\frac{1}{2}} (\hat{a}e^{i\theta} - \hat{a}^\dagger e^{-i\theta}). \tag{4.23a}$$

$$\hat{p} = \frac{1}{\sqrt{2}}(\hbar m\omega)^{\frac{1}{2}} \left(\hat{a}e^{i\theta} + \hat{a}^\dagger e^{-i\theta}\right). \tag{4.23b}$$

Calculating $[\hat{p}, \hat{x}]_-$ from Eq. (4.23), we have

$$[\hat{p}, \hat{x}]_- = \frac{i\hbar}{2}\left(-[\hat{a}, \hat{a}^\dagger]_{\mp}[\hat{a}^\dagger, \hat{a}]_-\right) = \frac{\hbar}{i}, \tag{4.24}$$

which shows that

$$[\hat{a}, \hat{a}^\dagger]_- = 1. \tag{4.25}$$

Then, from Eq. (4.22) we have

$$\hat{a}^\dagger\hat{a} = \frac{1}{2\hbar m\omega}\left[\hat{p}^2 + m^2\omega^2\hat{x}^2 - im\omega(\hat{p}\hat{x} - \hat{x}\hat{p})\right], \tag{4.26}$$

from which we easily obtain

$$\hat{H} = \hbar\omega\left(\hat{a}^\dagger\hat{a} + \frac{1}{2}\right). \tag{4.27}$$

We have repeated some of the material from Chapter 2 here, since the reasoning in Section 4.2 will be used in our development of the quantum field theory of spinless particles, which we begin in the next section.

4.3. Classical Klein-Gordon Field Theory

We generalize Lagrangian mechanics by replacing the coördinate x by a scalar field $\varphi(x)$; that is, a scalar function of the coördinate. The time derivative $\dot{x}$ is similarly replaced by a derivative of $\varphi(x)$. But since we are developing a general theory of possibly relativistic scalar particles, we take x to be a 4-vector in space-time and generalize the time derivative to a 4-vector of derivatives with respect to each of the components of the 4-vector x. Let us review the notation rules for vectors in space-time.

4.3.1. *Vectors and Derivatives in Space-time*

The 4-vector x has components x^μ with $x^0 = ct$, $x^1 = x$, $x^2 = y$, $x^3 = z$, where c is the speed of light; or we may write

$$x^\mu = (ct, \boldsymbol{r}). \tag{4.28}$$

When the index is lowered, we have

$$x_\mu = (ct, -\boldsymbol{r}). \tag{4.29}$$

To minimize confusion, we will use x to mean the 4-vector and indicate the 3-vector (x, y, z) by $\boldsymbol{r}$.

Greek indices are used for components of 4-vectors and Latin indices for those of 3-vectors. Repeated indices are assumed to be summed. Thus, a scalar product of 4-vectors A and B is indicated by

$$\begin{aligned} A^\mu B_\mu = A_\mu B^\mu &= A_0 B_0 - A_x B_x - A_y B_y - A_z B_z \\ &= A_0 B_0 - A_j B_j. \end{aligned} \tag{4.30}$$

The derivative ∂x^μ is indicated by ∂_μ that is,

$$\partial_\mu = \left(\frac{1}{c}\frac{\partial}{\partial t}, \frac{\partial}{\partial x}, \frac{\partial}{\partial y}, \frac{\partial}{\partial z}\right) = \left(\frac{1}{c}\frac{\partial}{\partial t}, \nabla\right) \tag{4.31a}$$

$$\partial^\mu = \left(\frac{1}{c}\frac{\partial}{\partial t}, -\frac{\partial}{\partial x}, -\frac{\partial}{\partial y}, -\frac{\partial}{\partial z}\right) = \left(\frac{1}{c}\frac{\partial}{\partial t}, -\nabla\right). \tag{4.31b}$$

4.3.2. *The Klein-Gordon Lagrangian*

We are now ready to implement the field-theoretic generalization of the harmonic oscillator as described in the first paragraph of Section 4.3. The action is now given by an integral over a region of space-time:

$$S = \int d^4x \mathscr{L}[\varphi(x), \partial_\mu \varphi(x)], \tag{4.32}$$

where the Lagrangian *density* $\mathscr{L}$ is now a functional of the field $\varphi(x)$ and the derivative fields $\partial_\mu \varphi(x)$. The Lagrangian L is defined as the spatial integral of the Lagrangian density $\mathscr{L}$,

$$L(t) = c \int dr \mathscr{L}[\varphi(x), \partial_\mu \varphi(x)], \tag{4.33}$$

so that

$$S = \int dt L(t). \tag{4.34}$$

The field-theoretic analog to the Lagrangian for the harmonic oscillator is the Lagrangian density

$$\mathscr{L} = \frac{1}{2}\left(\partial^\mu \varphi(x) \partial_\mu \varphi(x) - \frac{m^2c^2}{\hbar^2}\varphi^2(x)\right). \tag{4.35}$$

The reason for the choice of the "force" constant $\mathrm{m}^2c^2/\hbar^2$ will soon become clear.

4.3.3. *The Klein-Gordon Equation*

We now derive an equation of motion analogous to Eq. (4.6) using the Principle of Least Action. The variation in the action S within the region of integration in space-time is

$$\delta S = \int d^4x \delta \mathscr{L}[\varphi(x), \partial_\mu \varphi(x)]. \tag{4.36}$$

The variation in the Lagrangian density $\mathscr{L}$ is

$$\delta \mathscr{L} = \frac{\partial \mathscr{L}}{\partial \varphi}\delta\varphi + \frac{\partial \mathscr{L}}{\partial(\partial_\mu \varphi)}\delta(\partial_\mu \varphi). \tag{4.37}$$

The last term on the right-hand-side of Eq. (4.37) can be written as

$$\frac{\partial \mathscr{L}}{\partial(\partial_\mu \varphi)}\delta(\partial_\mu \varphi) = \partial_\mu \left[\frac{\partial \mathscr{L}}{\partial(\partial_\mu \varphi)}\delta\varphi\right] - \left[\partial_\mu \frac{\partial \mathscr{L}}{\partial(\partial_\mu \varphi)}\right]\delta\varphi. \tag{4.38}$$

The first term on the right-hand-side of Eq. (4.38) is a 4-divergence; accordingly, its integral over volume in space-time becomes an integral over the surface of that volume.

$$\int_{\text{vol}} d^4x \partial_\mu \left[\frac{\partial \mathscr{L}}{\partial(\partial_\mu \varphi)}\delta\varphi\right] = \int_{\text{surf}} d\sigma_\mu \frac{\partial \mathscr{L}}{\partial(\partial_\mu \varphi)}\delta\varphi = 0, \tag{4.39}$$

where $d\sigma_\mu$ is an infinitesimal element of surface centered at x^μ. But since the field φ is assumed to be fixed on this surface, the variation $\delta\varphi$ vanishes there; that is

$$\delta S = \int d^4x \left[\frac{\partial \mathscr{L}}{\partial \varphi} - \partial_\mu \frac{\partial \mathscr{L}}{\partial(\partial_\mu \varphi)}\right]\delta\varphi = 0. \tag{4.40}$$

Taking into account Eqs. (4.37)–(4.39), we see that Eq. (4.36) becomes

$$\frac{\partial \mathscr{L}}{\partial \varphi} - \partial_\mu \frac{\partial \mathscr{L}}{\partial(\partial_\mu \varphi)} = 0. \tag{4.41}$$

Since $\delta\varphi$ is arbitrary within the volume of integration, Eq. (4.40) requires that

$$\frac{\partial \mathscr{L}}{\partial \varphi} = -\frac{m^2c^2}{\hbar}\varphi. \tag{4.42}$$

Using Eq. (4.35) for $\mathscr{L}$, we obtain

$$\partial_\mu \frac{\partial \mathscr{L}}{\partial(\partial_\mu \varphi)} = \partial_\mu \partial^\mu \varphi = \left[\left(\frac{1}{c}\frac{\partial}{\partial t}\right)^2 - \nabla^2\right]\varphi = \Box^2\varphi. \tag{4.43}$$

The symbol $\Box^2$ for the 4-dimensional Laplacian operator in space-time is defined as shown in Eq. (4.43).

Combining Eqs. (4.41)–(4.43), we obtain the Klein-Gordon equation,

$$\left(\Box^2 + \frac{m^2c^2}{\hbar^2}\right)\varphi = 0. \tag{4.44}$$

In quantum mechanics $\partial/\partial t$ is associated with $(-i/\hbar)E$ and ∇ with $(i/\hbar)\boldsymbol{p}$, where E and $\boldsymbol{p}$ are the energy and momentum of the system, respectively.

Thus, Eq. (4.44) can be interpreted as stating that $\mathrm{E}^2 = c^2(m^2c^2 + p^2)$, in agreement with the relativistic energy of a free particle of mass m and momentum $\boldsymbol{p}$. This shows that we are on the right track. But as it stands, our theory is that of a *classical* field $\varphi(x)$. We must now proceed to quantize it and find creation and annihilation operators for the particles that are its quanta. As in the case of the harmonic oscillator, first we identify the momentum and the Hamiltonian.

4.3.4. *Canonical Momentum and Hamiltonian*

The energy relation suggested by Eq. (4.44) is preserved in a Hamiltonian treatment of the Klein-Gordon field if we define the canonical momentum as

$$\mathscr{P}(x) = \frac{\partial\mathscr{L}}{\partial(\partial_0\varphi)} = \frac{1}{c}\dot{\varphi}(x). \tag{4.45}$$

Then we have for the energy density

$$\mathcal{H}(x)/c = \mathscr{P}\partial_0\varphi - \mathscr{L} = \frac{1}{2}\left[\mathscr{P}^2 + (\nabla_\varphi)^2 + \frac{m^2c^2}{\hbar^2}\varphi^2\right]. \tag{4.46}$$

As we have defined it, $\mathcal{H}(x)$ is the Hamiltonian *density*; the Hamiltonian H is the 3-space integral

$$H(t) = \int d\boldsymbol{r}\mathcal{H}(x). \tag{4.47}$$

To bring out the *particle* momentum $\boldsymbol{p}$ explicitly, we express both $\varphi(x)$ and $\mathscr{P}(x)$ as Fourier integrals over $\boldsymbol{p}$:

$$\varphi(x) = \frac{1}{h^{\frac{3}{2}}} \int d\boldsymbol{p} \left[\chi(\boldsymbol{p}) e^{\frac{i}{\hbar}\boldsymbol{p}\cdot\boldsymbol{r}} + \chi^*(\boldsymbol{p}) e^{-\frac{i}{\hbar}\boldsymbol{p}\cdot\boldsymbol{r}}\right] \tag{4.48a}$$

$$\mathscr{P}(x) = -\frac{i}{h^{\frac{3}{2}}} \int d\boldsymbol{p} \left[\gamma(\boldsymbol{p}) e^{\frac{i}{\hbar}\boldsymbol{p}\cdot\boldsymbol{r}} - \gamma^*(\boldsymbol{p}) e^{-\frac{i}{\hbar}\boldsymbol{p}\cdot\boldsymbol{r}}\right]. \tag{4.48b}$$

To get an expression for H in terms of $\boldsymbol{p}$, we use Eq. (4.48) to obtain

$$\begin{aligned}\int d\boldsymbol{r}\mathscr{P}^2(x) &= -\frac{1}{\hbar^3} \int d\boldsymbol{r} \int d\boldsymbol{p} \int d\boldsymbol{p}' \left[-\gamma(\boldsymbol{p})\gamma^*(\boldsymbol{p}) e^{\frac{i}{\hbar}(\boldsymbol{p}-\boldsymbol{p}')\cdot\boldsymbol{r}}\right. \\ &\quad - \gamma^*(\boldsymbol{p})\gamma(\boldsymbol{p}') e^{-\frac{i}{\hbar}(\boldsymbol{p}-\boldsymbol{p}')\cdot\boldsymbol{r}} + \gamma(\boldsymbol{p})\gamma(\boldsymbol{p}') e^{\frac{i}{\hbar}(\boldsymbol{p}+\boldsymbol{p}')\cdot\boldsymbol{r}} \\ &\quad + \gamma^*(\boldsymbol{p})\gamma^*(\boldsymbol{p}') e^{-\frac{i}{\hbar}(\boldsymbol{p}+\boldsymbol{p}')\cdot\boldsymbol{r}} \\ &= 2\int d\boldsymbol{p} \left\{|\gamma(\boldsymbol{p})|^2 - \frac{1}{2}[\gamma(\boldsymbol{p})\gamma(-\boldsymbol{p}) + \gamma^*(\boldsymbol{p})\gamma^*(-\boldsymbol{p})]\right\}\end{aligned} \tag{4.49}$$

$$\nabla\varphi(x) = \left(\frac{i}{\hbar}\right) \frac{1}{h^{\frac{3}{2}}} \int d\boldsymbol{p}p^2 \left[\chi(\boldsymbol{p}) e^{\frac{i}{\hbar}\boldsymbol{p}\cdot\boldsymbol{r}} - \chi^*(\boldsymbol{p}) e^{-\frac{i}{\hbar}\boldsymbol{p}\cdot\boldsymbol{r}}\right] \tag{4.50}$$

$$\begin{aligned}\int d\boldsymbol{r}[\nabla\varphi(x)]^2 &= \frac{2}{\hbar^2} \int d\boldsymbol{p}p^2 \left\{|\chi(\boldsymbol{p})|^2\right. \\ &\quad \left. + \frac{1}{2}\left[\chi(\boldsymbol{p})\chi(-\boldsymbol{p}) + \chi^*(\boldsymbol{p})\chi^*(-\boldsymbol{p})\right]\right\}\end{aligned} \tag{4.51}$$

From Eqs. (4.51) and (4.52), we see that

$$\int d\boldsymbol{r}\varphi^2(x) = 2\int d\boldsymbol{p} \left\{|\chi(\boldsymbol{p})|^2 + \frac{1}{2}\left[\chi(\boldsymbol{p})\chi(-\boldsymbol{p}) + \chi^*(\boldsymbol{p})\chi^*(-\boldsymbol{p})\right]\right\}. \tag{4.52}$$

From Eqs. (4.51) and (4.52), we see that

$$\int d\boldsymbol{r}\left[(\nabla\varphi)^2+\frac{m^2c^2}{\hbar_2}\varphi^2\right]=2\int d\boldsymbol{p}\frac{(m^2c^2+\boldsymbol{p}_2)}{\hbar^2}$$
$$\times\left\{|\chi(\boldsymbol{p})|^2+\frac{1}{2}\left[\chi(\boldsymbol{p})\chi(-\boldsymbol{p})+\chi^*(\boldsymbol{p})\chi^*(-\boldsymbol{p})\right]\right\}. \quad (4.53)$$

The "diagonal" terms with the negative $\boldsymbol{p}$ arguments will cancel in the addition of Eqs. (4.49) and (4.53) to obtain the Hamiltonian H if we set

$$\gamma(\boldsymbol{p})=\frac{\sqrt{m^2c^2+\boldsymbol{p}^2}}{\hbar}\chi(\boldsymbol{p}). \quad (4.54)$$

We then get

$$H=2c\frac{(m^2c^2+\boldsymbol{p}_2)}{\hbar^2}|\chi(\boldsymbol{p})|^2. \quad (4.55)$$

If we let

$$\chi(\boldsymbol{p})=\frac{\hbar}{\sqrt{2}(m^2c^2+\boldsymbol{p}^2)^{\frac{1}{4}}}a(\boldsymbol{p}),\quad \gamma(\boldsymbol{p})=\frac{1}{\sqrt{2}}(m^2c^2+\boldsymbol{p}^2)^{\frac{1}{4}}a(\boldsymbol{p}), \quad (4.56)$$

then the Klein-Gordon fields and Hamiltonian take the forms

$$\varphi(x)=\frac{1}{h^{\frac{3}{2}}}\frac{\hbar}{\sqrt{2}}\int d\boldsymbol{p}(m^2c^2+\boldsymbol{p}^2)^{-\frac{1}{4}}\left[a(\boldsymbol{p})e^{\frac{i}{\hbar}\boldsymbol{p}\cdot\boldsymbol{r}}+a^*(\boldsymbol{p})e^{-\frac{i}{\hbar}\boldsymbol{p}\cdot\boldsymbol{r}}\right] \quad (4.57)$$

$$\mathscr{P}(x)=-\frac{i}{h^{\frac{3}{2}}}\frac{1}{\sqrt{2}}\int d\boldsymbol{p}(m^2c^2+\boldsymbol{p}^2)^{\frac{1}{4}}\left[a(\boldsymbol{p})e^{\frac{i}{\hbar}\boldsymbol{p}\cdot\boldsymbol{r}}-a^*(\boldsymbol{p})e^{-\frac{i}{\hbar}\boldsymbol{p}\cdot\boldsymbol{r}}\right] \quad (4.58)$$

$$H=c\int d\boldsymbol{p}\sqrt{m^2c^2+\boldsymbol{p}_2}|a(\boldsymbol{p}|^2. \quad (4.59)$$

4.3.5. *Constants of Motion*[2]

We may find the constants of motion for the Klein-Gordon field by a generalization of the method we used in Section 4.2.2 for the

---

[2]See the footnote on p. 95.

harmonic oscillator. If $\mathscr{L}$ depends on x only through φ and $\partial_\mu\varphi$, the partial derivative of $\mathscr{L}$ with respect to x^ν is

$$\begin{aligned}\frac{\partial\mathscr{L}}{\partial x^\nu} &= \frac{\partial\mathscr{L}}{\partial\varphi}\partial_\nu\varphi + \frac{\partial\mathscr{L}}{\partial(\partial_\mu\varphi)}\partial_\nu(\partial_\mu\varphi)\\ &= \frac{\partial\mathscr{L}}{\partial\varphi}\partial_\nu\varphi + \partial_\mu\left[\frac{\partial\mathscr{L}}{\partial(\partial_\mu\varphi)}\partial_\nu\varphi\right] - \left[\partial_\mu\frac{\partial\mathscr{L}}{\partial(\partial_\mu\varphi)}\right]\partial_\nu\varphi. \quad (4.60)\end{aligned}$$

The first and third terms on the right-hand side of Eq. (4.60) cancel as a result of the Principle of Least Action, Eq. (4.41), so that this equation can be rearranged to read

$$\partial_\mu\left[\frac{\partial\mathscr{L}}{\partial(\partial_\mu\varphi)}\partial_\nu\varphi - \mathscr{L}\delta^\mu_\nu\right] = 0. \quad (4.61)$$

Eq. (4.61) identifies the energy-momentum 4-tensor, T^μ_ν,

$$\mathrm{T}^\mu_\nu = \frac{\partial\mathscr{L}}{\partial(\partial_\mu\varphi)}\partial_\nu\varphi - \mathscr{L}\delta^\mu_\nu, \quad (4.62)$$

such that

$$\partial_\mu\mathrm{T}^\mu_\nu = 0. \quad (4.63)$$

From Eq. (4.63), we see that for each value of ν, the 4-tensor T^μ_ν is a 4-vector current density. The 4-vector T^0_ν is of particular interest. If we integrate Eq. (4.63) over spatial coördinates, we have

$$c\int d\boldsymbol{r}\partial_\mu\mathrm{T}^\mu_\nu = 0 = c\int d\boldsymbol{r}\partial_0\mathrm{T}^0_\nu - c\int d\boldsymbol{r}\partial_j\mathrm{T}^j_\nu. \quad (4.64)$$

But the second term on the right-hand side of Eq. (4.64) can be written as a surface integral:

$$c\int d\boldsymbol{r}\partial_j\mathrm{T}^j_\nu = c\int d\boldsymbol{r}\nabla\cdot\mathbf{T}_\nu = c\int_{\text{surf}} d\boldsymbol{\sigma}\cdot\mathbf{T}_\nu = 0, \quad (4.65)$$

where $\mathbf{T}_\nu$ is the 3-vector with components $\mathrm{T}^j_\nu, j = 1, 2, 3$. We have set the surface integral equal to zero, since the surface of integration is the spatial boundary of the system on which the fields, and therefore the tensor T, vanishes. Thus, Eqs. (4.64) and (4.65) give the result

$$c\int d\boldsymbol{r}\partial_0 \mathrm{T}^0_\nu = \frac{d}{dt}\int d\boldsymbol{r}\mathrm{T}^0_\nu = 0; \tag{4.66}$$

that is, the four quantities $\int d\boldsymbol{r}\mathrm{T}^0_\nu$ are conserved. Using Eq. (4.35) for the Klein-Gordon Lagrangian density, we obtain

$$\mathrm{T}^0_0 = \frac{\partial \mathscr{L}}{\partial(\partial_0\varphi)}\partial_0\varphi - \mathscr{L} = \mathscr{P}\partial_0\varphi - \mathscr{L} = \mathcal{H}/c \tag{4.67a}$$

$$\mathrm{T}^0_j = -\frac{\partial \mathscr{L}}{\partial(\partial_0\varphi)}\nabla_\varphi = -\mathscr{P}\nabla\varphi. \tag{4.67b}$$

We see that the time component of the 4-vector T^0_ν gives the Hamiltonian density; the spatial components form a 3-vector that will turn out to be the total momentum of the particles that are the quanta in the quantized form of the theory. We have obtained an expression for $H = c\int d\boldsymbol{r}\mathcal{H}$ in terms of the Fourier coefficient $a(\boldsymbol{p})$ in Eq. (4.59); we now use Eqs. (4.57) and (4.58) to derive the corresponding expression:

$$\begin{aligned}\boldsymbol{P} &= \int d\boldsymbol{r}[-\mathscr{P}(\nabla\varphi)] = -\int d\boldsymbol{r}\frac{1}{h^3}\int d\boldsymbol{p}\int d\boldsymbol{p}' \\ &\quad\times\left[a(\boldsymbol{p})e^{\frac{i}{\hbar}\boldsymbol{p}\cdot\boldsymbol{r}} - a^*(\boldsymbol{p})e^{-\frac{i}{\hbar}\boldsymbol{p}\cdot\boldsymbol{r}}\right]\boldsymbol{p}\left[a(\boldsymbol{p})e^{\frac{i}{\hbar}\boldsymbol{p}\cdot\boldsymbol{r}} - a^*(\boldsymbol{p})e^{-\frac{i}{\hbar}\boldsymbol{p}\cdot\boldsymbol{r}}\right] \\ &= \int d\boldsymbol{p}\boldsymbol{p}[|a(\boldsymbol{p}|^2 - a(\boldsymbol{p}))a(-\boldsymbol{p}) - a^*(\boldsymbol{p})a^*(-\boldsymbol{p})].\end{aligned} \tag{4.68}$$

The last two terms in brackets in the last line of Eq. (4.68) may be neglected, since they are both even with respect to $\boldsymbol{p}$, whereas the integrand contains the (odd) factor $\boldsymbol{p}$ before the bracket; thus, we have

$$\boldsymbol{P} = \int d\boldsymbol{p}\boldsymbol{p}|a(\boldsymbol{p})|^2, \tag{4.69}$$

which is consistent with our identification of T_j^0 with the total momentum density.

4.4. The Quantized Klein-Gordon Field

In the previous section, we have seen the analogy between the harmonic oscillator and the classical Klein-Gordon field as the basis for a theory of relativistic spinless particles. It is now a conceptually simple matter to quantize the Klein-Gordon field and identify the excitation quanta with spinless particles.

4.4.1. *Quantization of the Fields*

Let the fields $\varphi(x)$ and $\mathscr{P}(x)$ be replaced in all the equations of Section 4.3.4 by the operators

$$\hat{\varphi}(\boldsymbol{r}) = \frac{1}{h^{\frac{3}{2}}}\frac{\hbar}{\sqrt{2}}\int d\boldsymbol{p}(m^2c^2+\boldsymbol{p}^2)^{-\frac{1}{4}}\left[\hat{a}(\boldsymbol{p})e^{\frac{i}{\hbar}\boldsymbol{p}\cdot\boldsymbol{r}} + \hat{a}^{\dagger}(\boldsymbol{p})e^{-\frac{i}{\hbar}\boldsymbol{p}\cdot\boldsymbol{r}}\right], \tag{4.70}$$

$$\hat{\mathscr{P}}(\boldsymbol{r}) = -\frac{i}{h^{\frac{3}{2}}}\frac{1}{\sqrt{2}}\int d\boldsymbol{p}(m^2c^2+\boldsymbol{p}^2)^{\frac{1}{4}}\left[\hat{a}(\boldsymbol{p})e^{\frac{i}{\hbar}\boldsymbol{p}\cdot\boldsymbol{r}} - \hat{a}^{\dagger}(\boldsymbol{p})e^{-\frac{i}{\hbar}\boldsymbol{p}\cdot\boldsymbol{r}}\right]. \tag{4.71}$$

In writing these operators as functions of the spatial position vector $\boldsymbol{r}$ rather than the 4-vector x, we have assumed that they are evaluated at a given instant of time, namely $t = 0$. Time dependence is treated in Section 4.4.

4.4.2. *Commutation Relations in Quantum Field Theory*

By analogy with the quantization of the harmonic oscillator, we assume the commutation relation to be a generalization of Eq. (4.21):

$$[\mathscr{P}(\boldsymbol{r}), \hat{\varphi}(\boldsymbol{r})]_- = \frac{\hbar}{i}\delta(\boldsymbol{r}-\boldsymbol{r}'). \tag{4.72}$$

The commutation rules for $\hat{a}(\boldsymbol{p})$ and $\hat{a}^\dagger(\boldsymbol{p})$ are found by evaluating Eq. (4.72) from Eqs. (4.70) and (4.71):

$$\begin{aligned}
&[\hat{\mathscr{P}}(\boldsymbol{r}), \hat{\varphi}(\boldsymbol{r}')]_- \\
&= \frac{\hbar}{i}\frac{1}{2}\frac{1}{h^3}\int d\boldsymbol{p}\int d\boldsymbol{p}'\left\{[\hat{a}(\boldsymbol{p}), \hat{a}(\boldsymbol{p}')]_- e^{\frac{i}{\hbar}(\boldsymbol{p}\cdot\boldsymbol{r}+\boldsymbol{p}'\cdot\boldsymbol{r}')}\right. \\
&\quad +[\hat{a}(\boldsymbol{p}), \hat{a}^\dagger(\boldsymbol{p}')]_- e^{\frac{i}{\hbar}(\boldsymbol{p}\cdot\boldsymbol{r}-\boldsymbol{p}'\cdot\boldsymbol{r}')} - [\hat{a}^\dagger(\boldsymbol{p}), \hat{a}(\boldsymbol{p}')]_- e^{-\frac{i}{\hbar}(\boldsymbol{p}\cdot\boldsymbol{r}-\boldsymbol{p}'\cdot\boldsymbol{r}')} \\
&\quad \left. -[\hat{a}^\dagger(\boldsymbol{p}), \hat{a}^\dagger(\boldsymbol{p}')]_- e^{-\frac{i}{\hbar}(\boldsymbol{p}\cdot\boldsymbol{r}+\boldsymbol{p}'\cdot\boldsymbol{r}')}\right\}. \qquad (4.73)
\end{aligned}$$

Eq. (4.73) will be consistent with Eq. (4.72) if

$$[\hat{a}(\boldsymbol{p}), \hat{a}(\boldsymbol{p})]_- = [\hat{a}^\dagger(\boldsymbol{p}), \hat{a}^\dagger(\boldsymbol{p}')]_- = 0 \qquad (4.74a)$$

$$[\hat{a}(\boldsymbol{p}), \hat{a}^\dagger\boldsymbol{p}']_- = -[\hat{a}^\dagger(\boldsymbol{p}), \hat{a}(\boldsymbol{p}')]_- = \delta(\boldsymbol{p}-\boldsymbol{p}'). \qquad (4.74b)$$

As we shall see in some detail in Chapter 6, the commutation rules of Eq. (4.74) lead to Bose-Einstein statistics for the spinless particles created and annihilated by $\hat{a}^\dagger(\boldsymbol{p})$ and $\hat{a}(\boldsymbol{p})$, respectively.

A crucial question is how to deal with the apparent infinity $\delta(\boldsymbol{p} - \boldsymbol{p})$? In terms of the Fourier integral representation of the Dirac delta function,

$$\delta(\boldsymbol{p}-\boldsymbol{p}) = \frac{1}{h^3}\int d\boldsymbol{r} e^{\frac{i}{\hbar}(\boldsymbol{p}-\boldsymbol{p})\cdot\boldsymbol{r}} = \frac{1}{h^3}\int d\boldsymbol{r} = \frac{V}{h^3}, \qquad (4.75)$$

where V is the total volume of the system. For many experiments in which there are a few particles in collision with one another, this interpretation makes little or no sense. As it turns out, however, when we use quantum field theory to treat a macroscopic sample of molecules, the interpretation given by Eq. (4.75) introduces the volume in an entirely consistent way. We shall return to this point in Chapter 6. The indicated replacements give

$$\hat{H} = c\int d\boldsymbol{p}\sqrt{m^2c^2+p^2}\left[\hat{a}^\dagger(\boldsymbol{p})\hat{a}(\boldsymbol{p}) + \frac{V}{2h^3}\right] \qquad (4.76)$$

$$\hat{P} = \int d\boldsymbol{p}\boldsymbol{p}\left[\hat{a}^{\dagger}(\boldsymbol{p})\hat{a}(\boldsymbol{p}) + \frac{V}{2h^3}\right] \tag{4.77}$$

$$\hat{H} = c\int d\boldsymbol{p}\sqrt{m^2c^2 + \boldsymbol{p}^2}\hat{a}^{\dagger}(\boldsymbol{p})\hat{a}(\boldsymbol{p}). \tag{4.78}$$

We see that the "volume" terms in these equations lead to both an infinite energy and an infinite momentum. So here we just ignore these infinite terms and use

$$\hat{\boldsymbol{P}} = \int d\boldsymbol{p}\boldsymbol{p}\hat{a}^{\dagger}(\boldsymbol{p})\hat{a}(\boldsymbol{p}). \tag{4.79}$$

The forms of these operators for the energy and momentum of the particles corresponding to the excitations of the quantized Klein-Gordon field indicate that $\hat{a}^{\dagger}(\boldsymbol{p})\hat{a}(\boldsymbol{p})$ is the operator for the number of particles with momentum between $\boldsymbol{p}$ and $\boldsymbol{p} + d\boldsymbol{p}$. The operator for the total number of particles is therefore

$$\hat{N} = \int d\boldsymbol{p}\hat{N}(\boldsymbol{p}) = \int d\boldsymbol{p}\hat{a}^{\dagger}(\boldsymbol{p})\hat{a}(\boldsymbol{p}). \tag{4.80}$$

The question now arises: what is the operator $\hat{\boldsymbol{R}}$ that is conjugate to the total particle momentum operator $\hat{\boldsymbol{P}}$ and what is the form of the commutation relation $[\hat{\boldsymbol{P}}, \hat{\boldsymbol{R}}]_{-}$? An obvious candidate for $\hat{\boldsymbol{R}}$ is

$$\hat{\boldsymbol{R}} = \int d\boldsymbol{r}\boldsymbol{r}\hat{\psi}^{\dagger}(\boldsymbol{r})\hat{\psi}(\boldsymbol{r}), \tag{4.81}$$

where

$$\hat{\psi}(\boldsymbol{r}) = \frac{1}{h^{\frac{3}{2}}}\int d\boldsymbol{p}e^{\frac{i}{\hbar}\boldsymbol{p}\cdot\boldsymbol{r}}\hat{a}(\boldsymbol{p}). \tag{4.82}$$

With a little effort, it can be shown that for this definition of $\hat{\boldsymbol{R}}$ the commutation relation is

$$[\hat{P}_j, \hat{R}_k]_{-} = \frac{\hbar}{i}\hat{N}\delta_{jk} \tag{4.83}$$

and that

$$\int d\boldsymbol{r}\hat{\psi}^{\dagger}(\boldsymbol{r})\hat{\psi}(\boldsymbol{r}) = \hat{N}. \tag{4.84}$$

The commutation relation is quite satisfactory, since the replacement of 1 by the operator $\hat{N}$ is appropriate for the generalization from 1 particle to a system of multiple particles. Furthermore, Eq. (4.84) indicates that $\hat{\psi}^{\dagger}(\boldsymbol{r})\hat{\psi}(\boldsymbol{r})$ is the operator for the number of particles with position between $\boldsymbol{r}$ and $\boldsymbol{r} + d\boldsymbol{r}$, in analogy with the corresponding result for momentum. Thus, we interpret $\hat{a}^{\dagger}(\boldsymbol{p})$ to be the creation operator for a particle with momentum $\boldsymbol{p}$, and $\hat{\psi}^{\dagger}(\boldsymbol{r})$ to be the creation operator for a particle at point $\boldsymbol{r}$; the corresponding annihilation operators are the respective Hermitian conjugates.

4.4.3. *Excitation States of the Quantized Klein-Gordon Field*

The "ground state" of the system is the vacuum *ket*, $|0\rangle$, such that the vacuum bracket is normalized to 1; that is, $\langle 0|0\rangle = 1$. The excited *ket* states are indicated by left-multiplication of $|0\rangle$ by a string of n particle creation operators, $\hat{a}^{\dagger}(\boldsymbol{p})$; for example,

$$\hat{a}^{\dagger}(\boldsymbol{p}_n)\cdots\hat{a}^{\dagger}(\boldsymbol{p}_2)\hat{a}^{\dagger}(\boldsymbol{p}_1)|0\rangle. \tag{4.85a}$$

This n-particle *ket* has the corresponding *bra*

$$\langle 0|\hat{a}(\boldsymbol{p}_1)\hat{a}(\boldsymbol{p}_2)\cdots\hat{a}(\boldsymbol{p}_n). \tag{4.85b}$$

The states are normalized to $(V/\hbar^3)^n$, as can be shown by evaluating

$$\langle 0|\hat{a}(\boldsymbol{p}_1)\hat{a}(\boldsymbol{p}_2)\cdots\hat{a}(\boldsymbol{p}_n)|\hat{a}^{\dagger}(\boldsymbol{p}_n)\cdots\hat{a}^{\dagger}(\boldsymbol{p}_2)\hat{a}^{\dagger}(\boldsymbol{p}_1)|0\rangle \tag{4.86}$$

according to the commutation rules of Eq. (4.74) and using the fact that

$$\hat{a}(\boldsymbol{p})|0\rangle = 0 \tag{4.87}$$

for all values of $\boldsymbol{p}$, since this represents annihilation of the vacuum. A more systematic method of evaluating vacuum expectations of mixed strings of creation and annihilation operators is Wick's theorem, which we have introduced in Chapter 2 and will discuss in some detail in Chapter 6.

As an illustration of the calculation of expectation values of operators in a single-particle state of the quantized Klein-Gordon field, consider

$$\langle 0|\hat{a}(\boldsymbol{p})|\hat{H}|\hat{a}^\dagger(\boldsymbol{p})|0\rangle. \tag{4.88}$$

From the commutation rules, Eq. (4.78), and making use of the normalization, we get for the expectation of $\hat{H}$ in a field containing a single particle with momentum $\boldsymbol{p}$

$$\frac{\langle 0|\hat{a}(\boldsymbol{p})|\hat{H}|\hat{a}^\dagger(\boldsymbol{p})|0\rangle}{\langle 0|\hat{a}(\boldsymbol{p})\hat{a}^\dagger(\boldsymbol{p})|0\rangle} = c\sqrt{m^2c^2 + \boldsymbol{p}^2}, \tag{4.89}$$

as we would expect. Similarly, we get

$$\frac{\langle 0|\hat{a}(\boldsymbol{p})|\hat{P}|\hat{a}^\dagger(\boldsymbol{p})|0\rangle}{\langle 0|\hat{a}(\boldsymbol{p})\hat{a}^\dagger(\boldsymbol{p})|0\rangle} = \boldsymbol{p} \tag{4.90}$$

and

$$\frac{\langle 0|\hat{a}(\boldsymbol{p})|\hat{N}|\hat{a}^\dagger(\boldsymbol{p})|0\rangle}{\langle 0|\hat{a}(\boldsymbol{p})\hat{a}^\dagger(\boldsymbol{p})|0\rangle} = 1. \tag{4.91}$$

The reader may work out the corresponding expectation values for 2- and 3-particle states as a useful exercise to gain familiarity with the formalism and the use of the commutation rules for Klein-Gordon particles.

The final example in our discussion of the states of the quantized Klein-Gordon field makes contact with the wave functions of the Schrödinger quantum theory. Consider the coördinate representation of an eigenstate of momentum. In our formalism this works out to be

$$\langle 0|\hat{\psi}(\boldsymbol{r})\hat{a}^\dagger(\boldsymbol{p})|0\rangle = h^{-\frac{3}{2}}e^{\frac{i}{\hbar}\boldsymbol{p}\cdot\boldsymbol{r}}, \tag{4.92}$$

which is the delta-function normalized momentum eigenfunction.

4.5. Time Dependence: The Schrödinger and Heisenberg Pictures

The treatment of time dependence in quantum field theory is essentially the same as in quantum mechanics. Although we have discussed time dependence in some length in Chapter 3, for completeness we include here a brief discussion in the language of quantum field theory of multi-particle states.

4.5.1. *The Schrödinger Picture*

In the Schrödinger picture, the wave function depends upon time according to the equation

$$i\hbar\frac{\partial}{\partial t}\Psi_{\rm S}(\boldsymbol{r},t) = \hat{H}\Psi_{\rm S}(\boldsymbol{r},t), \tag{4.93}$$

where the Hamiltonian operator $\hat{H}$ is independent of t, as are the operators for all observables in the theory; we indicate the Schrödinger picture by the subscript S on the wave function. Integration of Eq. (4.93) gives

$$\Psi_{\rm S}(\boldsymbol{r},t_2) = e^{-\frac{i}{\hbar}\hat{H}(t_2-t_1)}\boldsymbol{\Psi}_{\rm S}(\boldsymbol{r},t_1) = \hat{U}(t_2-t_1)\boldsymbol{\Psi}_{\rm S}(\boldsymbol{r},t_1), \tag{4.94}$$

which defines the time-evolution operator $\hat{U}(t)$. For a single-particle state in quantum field theory, this translates to

$$\hat{a}^\dagger_{\rm S}(\boldsymbol{p},t)|0\rangle = e^{-\frac{i}{\hbar}\hat{H}t}\hat{a}^\dagger(\boldsymbol{p})|0\rangle, \tag{4.95}$$

with $\hat{H}$ given by Eq. (4.78). Using the techniques introduced in Chapter 3, we find that

$$\hat{a}^\dagger_{\rm S}(\boldsymbol{p},t) = e^{-\frac{i}{\hbar}\hat{H}t}\,\hat{a}^\dagger(\boldsymbol{p})e^{\frac{i}{\hbar}\hat{H}t} = e^{-\frac{i}{\hbar}\varepsilon(\boldsymbol{p})t}\,\hat{a}^\dagger(\boldsymbol{p}), \tag{4.96}$$

where $\varepsilon(\boldsymbol{p}) = c\sqrt{m^2c^2+\boldsymbol{p}^2}$. For a multi-particle state, we have

$$\begin{aligned} &e^{-\frac{i}{\hbar}\hat{H}t}\hat{a}^\dagger(\boldsymbol{p}_n)\cdots\hat{a}^\dagger(\boldsymbol{p}_2)\hat{a}^\dagger(\boldsymbol{p}_1)|0\rangle \\ &\quad= \hat{a}^\dagger_{\rm S}(\boldsymbol{p}_n,t)\cdots\hat{a}^\dagger_{\rm S}(\boldsymbol{p}_2 t)\hat{a}^\dagger_{\rm S}(\boldsymbol{p}_1,t)|0\rangle \\ &\quad= e^{-\frac{i}{\hbar}\sum_{j=1}^n\varepsilon(\boldsymbol{p}_j)t}\,\hat{a}^\dagger(\boldsymbol{p}_n)\cdots\hat{a}^\dagger(\boldsymbol{p}_2)\hat{a}^\dagger(\boldsymbol{p}_1)|0\rangle. \end{aligned} \tag{4.97}$$

Note that the normalization is independent of time, since the *bra* for the state is the Hermitian conjugate of the *ket*, and $\hat{U}^\dagger(t)\hat{U}(t) = 1$.

The time-dependent expectation value of the time-independent Schrödinger-picture operator $\hat{O}_S$ is

$$\frac{\langle 0|\hat{a}(\boldsymbol{p}_1)\hat{a}(\boldsymbol{p}2)\cdots\hat{a}(\boldsymbol{p}_n)e^{\frac{i}{\hbar}\hat{H}t}|\hat{O}_{\mathrm{S}}|e^{-\frac{i}{\hbar}\hat{H}t}\hat{a}^\dagger(\boldsymbol{p}_n)\cdots\hat{a}^\dagger(\boldsymbol{p}_2)\hat{a}^\dagger(\boldsymbol{p}_1)|0\rangle}{\langle 0|\hat{a}(\boldsymbol{p}_1)\hat{a}(\boldsymbol{p}_2)\cdots\hat{a}(\boldsymbol{p}_n)|\hat{a}^\dagger(\boldsymbol{p}_n)\cdots\hat{a}^\dagger(\boldsymbol{p}_2)\hat{a}^\dagger(\boldsymbol{p}_1)|0\rangle}. \tag{4.98}$$

4.5.2. *The Heisenberg Picture*

In the Heisenberg picture, expectation values are still given by Eq. (4.98), but the states are taken to be time-independent and the operators to be time-dependent. Thus, the Heisenberg operator $\hat{O}_{\mathrm{H}}(t)$ is defined as

$$\hat{O}_{\mathrm{H}}(t) = e^{\frac{i}{\hbar}\hat{H}t}\,\hat{O}_{\mathrm{S}}e^{-\frac{i}{\hbar}\hat{H}t} \tag{4.99}$$

and expectation values are given by

$$\frac{\langle 0|\hat{a}(\boldsymbol{p}_1)\hat{a}(\boldsymbol{p}_2)\cdots\hat{a}(\boldsymbol{p}_n)|\hat{O}_{\mathrm{H}}(t)|\hat{a}^\dagger(\boldsymbol{p}_n)\cdots\hat{a}^\dagger(\boldsymbol{p}_2)\hat{a}^\dagger(\boldsymbol{p}_1)|0\rangle}{\langle 0|\hat{a}(\boldsymbol{p}_1)\hat{a}(\boldsymbol{p}_2)\cdots\hat{a}(\boldsymbol{p}_n)|\hat{a}^\dagger(\boldsymbol{p}_n)\cdots\hat{a}^\dagger(\boldsymbol{p}_2)\hat{a}(\boldsymbol{p}_1)|0\rangle}. \tag{4.100}$$

If the operator $\hat{O}$ is composed of a string of operators $\hat{a}(\boldsymbol{p})$ and $\hat{a}^\dagger(\boldsymbol{p})$, their Heisenberg time-dependent forms are

$$\hat{a}_{\mathrm{H}}(\boldsymbol{p},t) = e^{-\frac{i}{\hbar}\varepsilon(\boldsymbol{p})t}\,\hat{a}(\boldsymbol{p}), \quad \hat{a}_{\mathrm{H}}^\dagger(\boldsymbol{p},t) = e^{\frac{i}{\hbar}\varepsilon(\boldsymbol{p})t}\,\hat{a}(\boldsymbol{p})^\dagger. \tag{4.101}$$

This contrasts with the Schrödinger form for these operators, as indicated in Eq. (4.98).

As an illustration of calculations in the Heisenberg picture, we may obtain the Hamiltonian equations of motion. The time derivative of an operator is its commutator with the Hamiltonian operator:

$$\frac{\partial}{\partial t}\hat{O}(\boldsymbol{r},t) = \frac{\partial}{\partial t}\left[e^{\frac{i}{\hbar}\hat{H}t}\hat{O}(\boldsymbol{r},0)e^{-\frac{i}{\hbar}\hat{H}t}\right] = \frac{i}{\hbar}[\hat{H},\hat{O}(\boldsymbol{r},t)]_-. \tag{4.102}$$

Thus, from the quantized version of Eq. (4.46) and the canonical commutation rule, Eq. (4.71), we have the time derivatives of the fields at $t = 0$ as follows:

$$\frac{\partial}{\partial t}\hat{\varphi}(\boldsymbol{r},t)\Big|_{t=0} = \frac{ic}{2\hbar}\int d\boldsymbol{r}' \left\{[\hat{\mathscr{P}}^2(\boldsymbol{r}'),\hat{\varphi}(\boldsymbol{r})]_-\right\}$$
$$= \frac{ic}{2\hbar}\int d\boldsymbol{r}' \left\{2\hat{\mathscr{P}}(\boldsymbol{r})[\hat{\mathscr{P}}(\boldsymbol{r}'),\hat{\varphi}(\boldsymbol{r})]_-\right\} = c\hat{\mathscr{P}}(\boldsymbol{r}), \quad (4.103)$$

$$\frac{\partial}{\partial t}\hat{\mathscr{P}}(\boldsymbol{r},t)\Big|_{t=0} = c\left(\nabla^2 - \frac{m^2c^2}{\hbar^2}\right)\hat{\varphi}(\boldsymbol{r}). \quad (4.104)$$

Notice that Eq. (4.103) is in agreement with Eq. (4.45). Then, combining Eqs. (4.103) and (4.104), we obtain the Klein-Gordon equation:

$$\frac{1}{c^2}\frac{\partial}{\partial t^2}\hat{\varphi}(\boldsymbol{r},t) = \frac{1}{c}\frac{\partial}{\partial t}\hat{\mathscr{P}}(\boldsymbol{r},t) = \left(\nabla^2 - \frac{m^2c^2}{\hbar^2}\right)\hat{\varphi}(\boldsymbol{r},t), \quad (4.105)$$

or, upon rearrangement,

$$\left(\frac{1}{c^2}\frac{\partial}{\partial t^2} - \nabla^2 + \frac{m^2c^2}{\hbar^2}\right)\hat{\varphi}(\boldsymbol{r},t) = 0. \quad (4.106)$$

Eqs. (4.103)–(4.106) show that our definition of the canonical momentum conjugate to $\varphi(x)$ given in Eq. (4.45) is consistent with the Hamiltonian equations of motion.

4.6. A Note on Relativistic Invariance

In anticipation of our applications of quantum field theory to molecules at non-relativistic velocities in later chapters, we have not paid strict attention to the requirements of relativistic invariance. In particular, the integrals over the 3-vector $\boldsymbol{p}$ in this chapter should have been over the 4-vector p. For example, the Fourier transform of Eq. (4.83) might properly take the form

$$\hat{\psi}(\boldsymbol{r},t) = \frac{1}{h^2}\int d^4p e^{-\frac{i}{\hbar}[-\varepsilon(\boldsymbol{p})t+\boldsymbol{p}\cdot\boldsymbol{r}]}\hat{a}(\boldsymbol{p}),$$

or in space-time notation,

$$\hat{\psi}(x) = \frac{1}{h^2}\int d^4p e^{-\frac{i}{\hbar}p_\mu x^\mu}\hat{a}(\boldsymbol{p}).$$

But this does not restrict the integration variables to the "energy shell"; *i.e.*, to positive values of p_0 satisfying the Klein-Gordon equation:

$$p_0^2 = m^2c^2 + \boldsymbol{p}^2,\ p_0 \leq 0.$$

This restriction is accomplished by inserting a relativistically invariant delta function into the integrand:

$$\begin{aligned}\hat{\psi}(x) &= \frac{1}{h^2}\int d^4pe^{-\frac{i}{\hbar}p_\mu x^\mu}\delta(p^2 - m^2c^2)\hat{a}(\boldsymbol{p}) \\ &= \frac{1}{h^2}\int dp_0 \int d\boldsymbol{p}\delta(p_0^2 - m^2c^2 - \boldsymbol{p}^2)e^{-\frac{i}{\hbar}p_\mu x^\mu}\hat{a}(\boldsymbol{p}).\end{aligned} \tag{4.107}$$

Now we must convert the delta function to one whose argument is a linear function of the variable of integration. This conversion is facilitated by the identity

$$\delta[f(y) - f(y_0)]df(y) = \delta(y - y_0)dy, \tag{4.108}$$

from which we obtain

$$\delta[f(y) - f(y_0)] = \left(\frac{df}{dy}\right)^{-1}\delta(y - y_0).$$

In the present case, this gives

$$\delta(p_0^2 - m^2c^2 - \boldsymbol{p}^2) = \frac{1}{2p_0}\delta\left[p_0 - \frac{\varepsilon(\boldsymbol{p})}{c}\right].$$

After inserting this expression into Eq. (4.107) and integrating over p_0 we obtain

$$\hat{\psi}(\boldsymbol{r}, t) = \frac{c}{h^2}\int \frac{d\boldsymbol{p}}{2\varepsilon(\boldsymbol{p})}e^{-\frac{i}{\hbar}\varepsilon(\boldsymbol{p})t}e^{\frac{i}{\hbar}\boldsymbol{p}\cdot\boldsymbol{r}}\hat{a}(\boldsymbol{p}). \tag{4.109}$$

We will similarly obtain for the relavistic commutator

$$[\hat{a}(\boldsymbol{p}), \hat{a}^\dagger(\boldsymbol{p}')]_- = \frac{2\varepsilon(\boldsymbol{p})}{c}\delta(\boldsymbol{p} - \boldsymbol{p}')$$

instead of Eq. (4.74b). In the non-relativistic regime, $\varepsilon(\boldsymbol{p})/c$ is essentially mc, so that the effect of relativistic invariance is to change normalization constants. For energies *not* infinitesimally close to mc^2 the appearance of $\varepsilon(\boldsymbol{p})/c$ in the 3-momentum integrals presents an additional complication. Fortunately, we may ignore it in applications to molecular ensembles under normally observed conditions.

4.7. Summary for Chapter 4

In this chapter, we have shown how the theory of a classical field obeying the Klein-Gordon equation can be developed from Lagrangian mechanics and how the theory is quantized by requiring that the field and its canonically conjugate momentum obey a natural generalization of the coördinate-momentum commutation rules of quantum mechanics. The result is an extension of the Fock-operator treatment of the harmonic oscillator to a quantized field whose quanta are relativistic spinless particles of a specified momentum $\boldsymbol{p}$ and energy ε_0. The commutation rules for the creation and annihilation operators for these particles, $\hat{a}^\dagger(\boldsymbol{p})$ and $\hat{a}(\boldsymbol{p})$, respectively, are worked out and found to be a straightforward generalization of those for the corresponding harmonic-oscillator Fock operators; the particles therefore obey Bose-Einstein statistics. The states of the field are described by both the Schrödinger and Heisenberg pictures of time-dependent calculations. The Heisenberg picture is used to derive the Hamiltonian equations of motion. In the next chapter, charged particles and particles with non-zero spin are considered and elements of the theory of the Dirac field are presented as a model for spin-$\frac{1}{2}$ particles that are seen to be obeying Fermi-Dirac statistics.

Chapter 5

Charge and Spin

5.1. Introduction

Now that we have outlined the basic features of the quantum field theory of spinless particles in the preceding chapter, we extend the theory to charged particles and to those with spin. To imbue particles with charge, we take the fields to be complex, with $\varphi(x)$ and $\varphi^*(x)$ independent of one another. The charge is the spatial integral of the time-component of the conserved current density when the Lagrangian is invariant with respect to variations in the global phase angle of the field.

The intrinsic angular momentum, or spin, of particles results from taking the field $\varphi(x)$ to be an n-component vector in space-time (the spatial component of which is $(n-1)$-dimensional. If n is odd, the field obeys the Klein-Gordon equation; if n is even, the field obeys the Dirac equation. The meaning of these statements will become clear as this chapter develops.

5.2. Charged Particles

Quantum field theory treats *charged* particles as the quanta of a *complex* field. We first treat the case of spinless charged particles, for which the complex field is a scalar with respect to Lorentz transformations.

5.2.1. *Complex Classical Scalar Fields*

If the classical scalar field $\varphi(x)$ is complex, there are two independent numbers, $\varphi(x)$ and its complex conjugate $\varphi^*(x)$, associated with each point x in space-time. As we shall see, each of these fields obeys the Klein-Gordon equation. The Lagrangian density which assures this is

$$\begin{aligned}\mathscr{L}&[\varphi(x), \varphi^*(x), \partial_\mu\varphi(x), \partial_\mu\varphi^*(x)]\\ &= \partial^\mu\varphi^*(x)\partial_\mu\varphi(x) - \frac{m^2c^2}{\hbar^2}\varphi^*(x)\varphi(x). \end{aligned} \tag{5.1}$$

Application of the Principle of Least Action gives

$$\delta S = \int d^4x\delta\mathscr{L} = 0, \tag{5.2}$$

where

$$\begin{aligned}\delta\mathscr{L} &= \frac{\partial\mathscr{L}}{\partial(\partial_\mu\varphi)}\delta(\partial_\mu\varphi) + \frac{\partial\mathscr{L}}{\partial(\partial^\mu\varphi^*)}\delta(\partial^\mu\varphi^*)\\ &\quad + \frac{\partial\mathscr{L}}{\partial\varphi}\delta\varphi + \frac{\partial\mathscr{L}}{\partial\varphi^*}\delta\varphi^* = 0. \end{aligned} \tag{5.3}$$

If we rewrite the first two terms in Eq. (5.3) as

$$\begin{aligned}\frac{\partial\mathscr{L}}{\partial(\partial_\mu\varphi)}\delta(\partial_\mu\varphi) &= \partial_\mu\left[\frac{\partial\mathscr{L}}{\partial(\partial_\mu\varphi)}\delta\varphi\right] - \partial_\mu\left(\frac{\partial\mathscr{L}}{\partial_\mu\varphi}\right)\delta\varphi\\ \frac{\partial\mathscr{L}}{\partial(\partial_\mu\varphi^*)}\delta(\partial_\mu\varphi^*) &= \partial_\mu\left[\frac{\partial\mathscr{L}}{\partial(\partial^\mu\varphi^*)}\delta\varphi^*\right] - \partial_\mu\left(\frac{\partial\mathscr{L}}{\partial^\mu\varphi^*}\right)\delta\varphi^* \end{aligned} \tag{5.4}$$

and perform the integral in Eq. (5.2), using the fact that both $\delta\varphi$ and $\delta\varphi^*$ vanish on the boundary of the integration volume,[1] we have

$$\begin{aligned}\delta S = \int d^4x\Bigg\{&\left[\frac{\partial\mathscr{L}}{\partial\varphi} - \partial_\mu\left(\frac{\partial\mathscr{L}}{\partial(\partial_\mu\varphi)}\right)\right]\delta\varphi\\ &+ \left[\frac{\partial\mathscr{L}}{\partial\varphi^*} - \partial_\mu\left(\frac{\partial\mathscr{L}}{\partial(\partial_\mu\varphi^*)}\right)\right]\delta\varphi^*\Bigg] = 0. \end{aligned} \tag{5.5}$$

[1]See the discussion in Chapter 4 between Eqs. (4.38) and (4.39).

Since Eq. (5.5) is assumed to hold for arbitrary values of both $\delta\varphi$ and $\delta\varphi^*$ within the integration volume, the two terms in brackets must vanish independently:

$$\partial_\mu\left[\frac{\partial\mathscr{L}}{\partial(\partial_\mu\varphi)}\right]-\frac{\partial\mathscr{L}}{\partial\varphi}=0,\quad \partial_\mu\left[\frac{\partial\mathscr{L}}{\partial(\partial^\mu\varphi^*)}\right]-\frac{\partial\mathscr{L}}{\partial\varphi^*}=0. \tag{5.6}$$

Using Eq. (5.1) for $\mathscr{L}$, we have

$$\begin{aligned}\frac{\partial\mathscr{L}}{\partial(\partial_\mu\varphi)}&=\partial^\mu\varphi^*, & \frac{\partial\mathscr{L}}{\partial\varphi}&=-\frac{m^2c^2}{\hbar^2}\varphi^*,\\ \frac{\partial\mathscr{L}}{\partial(\partial^\mu\varphi^*)}&=\partial_\mu\varphi^*, & \frac{\partial\mathscr{L}}{\partial\varphi^*}&=-\frac{m^2c^2}{\hbar^2}\varphi.\end{aligned} \tag{5.7}$$

When these expressions are inserted into Eqs. (5.6), the result is

$$\partial_\mu\partial^\mu\varphi^*+\frac{m^2c^2}{\hbar^2}\varphi^*=0,\quad \partial^\mu\partial_\mu\varphi+\frac{m^2c^2}{\hbar^2}\varphi=0. \tag{5.8}$$

Thus, the Lagrangian density given in Eq. (5.1) requires that both φ and φ^* obey the Klein-Gordon equation. Our next step is to find the effect upon $\mathscr{L}$ when φ and φ^* are modified by a global (*i.e.*, a constant) change in phase, $\varphi\to\varphi e^{i\alpha}$; $\varphi^*\to\varphi^*e^{-i\alpha}$. The total derivative of $\mathscr{L}$ with respect to the phase angle α is

$$\begin{aligned}\frac{d\mathscr{L}}{d\alpha}&=\frac{\partial\mathscr{L}}{\partial(\partial^\mu\varphi^*)}\frac{\partial(\partial^\mu\varphi^*)}{\partial\alpha}+\frac{\partial\mathscr{L}}{\partial\varphi^*}\frac{\partial\varphi^*}{\partial\alpha}+\frac{\partial\mathscr{L}}{\partial(\partial_\mu\varphi)}\frac{\partial(\partial_\mu\varphi)}{\partial\alpha}+\frac{\partial\mathscr{L}}{\partial\varphi}\frac{\partial\varphi}{\partial\alpha}\\ &=\partial^\mu\left[\frac{\partial\mathscr{L}}{\partial(\partial^\mu\varphi^*)}\frac{\partial\varphi^*}{\partial\alpha}\right]-\partial^\mu\left[\frac{\partial\mathscr{L}}{\partial(\partial^\mu\varphi^*)}\right]\frac{\partial\varphi^*}{\partial\alpha}+\frac{\partial\mathscr{L}}{\partial\varphi^*}\frac{\partial\varphi^*}{\partial\alpha}\\ &\quad+\partial_\mu\left[\frac{\partial\mathscr{L}}{\partial(\partial_\mu\varphi)}\frac{\partial\varphi}{\partial\alpha}\right]-\partial_\mu\left[\frac{\partial\mathscr{L}}{\partial(\partial_\mu\varphi)}\right]\frac{\partial\varphi}{\partial\alpha}+\frac{\partial\mathscr{L}}{\partial\varphi}\frac{\partial\varphi}{\partial\alpha}.\end{aligned} \tag{5.9}$$

The cancellation of terms indicated in Eq. (5.9) is the result of insertion of the equations of motion, Eqs. (5.6).

If $\mathscr{L}$ is invariant with respect to changes of phase angle,

$$\frac{d\mathscr{L}}{d\alpha}=\partial^\mu\left[\frac{\partial\mathscr{L}}{\partial(\partial^\mu\varphi^*)}\frac{\partial\varphi^*}{\partial\alpha}\right]+\partial_\mu\left[\frac{\partial\mathscr{L}}{\partial(\partial_\mu\varphi)}\frac{\partial\varphi}{\partial\alpha}\right]=0. \tag{5.10}$$

Insertion into Eq. (5.10) of expressions for the derivatives of $\mathscr{L}$ from Eqs. (5.7) and use of

$$\frac{\partial\varphi}{\partial\alpha} = i\varphi, \quad \frac{\partial\varphi^*}{\partial\alpha} = -i\varphi^* \tag{5.11}$$

gives

$$-i\partial_\mu[\varphi^*\partial^\mu\varphi - (\partial^\mu\varphi^*)\varphi] = 0. \tag{5.12}$$

This equation identifies the conserved current density

$$j^\mu(x) = -i\{\varphi^*(x)\partial^\mu\varphi(x) - [\partial^\mu\varphi^*(x)]\varphi(x)\}, \tag{5.13}$$

such that

$$\partial_\mu j^\mu(x) = 0, \tag{5.14}$$

$$c\int d\boldsymbol{r}\,\partial_\mu j^\mu(x) = c\int d\boldsymbol{r}\,\partial_0 j^0(x) - c\int d\boldsymbol{r}\nabla\cdot\boldsymbol{j}(x) = 0. \tag{5.15}$$

As we shall see, we may identify $j^\mu(x)$ with the *charge current density* for the field quanta. Integration of Eq. (5.14) over space gives where the second term on the right-hand side is equal to zero because the field vanishes on the surface of the integration volume.[2] Thus we have

$$c\partial_0\int d\boldsymbol{r}j^0(x) = \frac{d}{dt}\int d\boldsymbol{r}j^0(x) = 0. \tag{5.16}$$

5.2.2. *Quantum Field Theory of Charged Particles*

At this point, we quantize the field by finding the canonical momenta $\wp$ and $\wp^*$ conjugate to φ and φ^*, respectively, impose the commutation relations, and introduce the Fock creation and annihilation operators. From Eq. (5.1), we see that

$$\wp(x) = \frac{\partial\mathscr{L}}{\partial(\partial_0\varphi)} = \partial^0\varphi^*(x), \quad \wp^*(x) = \frac{\partial\mathscr{L}}{\partial(\partial^0\varphi^*)} = \partial_0\varphi(x), \tag{5.17}$$

[2]See Eq. (4.65) and the discussion following it.

so that the Hamiltonian is

$$\begin{aligned} H &= c\int d\boldsymbol{r}\{\wp^*(x)\partial^0\varphi^*(x) + \wp(x)\partial_0\varphi(x) - \mathscr{L}\} \\ &= c\int d\boldsymbol{r}\{2\partial^0\varphi^*\partial_0\varphi - \mathscr{L}\} \\ &= c\int d\boldsymbol{r}\left\{\wp^*(x)\wp(x) + \nabla\varphi^*(x)\cdot\nabla\varphi(x) + \frac{m^2c^2}{\hbar^2}\varphi^*(x)\varphi(x)\right\}. \end{aligned} \tag{5.18}$$

Since φ and φ^* are independent fields, the canonical commutation relations for the corresponding operators and their conjugate momenta at time $t = 0$ are

$$\begin{aligned} [\hat{\wp}(\boldsymbol{r}), \hat{\varphi}(\boldsymbol{r}')]_- &= [\hat{\wp}^\dagger(\boldsymbol{r}), \hat{\varphi}^\dagger(\boldsymbol{r}')]_- = \delta(\boldsymbol{r}-\boldsymbol{r}') \\ [\hat{\varphi}^\dagger(\boldsymbol{r}), \hat{\varphi}(\boldsymbol{r}')]_- &= [\hat{\wp}^\dagger(\boldsymbol{r}), \hat{\wp}(\boldsymbol{r}')]_- = 0 \\ [\hat{\wp}^\dagger(\boldsymbol{r}), \hat{\varphi}(\boldsymbol{r}')]_- &= [\hat{\wp}(\boldsymbol{r}), \hat{\varphi}^\dagger(\boldsymbol{r}')]_- = 0. \end{aligned} \tag{5.19}$$

We may write the operators for the fields and canonical momenta in terms of the Fock operators as follows:

$$\begin{aligned} \hat{\varphi}(\boldsymbol{r}) &= \frac{\hbar}{\sqrt{2h^3}}\int d\boldsymbol{p}(m^2c^2+\boldsymbol{p}^2)^{-\frac{1}{4}}\left[\hat{a}(\boldsymbol{p})e^{\frac{i}{\hbar}\boldsymbol{p}\cdot\boldsymbol{r}} + \hat{b}^\dagger(\boldsymbol{p})e^{-\frac{i}{\hbar}\boldsymbol{p}\cdot\boldsymbol{r}}\right] \\ \hat{\varphi}^\dagger(\boldsymbol{r}) &= \frac{\hbar}{\sqrt{2h^3}}\int d\boldsymbol{p}(m^2c^2+\boldsymbol{p}^2)^{-\frac{1}{4}}\left[\hat{b}(\boldsymbol{p})e^{\frac{i}{\hbar}\boldsymbol{p}\cdot\boldsymbol{r}} + \hat{a}^\dagger(\boldsymbol{p})e^{-\frac{i}{\hbar}\boldsymbol{p}\cdot\boldsymbol{r}}\right] \\ \hat{\wp}(\boldsymbol{r}) &= \frac{(-i)}{\sqrt{2h^3}}\int d\boldsymbol{p}(m^2c^2+\boldsymbol{p}^2)^{-\frac{1}{4}}\left[\hat{b}(\boldsymbol{p})e^{\frac{i}{\hbar}\boldsymbol{p}\cdot\boldsymbol{r}} - \hat{a}^\dagger(\boldsymbol{p})e^{-\frac{i}{\hbar}\boldsymbol{p}\cdot\boldsymbol{r}}\right] \\ \hat{\wp}^\dagger(\boldsymbol{r}) &= \frac{(-i)}{\sqrt{2h^3}}\int d\boldsymbol{p}(m^2c^2+\boldsymbol{p}^2)^{-\frac{1}{4}}\left[\hat{a}(\boldsymbol{p})e^{\frac{i}{\hbar}\boldsymbol{p}\cdot\boldsymbol{r}} - \hat{b}^\dagger(\boldsymbol{p})e^{-\frac{i}{\hbar}\boldsymbol{p}\cdot\boldsymbol{r}}\right]. \end{aligned} \tag{5.20}$$

Note that the expressions in Eq. (5.20) are similar to those in Eqs. (4.70) and (4.71) for real fields, except that the quantized operators corresponding to complex fields are no longer Hermitian.

The canonical commutation relations of Eqs. (5.19) are satisfied if

$$[\hat{a}(\boldsymbol{p}), \hat{a}^\dagger(\boldsymbol{p}')]_- = [\hat{b}(\boldsymbol{p}), \hat{b}^\dagger(\boldsymbol{p}')]_- = \delta(\boldsymbol{p} - \boldsymbol{p}') \tag{5.21}$$

and all the commutators between an $\hat{a}$ or $\hat{a}^\dagger$ and a $\hat{b}$ or $\hat{b}^\dagger$ are equal to zero. To demonstrate this, we use Eqs. (5.20) to evaluate the commutation relations of Eq. (5.19):

$$\begin{aligned}[\hat{\wp}(\boldsymbol{r}), \hat{\varphi}(\boldsymbol{r}')]_- = \frac{\hbar}{i}\frac{1}{h^3}\int d\boldsymbol{p}\int d\boldsymbol{p}' \left(\frac{m^2c^2 + \boldsymbol{p}^2}{m^2c^2 + \boldsymbol{p}'^2}\right)^{\frac{1}{4}} \\ \times \frac{1}{2}\{[\hat{b}(\boldsymbol{p}), \hat{b}^\dagger(\boldsymbol{p}')]_- e^{\frac{i}{\hbar}(\boldsymbol{p}\cdot\boldsymbol{r} - \boldsymbol{p}'\cdot\boldsymbol{r}')} \\ - [\hat{a}^\dagger(\boldsymbol{p}), \hat{a}(\boldsymbol{p})]_- e^{-\frac{i}{\hbar}(\boldsymbol{p}\cdot\boldsymbol{r} - \boldsymbol{p}'\cdot\boldsymbol{r}')} \\ + [\hat{b}(\boldsymbol{p}), \hat{a}(\boldsymbol{p}')]_- e^{-\frac{i}{\hbar}(\boldsymbol{p}\cdot\boldsymbol{r} - \boldsymbol{p}'\cdot\boldsymbol{r}')} \\ - [\hat{a}^\dagger(\boldsymbol{p}), \hat{b}^\dagger(\boldsymbol{p}')]_- e^{-\frac{i}{\hbar}(\boldsymbol{p}\cdot\boldsymbol{r} - \boldsymbol{p}'\cdot\boldsymbol{r}')}\}\end{aligned} \tag{5.22a}$$

$$\begin{aligned}[\hat{\wp}^\dagger(\boldsymbol{r}), \hat{\varphi}^\dagger(\boldsymbol{r}')]_- = \frac{\hbar}{i}\frac{1}{h^3}\int d\boldsymbol{p}\int d\boldsymbol{p}' \left(\frac{m^2c^2 + \boldsymbol{p}^2}{m^2c^2 + \boldsymbol{p}'^2}\right)^{\frac{1}{4}} \\ \times \frac{1}{2}\{[\hat{a}(\boldsymbol{p}), \hat{a}^\dagger(\boldsymbol{p}')]_- e^{\frac{i}{\hbar}(\boldsymbol{p}\cdot\boldsymbol{r} - \boldsymbol{p}'\cdot\boldsymbol{r}')} \\ - [\hat{b}^\dagger(\boldsymbol{p}), \hat{b}(\boldsymbol{p}')]_- e^{-\frac{i}{\hbar}(\boldsymbol{p}\cdot\boldsymbol{r} - \boldsymbol{p}'\cdot\boldsymbol{r}')} \\ + [\hat{a}(\boldsymbol{p}), \hat{b}(\boldsymbol{p}')]_- e^{-\frac{i}{\hbar}(\boldsymbol{p}\cdot\boldsymbol{r} - \boldsymbol{p}'\cdot\boldsymbol{r}')} \\ - [\hat{b}^\dagger(\boldsymbol{p}), \hat{a}^\dagger(\boldsymbol{p}')]_- e^{-\frac{i}{\hbar}(\boldsymbol{p}\cdot\boldsymbol{r} - \boldsymbol{p}'\cdot\boldsymbol{r}')}\}\end{aligned} \tag{5.22b}$$

$$\begin{aligned}[\hat{\wp}^\dagger(\boldsymbol{r}), \hat{\varphi}^\dagger(\boldsymbol{r}')]_- = \frac{\hbar}{i}\frac{1}{h^3}\int d\boldsymbol{p}\int d\boldsymbol{p}' \left(\frac{m^2c^2 + \boldsymbol{p}^2}{m^2c^2 + \boldsymbol{p}'^2}\right)^{\frac{1}{4}} \\ \times \frac{1}{2}\{[\hat{b}(\boldsymbol{p}), \hat{b}^\dagger(\boldsymbol{p}')]_- e^{\frac{i}{\hbar}(\boldsymbol{p}\cdot\boldsymbol{r} - \boldsymbol{p}'\cdot\boldsymbol{r}')} \\ - [\hat{a}^\dagger(\boldsymbol{p}), \hat{a}(\boldsymbol{p}')]_- e^{-\frac{i}{\hbar}(\boldsymbol{p}\cdot\boldsymbol{r} - \boldsymbol{p}'\cdot\boldsymbol{r}')} \\ + [\hat{b}(\boldsymbol{p}), \hat{a}(\boldsymbol{p}')]_- e^{\frac{i}{\hbar}(\boldsymbol{p}\cdot\boldsymbol{r} - \boldsymbol{p}'\cdot\boldsymbol{r}')} \\ - [\hat{a}^\dagger(\boldsymbol{p}), \hat{b}^\dagger(\boldsymbol{p}')]_- e^{-\frac{i}{\hbar}(\boldsymbol{p}\cdot\boldsymbol{r} - \boldsymbol{p}'\cdot\boldsymbol{r}')}\}\end{aligned} \tag{5.22c}$$

$$\begin{aligned}[\hat{\wp}^{\dagger}(\boldsymbol{r}), \hat{\wp}(\boldsymbol{r}')]_- = \frac{\hbar}{i}\frac{1}{h^3}\int d\boldsymbol{p}\int d\boldsymbol{p}'\left(\frac{m^2c^2+\boldsymbol{p}^2}{m^2c^2+\boldsymbol{p}'^2}\right)^{\frac{1}{4}} \\ \times\frac{1}{2}\{-[\hat{a}(\boldsymbol{p}), \hat{a}^{\dagger}(\boldsymbol{p}')]_- e^{\frac{i}{\hbar}(\boldsymbol{p}\cdot\boldsymbol{r}-\boldsymbol{p}'\cdot\boldsymbol{r}')} \\ -[\hat{b}^{\dagger}(\boldsymbol{p}), \hat{b}(\boldsymbol{p}')]_- e^{-\frac{i}{\hbar}(\boldsymbol{p}\cdot\boldsymbol{r}-\boldsymbol{p}'\cdot\boldsymbol{r}')} \\ +[\hat{a}(\boldsymbol{p}), \hat{b}(\boldsymbol{p}')]_- e^{-\frac{i}{\hbar}(\boldsymbol{p}\cdot\boldsymbol{r}+\boldsymbol{p}'\cdot\boldsymbol{r}')} \\ \times[\hat{b}^{\dagger}(\boldsymbol{p}), \hat{a}^{\dagger}(\boldsymbol{p}')]_- e^{-\frac{i}{\hbar}(\boldsymbol{p}\cdot\boldsymbol{r}-\boldsymbol{p}'\cdot\boldsymbol{r}')}\}\end{aligned} \tag{5.22d}$$

$$\begin{aligned}[\hat{\varphi}^{\dagger}(\boldsymbol{r}), \hat{\varphi}(\boldsymbol{r}')]_- = \frac{\hbar}{i}\frac{1}{h^3}\int d\boldsymbol{p}\int d\boldsymbol{p}'\left(\frac{m^2c^2+\boldsymbol{p}^2}{m^2c^2+\boldsymbol{p}'^2}\right)^{\frac{1}{4}} \\ \times\frac{1}{2}\{[\hat{b}(\boldsymbol{p}), \hat{b}^{\dagger}(\boldsymbol{p}')]_- e^{\frac{i}{\hbar}(\boldsymbol{p}\cdot\boldsymbol{r}-\boldsymbol{p}'\cdot\boldsymbol{r}')} \\ +[\hat{a}^{\dagger}(\boldsymbol{p}), \hat{a}(\boldsymbol{p}')]_- e^{-\frac{i}{\hbar}(\boldsymbol{p}\cdot\boldsymbol{r}-\boldsymbol{p}'\cdot\boldsymbol{r}')} \\ +[\hat{b}(\boldsymbol{p}), \hat{a}(\boldsymbol{p}')]_- e^{\frac{i}{\hbar}(\boldsymbol{p}\cdot\boldsymbol{r}+\boldsymbol{p}'\cdot\boldsymbol{r}')} \\ +[\hat{a}^{\dagger}(\boldsymbol{p}), \hat{b}^{\dagger}(\boldsymbol{p}')]_- e^{-\frac{i}{\hbar}(\boldsymbol{p}\cdot\boldsymbol{r}+\boldsymbol{p}'\cdot\boldsymbol{r}')}\}\end{aligned} \tag{5.22e}$$

$$\begin{aligned}[\hat{\wp}^{\dagger}(\boldsymbol{r}), \hat{\varphi}(\boldsymbol{r}')]_- = \frac{\hbar}{i}\frac{1}{h^3}\int d\boldsymbol{p}\int d\boldsymbol{p}'\left(\frac{m^2c^2+\boldsymbol{p}^2}{m^2c^2+\boldsymbol{p}'^2}\right)^{\frac{1}{4}} \\ \times\frac{1}{2}\{[\hat{a}(\boldsymbol{p}), \hat{b}^{\dagger}(\boldsymbol{p}')]_- e^{\frac{i}{\hbar}(\boldsymbol{p}\cdot\boldsymbol{r}-\boldsymbol{p}'\cdot\boldsymbol{r}')} \\ -[\hat{b}^{\dagger}(\boldsymbol{p}), \hat{a}(\boldsymbol{p}')]_- e^{-\frac{i}{\hbar}(\boldsymbol{p}\cdot\boldsymbol{r}-\boldsymbol{p}'\cdot\boldsymbol{r}')} \\ +[\hat{a}(\boldsymbol{p}), \hat{a}(\boldsymbol{p}')]_- e^{\frac{i}{\hbar}(\boldsymbol{p}\cdot\boldsymbol{r}+\boldsymbol{p}'\cdot\boldsymbol{r}')} \\ -[\hat{b}^{\dagger}(\boldsymbol{p}), \hat{b}^{\dagger}(\boldsymbol{p}')]_- e^{-\frac{i}{\hbar}(\boldsymbol{p}\cdot\boldsymbol{r}+\boldsymbol{p}'\cdot\boldsymbol{r}')}\}\end{aligned} \tag{5.22f}$$

From Eqs. (5.22), it is easy to see that when the commutation rules of Eq. (5.21) are inserted and all other commutators are set to zero, the result is the canonical commutation rules of Eq. (5.19). The Hamiltonian of Eq. (5.18) is evaluated in terms of the Fock operators as follows:

$$c\int d\boldsymbol{r}\hat{\wp}^{\dagger}(\boldsymbol{r})\wp(\boldsymbol{r}) = \frac{c}{2h^3}\int d\boldsymbol{r}\int d\boldsymbol{p}\int d\boldsymbol{p}'(m^2c^2+\boldsymbol{p}^2)^{\frac{1}{4}}(m^2c^2+\boldsymbol{p}'^2)^{\frac{1}{4}}$$
$$\times\,[\hat{a}(\boldsymbol{p})\hat{a}^{\dagger}(\boldsymbol{p}')e^{\frac{i}{\hbar}(\boldsymbol{p}-\boldsymbol{p}')\cdot\boldsymbol{r}} + \hat{b}^{\dagger}(\boldsymbol{p})\hat{b}(\boldsymbol{p}')e^{-\frac{i}{\hbar}(\boldsymbol{p}-\boldsymbol{p}')\cdot\boldsymbol{r}}$$
$$-\,\hat{a}(\boldsymbol{p})\hat{b}(\boldsymbol{p}')e^{\frac{i}{\hbar}(\boldsymbol{p}+\boldsymbol{p}')\cdot\boldsymbol{r}} - \hat{b}^{\dagger}(\boldsymbol{p})\hat{a}^{\dagger}(\boldsymbol{p}')e^{-\frac{i}{\hbar}(\boldsymbol{p}+\boldsymbol{p}')\cdot\boldsymbol{r}}] \tag{5.23a}$$

$$c\int d\boldsymbol{r}\nabla\hat{\varphi}^{\dagger}(\boldsymbol{r})\cdot\nabla\hat{\varphi}(\boldsymbol{r})$$
$$= \frac{c}{2h^3}\int d\boldsymbol{p}\int d\boldsymbol{p}'(m^2c^2+\boldsymbol{p}^2)^{-\frac{1}{4}}(m^2c^2+\boldsymbol{p}'^2)^{-\frac{1}{4}}\boldsymbol{p}\cdot\boldsymbol{p}'$$
$$\times[+\hat{a}^{\dagger}(\boldsymbol{p})\hat{a}(\boldsymbol{p}')e^{-\frac{i}{\hbar}(\boldsymbol{p}-\boldsymbol{p}')\cdot\boldsymbol{r}} + \hat{b}(\boldsymbol{p})\hat{b}^{\dagger}(\boldsymbol{p}')e^{\frac{i}{\hbar}(\boldsymbol{p}-\boldsymbol{p}')\cdot\boldsymbol{r}}$$
$$-\,\hat{b}(\boldsymbol{p})\hat{a}(\boldsymbol{p}')e^{\frac{i}{\hbar}(\boldsymbol{p}+\boldsymbol{p}')\cdot\boldsymbol{r}} - \hat{a}^{\dagger}(\boldsymbol{p})\hat{b}^{\dagger}(\boldsymbol{p}')e^{-\frac{i}{\hbar}(\boldsymbol{p}+\boldsymbol{p}')\cdot\boldsymbol{r}}] \tag{5.23b}$$

$$c\left(\frac{m^2c^2}{\hbar^2}\right)\int d\boldsymbol{r}\hat{\varphi}^{\dagger}(\boldsymbol{r})\hat{\varphi}(\boldsymbol{r}')$$
$$= c\left(\frac{m^2c^2}{2\hbar^2}\right)\int d\boldsymbol{r}\int d\boldsymbol{p}\int d\boldsymbol{p}'$$
$$\times\,[\hat{a}^{\dagger}(\boldsymbol{p})\hat{a}(\boldsymbol{p}')e^{-\frac{i}{\hbar}(\boldsymbol{p}-\boldsymbol{p}')\cdot\boldsymbol{r}} + \hat{b}(\boldsymbol{p})\hat{b}^{\dagger}(\boldsymbol{p}')e^{\frac{i}{\hbar}(\boldsymbol{p}-\boldsymbol{p}')\cdot\boldsymbol{r}}$$
$$+\,\hat{b}(\boldsymbol{p})\hat{a}(\boldsymbol{p}')e^{\frac{i}{\hbar}(\boldsymbol{p}+\boldsymbol{p}')\cdot\boldsymbol{r}} + \hat{a}^{\dagger}(\boldsymbol{p})\hat{b}^{\dagger}(\boldsymbol{p})e^{-\frac{i}{\hbar}(\boldsymbol{p}+\boldsymbol{p}')\cdot\boldsymbol{r}}]. \tag{5.23c}$$

When the three expressions in Eqs. (5.23) are added and integrated over $\boldsymbol{r}$ and $\boldsymbol{p}'$ the result is

$$\hat{H} = \int d\boldsymbol{p}c\sqrt{m^2c^2+\boldsymbol{p}^2}[\hat{a}^{\dagger}(\boldsymbol{p})\hat{a}(\boldsymbol{p}) + \hat{b}^{\dagger}(\boldsymbol{p})\hat{b}(\boldsymbol{p})], \tag{5.24}$$

where we have ignored the term arising from $\delta(\boldsymbol{p}-\boldsymbol{p})$. Note that there are two distinct types of quanta (of the same mass) in this complex field. We now evaluate the quantized form of the conserved charge and current density 4-vector of Eqs. (5.13), beginning with $\hat{\jmath}^0(\boldsymbol{r})$.

In terms of the canonical momenta, this is

$$\begin{aligned}\hat{\jmath}^0(\boldsymbol{r}) &= -i\{\hat{\varphi}^\dagger(\boldsymbol{r})\partial^0\hat{\varphi}(\boldsymbol{r}) - [\partial^0\hat{\varphi}^\dagger(\boldsymbol{r})]\hat{\varphi}(\boldsymbol{r})\}\\ &= i[\hat{\wp}(\boldsymbol{r})\hat{\varphi}(\boldsymbol{r}) - \hat{\varphi}^\dagger(\boldsymbol{r})\hat{\wp}^\dagger(\boldsymbol{r})].\end{aligned} \tag{5.25}$$

Substitution of the expressions of Eqs. (5.20) for the field and canonical momentum operators in terms of the Fock operators gives

$$\begin{aligned}\hat{\jmath}^0(\boldsymbol{r}) = \frac{\hbar}{2h^3}\int d\boldsymbol{p}\int d\boldsymbol{p}'\Bigg\{&\left[\left(\frac{m^2c^2+\boldsymbol{p}^2}{m^2c^2+\boldsymbol{p}'^2}\right)^{\frac{1}{4}} + \left(\frac{m^2c^2+\boldsymbol{p}'^2}{m^2c^2+\boldsymbol{p}^2}\right)^{\frac{1}{4}}\right]\\ &\times[-\hat{a}^\dagger(\boldsymbol{p})\hat{a}(\boldsymbol{p})e^{-\frac{i}{\hbar}(\boldsymbol{p}-\boldsymbol{p}')\cdot\boldsymbol{r}} + \hat{b}^\dagger(\boldsymbol{p})\hat{b}(\boldsymbol{p})e^{-\frac{i}{\hbar}(\boldsymbol{p}-\boldsymbol{p}')\cdot\boldsymbol{r}}]\\ &+\left[\left(\frac{m^2c^2+\boldsymbol{p}^2}{m^2c^2+\boldsymbol{p}'^2}\right)^{\frac{1}{4}} - \left(\frac{m^2c^2+\boldsymbol{p}'^2}{m^2c^2-\boldsymbol{p}^2}\right)^{\frac{1}{4}}\right]\\ &\times[\hat{b}^\dagger(\boldsymbol{p})\hat{a}(\boldsymbol{p})e^{\frac{i}{\hbar}(\boldsymbol{p}+\boldsymbol{p}')\cdot\boldsymbol{r}} - \hat{a}^\dagger(\boldsymbol{p})\hat{b}^\dagger(\boldsymbol{p}')e^{-\frac{i}{\hbar}(\boldsymbol{p}+\boldsymbol{p}')\cdot\boldsymbol{r}}]\Bigg\}.\end{aligned} \tag{5.26}$$

This formidable expression for the charge density is considerably simpler after integration over 3-space, since this introduces delta functions that reduce $\boldsymbol{p}'$ to $\boldsymbol{p}$ or $-\boldsymbol{p}$ after integration over $\boldsymbol{p}'$. The result for the total charge operator, after ignoring the $\delta(\boldsymbol{p}-\boldsymbol{p})$ term when $\hat{b}(\boldsymbol{p})\hat{b}^\dagger(\boldsymbol{p})$ is replaced by $\hat{b}^\dagger(\boldsymbol{p})\hat{b}(\boldsymbol{p})$ and replacing the arbitrary "normalizing constant" $\hbar c$ by the unit charge e, is

$$\hat{Q} = c\int d\boldsymbol{r}\hat{\jmath}^0(\boldsymbol{r}) = e\int d\boldsymbol{p}[-\hat{a}^\dagger(\boldsymbol{p})\hat{a}(\boldsymbol{p}) + \hat{b}^\dagger(\boldsymbol{p})\hat{b}(\boldsymbol{p})]. \tag{5.27}$$

Thus, for a charged scalar quantum field, the number operators for positive and negative particles with momentum $\boldsymbol{p}$ and the same mass (*i.e.*, particles and antiparticles) are

$$\hat{N}_- = \hat{a}^\dagger(\boldsymbol{p})\hat{a}(\boldsymbol{p}), \quad \hat{N}_+ = \hat{b}^\dagger(\boldsymbol{p})\hat{b}(\boldsymbol{p}). \tag{5.28}$$

The total charge Q of a system composed of a mixture of such particles is a constant, as a result of the quantized equivalent of Eq. (5.16).

The charge current density $\hat{\boldsymbol{j}}(\boldsymbol{r})$ is given by

$$\begin{aligned}
\hat{\boldsymbol{j}}(\boldsymbol{r}) &= i\{\hat{\varphi}^{\dagger}(\boldsymbol{r})\nabla\hat{\varphi}(\boldsymbol{r}) - [\nabla\hat{\varphi}^{\dagger}(\boldsymbol{r})]\hat{\varphi}(\boldsymbol{r})\} \\
&= \frac{e}{2ch^3}\int d\boldsymbol{p}\int d\boldsymbol{p}'(m^2c^2+\boldsymbol{p}^2)^{-\frac{1}{4}}(m^2c^2+\boldsymbol{p}'^2)^{-\frac{1}{4}} \\
&\times\{(\boldsymbol{p}+\boldsymbol{p}')[-\hat{a}^{\dagger}(\boldsymbol{p})\hat{a}(\boldsymbol{p}')e^{-\frac{i}{\hbar}(\boldsymbol{p}-\boldsymbol{p}')\cdot\boldsymbol{r}} \\
&+\hat{b}^{\dagger}(\boldsymbol{p})\hat{b}^{\dagger}(\boldsymbol{p}')e^{\frac{i}{\hbar}(\boldsymbol{p}-\boldsymbol{p}')\cdot\boldsymbol{r}}] \\
&+(\boldsymbol{p}-\boldsymbol{p}')[\hat{b}(\boldsymbol{p})\hat{a}(\boldsymbol{p}')e^{-\frac{i}{\hbar}(\boldsymbol{p}+\boldsymbol{p}')\cdot\boldsymbol{r}} \\
&-\hat{a}^{\dagger}(\boldsymbol{p})\hat{b}^{\dagger}(\boldsymbol{p})e^{\frac{i}{\hbar}(\boldsymbol{p}+\boldsymbol{p}')\cdot\boldsymbol{r}}]\},
\end{aligned} \tag{5.29}$$

where we have again inserted e for $\hbar c$. It can be verified from Eqs. (5.26) and (5.29) that the charge density $\hat{j}^0(\boldsymbol{r})$ and current density $\hat{j}(\boldsymbol{r})$ satisfy the continuity equation

$$\frac{\partial}{\partial t}\hat{j}^0(\boldsymbol{r},t) = \nabla\cdot\hat{\boldsymbol{j}}(\boldsymbol{r},t), \tag{5.30}$$

where the time dependence of the operators is obtained by the insertions.[3]

$$\begin{aligned}
\hat{a}^{\dagger}(\boldsymbol{p},t) &= e^{\frac{i}{\hbar}\varepsilon(\boldsymbol{p})t}\hat{a}^{\dagger}(\boldsymbol{p}), \quad \hat{a}(\boldsymbol{p},t) = e^{-\frac{i}{\hbar}\varepsilon(\boldsymbol{p})t}\hat{a}(\boldsymbol{p}), \\
\hat{b}^{\dagger}(\boldsymbol{p},t) &= e^{\frac{i}{\hbar}\varepsilon(\boldsymbol{p})t}\hat{b}^{\dagger}(\boldsymbol{p}), \quad \hat{b}(\boldsymbol{p},t) = e^{-\frac{i}{\hbar}\varepsilon(\boldsymbol{p})t}\hat{b}(\boldsymbol{p}).
\end{aligned}$$

As in the case of the charge density, integration over $\boldsymbol{r}$ simplifies Eq. (5.29). The current is given by

$$\begin{aligned}
\hat{\boldsymbol{j}} &= c\int d\boldsymbol{r}\hat{j}(\boldsymbol{r}) \\
&= e\int d\boldsymbol{p}\boldsymbol{p}(m^2c^2+\boldsymbol{p}^2)^{-\frac{1}{2}}[-\hat{a}^{\dagger}(\boldsymbol{p})\hat{a}(\boldsymbol{p})+\hat{b}^{\dagger}(\boldsymbol{p})\hat{b}(\boldsymbol{p}) \\
&-\hat{b}(\boldsymbol{p})\hat{a}(-\boldsymbol{p})+\hat{a}^{\dagger}(\boldsymbol{p})\hat{b}^{\dagger}(-\boldsymbol{p})].
\end{aligned} \tag{5.31}$$

[3]See Section 4.5.2 for a discussion of the Heisenberg picture for time-dependent operators.

In contrast to the total charge, where the terms with two annihilation and two creation operators are eliminated by the integration over $\boldsymbol{r}$, these terms remain in the charge current. They correspond to the annihilation and creation of a particle–antiparticle pair with opposite momenta. The annihilation process requires the presence of one or more antiparticles in the system, while the creation process requires coupling to an energy source that can provide the total minimum energy of the pair, $2mc^2$. In the application to molecular ensembles at normal conditions, both of these processes are negligible.

At non-relativistic velocities the factor $c\boldsymbol{p}/(m^2c^2+\boldsymbol{p}^2)^{\frac{1}{2}}$ becomes the velocity $\boldsymbol{v}$; the current takes the classical form of velocity times density.

5.2.3. *Coupling to a Classical Electromagnetic Field*

Consider now a classical charged scalar field in which the field operators are subjected to a *local* phase change; that is, one for which the phase angle α depends on the position in space-time, x:

$$\varphi(x) \to e^{i\alpha(x)}\varphi(x), \quad \varphi^*(x) \to e^{-i\alpha(x)}\varphi^*(x). \tag{5.32}$$

Then the derivatives are modified to

$$\begin{aligned} \partial_\mu\varphi(x) &\to [\partial_\mu + ia^\mu(x)]\varphi(x), \\ \partial^\mu\varphi^*(x) &\to [\partial^\mu - ia_\mu(x)]\varphi^*(x), \end{aligned} \tag{5.33}$$

where $a^\mu(x)$ is defined by

$$a^\mu(x) = \partial_\mu\alpha(x). \tag{5.34}$$

The transformed Lagrangian is

$$\begin{aligned} \mathscr{L} = {} & [\partial^\mu - ia_\mu(x)]\varphi^*(x)[\partial_\mu + ia^\mu(x)]\varphi(x) \\ & -\frac{m^2c^2}{\hbar^2}\varphi^*(x)\varphi(x). \end{aligned} \tag{5.35}$$

The canonical momenta are

$$\begin{aligned} \wp(X) &= \frac{\partial\mathscr{L}}{\partial(\partial_0\varphi)} = [\partial^\mu - ia_\mu(x)]\varphi^*(x), \\ \wp^*(X) &= \frac{\partial\mathscr{L}}{\partial(\partial_0\varphi^*)} = [\partial_\mu - ia^\mu(x)]\varphi(x). \end{aligned} \tag{5.36}$$

The Hamiltonian density is

$$\begin{aligned}\mathcal{H} &= \wp(x)\partial_0\varphi(x) + [\partial^0\varphi^*(x)]\wp^*(x) - \mathscr{L} \\ &= \wp^*(x)\wp(x) + \nabla\varphi^*(x)\cdot\nabla\varphi(x) + \frac{m^2c^2}{\hbar^2}\varphi^*(x)\varphi(x) \\ &\quad - i\{\varphi^*(x)\nabla\varphi(x) - [\nabla\varphi^*(x)]\varphi(x)\}\cdot\boldsymbol{a}(x) \\ &\quad - i[\wp(x)\varphi(x) - \varphi^*(x)\wp^*(x)]a_0(x) \\ &\quad + \boldsymbol{a}(x)\cdot\boldsymbol{a}(x)\varphi^*(x)\varphi(x).\end{aligned} \tag{5.37}$$

The additions to the Hamiltonian density due to the field derived from the local phase change are in the last two lines of Eq. (5.37). The first of these lines is $\boldsymbol{j}(x)\cdot\boldsymbol{a}(x)$, which is the current density dotted into the vector potential; the last line is $-j_0(x)a_0(x)$, which is the negative of the charge density times the scalar potential, plus $\boldsymbol{a}^2$ an term. This interaction of charged particles with a 4-vector field a has been derived from a local phase shift. The field a looks like the electromagnetic 4-vector potential A, except that a does not obey Maxwell's equations. In particular,

$$\nabla\times\boldsymbol{a} = \nabla\times\nabla\alpha(x) = 0,$$

since the curl of a gradient vanishes. Furthermore, and of even greater consequence, the Lagrangian of Eq. (5.35) is not invariant under the local phase shifts of Eqs. (5.32). The question then is, how do we modify $\mathscr{L}$ so that it *is* invariant to local phase shifts. The answer is to modify the derivatives of the fields as follows:

$$\begin{aligned}\partial_\mu\varphi(x) &\to [\partial_\mu + iA^\mu(x)]\varphi(x), \\ \partial^\mu\varphi^*(x) &\to [\partial^\mu - iA_\mu(x)]\varphi^*(x),\end{aligned} \tag{5.38}$$

such that the local phase shifts of Eqs. (5.32) imply that

$$A_\mu(x) \to A_\mu(x) - a_\mu(x) = A_\mu(x) - \partial^\mu\alpha(x). \tag{5.39}$$

Then we have

$$\begin{aligned}
[\partial_\mu + iA^\mu(x)]\varphi(x) &\to \{\partial_\mu + ia^\mu(x) + i[A^\mu(x) - a^\mu(x)]\}\varphi(x) \\
&= [\partial_\mu + iA^\mu(x)]\varphi(x), \\
[\partial^\mu - iA_\mu(x)]\varphi^*(x) &\to \{\partial^\mu - ia_\mu(x) + i[A_\mu(x) - a_\mu(x)]\}\varphi^*(x) \\
&= [\partial^\mu - iA_\mu(x)]\varphi^*(x).
\end{aligned} \tag{5.40}$$

The 4-vector A is now free to satisfy Maxwell's equations and its gauge invariance is clear, since the Lagrangian, and therefore the dynamics, is unchanged by the transformation

$$\begin{aligned}
&A_0(x) \to A_0(x) - \partial_0\alpha(x), \quad \mathbf{A}(x) \to \mathbf{A}(x) - \nabla\alpha(x), \\
&\varphi(x) \to e^{i\alpha(x)}\varphi(x), \quad \varphi^*(x) \to e^{-i\alpha(x)}\varphi^*(x).
\end{aligned} \tag{5.41}$$

The Hamiltonian now takes the same form as Eq. (5.37), with $a_\mu(x)$ replaced by the electromagnetic potential $A_\mu(x)$. Now, the interpretation of the conserved 4-vector j as the charge-current density is justified. We shall return to the interaction of matter with a *quantized* electromagnetic field and the inclusion of a pure electromagnetic term in the Lagrangian in Chapter 9. Here we only want to show how the charge-current density of a particle field interacts with an electromagnetic field.

5.3. Particles with Spin[4]

At this writing, all material particles appear to be composed of elementary particles having spin 1/2, namely quarks, electrons, and their antiparticles. In particular, molecules consist of atoms, which in turn consist of electrons and nucleons, the latter consisting of three quarks each. We therefore limit our discussion of spin to particles that are quanta of fields obeying Fermi-Dirac statistics.

[4]See Appendix A for a discussion of Schwinger's harmonic-oscillator theory of spin.

5.3.1. *The Pauli Equation*

The first attempt at a theory of the empirically discovered intrinsic angular momentum, or spin, was by Wolfgang Pauli. He modified the Schrödinger equation for a particle with mass m and charge q (note: for an electron, $q = -e$) in an electromagnetic field $\mathbf{A}$, Φ as follows:

$$\begin{gathered}\left[\frac{1}{2m}\left(\hat{\boldsymbol{p}} - \frac{q}{c}\mathbf{A}\right)^2 + q\Phi\right]\psi = i\hbar\frac{\partial}{\partial t}\psi \rightarrow \\ \left\{\frac{1}{2m}\left[\boldsymbol{\sigma}\cdot\left(\hat{\boldsymbol{p}} - \frac{q}{c}\mathbf{A}\right)\right]^2 + q\Phi\right]\psi = i\hbar\frac{\partial}{\partial t}\psi.\end{gathered} \tag{5.42}$$

where $\boldsymbol{\sigma}$ is a 3-vector whose components are the 2×2 Pauli matrices

$$\sigma_1 = \begin{pmatrix} 0 & 1 \\ 1 & 0 \end{pmatrix} \quad \sigma_2 = \begin{pmatrix} 0 & -i \\ 1 & 0 \end{pmatrix} \quad \sigma_3 = \begin{pmatrix} 1 & 0 \\ 0 & -1 \end{pmatrix}. \tag{5.43}$$

It is easily verified that

$$\begin{aligned} \sigma_1^2 = \sigma_2^2 = \sigma_3^2 &= \begin{pmatrix} 1 & 0 \\ 0 & 1 \end{pmatrix} = \mathbf{1}^{(2)}, \\ \sigma_1\sigma_2 &= -\sigma_2\sigma_1 = i\sigma_3, \\ \sigma_2\sigma_3 &= -\sigma_3\sigma_2 = i\sigma_1, \\ \sigma_3\sigma_1 &= -\sigma_1\sigma_2 = i\sigma_2. \end{aligned} \tag{5.44}$$

When the squared term containing the momentum is expanded with the use of Eqs. (5.44), it becomes

$$\left[\boldsymbol{\sigma}\cdot\left(\hat{\boldsymbol{p}} - \frac{q}{c}\mathbf{A}\right)\right]^2 = \left(\hat{\boldsymbol{p}} - \frac{q}{c}\mathbf{A}\right)^2 \mathbf{1}^{(2)} - \frac{q\hbar}{c}\boldsymbol{\sigma}\cdot(\nabla\times\mathbf{A}), \tag{5.45}$$

where we have used relations such as

$$[\hat{p}_x, A_y]_- = \frac{\hbar}{i}\left[\frac{\partial}{\partial x}, A_y\right]_- = \frac{\hbar}{i}\frac{\partial}{\partial x}A_y.$$

Since the magnetic induction is given by $\mathbf{B} = \nabla\times\mathbf{A}$, the Pauli equation becomes

$$\left[\left(\hat{\boldsymbol{p}} - \frac{q}{c}\mathbf{A}\right)^2 + q\Phi - \frac{q\hbar}{2mc}\boldsymbol{\sigma}\cdot\mathbf{B}\right]\Psi = i\hbar\frac{\partial}{\partial t}\Psi. \tag{5.46}$$

If we define the spin vector $\boldsymbol{S}$ by

$$\boldsymbol{S} = \frac{\hbar}{2}\boldsymbol{\sigma}, \tag{5.47}$$

its commutation properties identify it as a quantum-mechanical angular momentum operator:

$$\boldsymbol{S} \times \boldsymbol{S} = i\hbar\boldsymbol{S}. \tag{5.48}$$

In terms of $\boldsymbol{S}$, the Pauli equation is

$$\left\{\left[\left(\hat{\boldsymbol{p}} - \frac{q}{c}\boldsymbol{A}\right)^2 + q\Phi\right]\mathbf{1}^{(2)} - \frac{q}{mc}\boldsymbol{S}\cdot\boldsymbol{B}\right\}\Psi = i\hbar\frac{\partial}{\partial t}\mathbf{1}^{(2)}\Psi. \tag{5.49}$$

Since the Hamiltonian operator in Eq. (5.49) is a 2×2 matrix, the wave function Ψ must be a 2-component column vector:

$$\Psi = \begin{pmatrix} \Psi_+ \\ \Psi_- \end{pmatrix} \tag{5.50}$$

so that

$$S_3\Psi = \frac{\hbar}{2}\begin{pmatrix} 1 & 0 \\ 0 & -1 \end{pmatrix}\begin{pmatrix} \Psi_+ \\ \Psi_- \end{pmatrix} = \begin{pmatrix} \frac{\hbar}{2}\Psi_+ \\ -\frac{\hbar}{2}\Psi_- \end{pmatrix}. \tag{5.51}$$

The Pauli equation thus includes spin, with a magnetic moment in approximate agreement with the observed value. But it is a *non-relativistic* equation and the Pauli matrices $\boldsymbol{\sigma}$ are introduced *ad hoc*.

5.3.2. *The Dirac Equation*

Dirac realized that a proper relativistic wave equation must treat time and position variables on the same footing. Although the Klein-Gordon equation does this, it has several failings when interpreted as a wave equation for a single particle and does not include spin. To be consistent with Schrödinger's approach, Dirac looked for an equation that included only the *first* derivatives $\partial/\partial t$ and ∇. In his derivation of a relativistic wave equation for the electron, Dirac started with the relativistic energy

$$E = c\sqrt{\boldsymbol{p}^2 + m^2c^2} \tag{5.52}$$

and sought a Hamiltonian operator of the form

$$\hat{h} = c(\boldsymbol{\alpha} \cdot \hat{\boldsymbol{p}} + \beta mc) = i\hbar\frac{\partial}{\partial t} \tag{5.53}$$

with α and β chosen so that

$$\hat{h}^2 = c^2(\hat{\boldsymbol{p}}^2 + m^2c^2) = -\hbar^2\frac{\partial^2}{\partial t^2}. \tag{5.54}$$

This requires that

$$\begin{aligned} \alpha_1^2 = \alpha_2^2 = \alpha_3^2 = \beta^2 = \mathbf{1}^{(2)}, \\ \alpha_j\alpha_k + \alpha_k\alpha_j = 0, \quad \boldsymbol{\alpha}\beta + \beta\boldsymbol{\alpha} = 0. \end{aligned} \tag{5.55}$$

Furthermore, since $\hat{h}^\dagger = \hat{h}$, these matrices must also be Hermitian:

$$\boldsymbol{\alpha}^\dagger = \boldsymbol{\alpha}, \quad \beta^\dagger = \beta. \tag{5.56}$$

These properties expressed in Eqs. (5.55) and (5.56) are realized by the 4×4 matrices.[5]

$$\beta = \begin{pmatrix} 0 & \mathbf{1}^{(2)} \\ \mathbf{1}^{(2)} & 0 \end{pmatrix}, \quad \boldsymbol{\alpha} = \begin{pmatrix} -\boldsymbol{\sigma} & 0 \\ 0 & \boldsymbol{\sigma} \end{pmatrix}, \tag{5.57}$$

where $\boldsymbol{\sigma}$ is the 3-vector whose components are the Pauli matrices given in Eq. (5.43). When the momentum operators are expressed in terms of spatial derivatives, namely

$$\hat{\boldsymbol{p}} = \frac{\hbar}{i}\nabla,$$

and substituted into the Dirac Hamiltonian in Eq. (5.53), we obtain the Dirac equation

$$c\left[\left(\frac{\hbar}{i}\right)\left(\frac{\partial}{c\partial t} + \boldsymbol{\alpha} \cdot \nabla\right) + \beta mc\right]\Psi = 0, \tag{5.58}$$

[5]The form of these matrices is not unique; here, we use the "chiral" representation.

where the wave function Ψ is the four-component column vector

$$\Psi = \begin{pmatrix} \Psi_a \\ \Psi_b \\ \Psi_c \\ \Psi_d \end{pmatrix}. \tag{5.59}$$

To put Eq. (5.58) into covariant form, we right-multiply by $-\beta$ to obtain

$$(i\hbar\gamma^\mu \partial_\mu - mc)\Psi = 0, \tag{5.60}$$

where

$$\gamma^0 = \beta, \quad \boldsymbol{\gamma} = \beta\boldsymbol{\alpha}. \tag{5.61}$$

Now we see that γ^0 is Hermitian, while the γ^j's are *anti-Hermitian*, since

$$(\gamma^j)^\dagger = (\gamma^0\alpha^j)^\dagger = (\alpha^j)^\dagger(\gamma^0)^\dagger = \alpha^j\gamma^0 = \alpha^j\beta = \beta\alpha^j = -\gamma^j. \tag{5.62}$$

5.3.2.1. *The Dirac Lagrangian*

To construct a Lagrangian that leads to the Dirac equation, Eq. (5.60), we might try

$$\mathscr{L} = \Psi^\dagger(i\hbar\gamma^\mu \partial_\mu - mc)\Psi,$$

treating Ψ and $\Psi^\dagger$ as independent fields. Since $\mathscr{L}$ is independent of $\partial_\mu \Psi^\dagger$ the application of Eq. (4.41) for each field gives

$$\frac{\delta\mathscr{L}}{\delta\Psi^\dagger} = (i\hbar\gamma^\mu\partial_\mu - mc)\Psi = 0,$$
$$\partial_\mu\left[\frac{\delta\mathscr{L}}{\delta(\partial_\mu\Psi)}\right] - \frac{\delta\mathscr{L}}{\delta\Psi} = i\hbar\partial_\mu\Psi^\dagger\gamma^\mu + mc\Psi^\dagger = 0.$$

The first line gives the Dirac equation as intended, but the second line gives a different equation which is *not* the Hermitian conjugate of the Dirac equation as might have been expected. In fact, if we take its

Hermitian conjugate, add it to the first equation, and use Eq. (5.62), we will obtain

$$2i\hbar\boldsymbol{\gamma}\cdot\nabla\Psi = -2\boldsymbol{\gamma}\cdot\hat{\boldsymbol{p}}\Psi = 0;$$

that is, this trial Lagrangian leads to a motionless particle. The solution is to take as the independent Dirac fields Ψ and $\Psi^\dagger\gamma^0$ and write

$$\mathscr{L} = \bar{\Psi}(i\hbar\gamma^\mu\partial_\mu - mc)\Psi, \qquad (5.63)$$

where we define the Dirac adjoint (or simply the adjoint) of the Dirac field as $\bar{\Psi} = \Psi^\dagger\gamma^0$.

Now, the two equations of motion we obtain from Eq. (5.63) are the Dirac equation and its Hermitian conjugate. To show this, we have

$$\frac{\delta\mathscr{L}}{\delta\bar{\Psi}} = (i\hbar\gamma^\mu\partial_\mu - mc)\Psi = 0, \qquad (5.64)$$

the Dirac equation, as before. But now

$$\partial_\mu\left[\frac{\partial\mathscr{L}}{\delta(\partial_\mu\Psi)}\right] - \frac{\delta\mathscr{L}}{\delta\Psi} = i\hbar\partial_\mu\bar{\Psi}\gamma^\mu + mc\bar{\Psi} = 0. \qquad (5.65)$$

Since

$$\bar{\Psi}\gamma^\mu = \Psi^\dagger\gamma^0\gamma^\mu = \Psi^\dagger(\gamma^0\gamma^\mu)^\dagger = \Psi^\dagger(\gamma^\mu)^\dagger(\gamma^0)^\dagger = \Psi^\dagger(\gamma^\mu)^\dagger\gamma^0$$

Eq. (5.65) can be written

$$i\hbar\partial_\mu\Psi^\dagger(\gamma^\mu)^\dagger\gamma^0 + mc\Psi^\dagger\gamma^0 = i\hbar\partial_\mu(\gamma^\mu\Psi)^\dagger\gamma^0 + mc\Psi^\dagger\gamma^0 = 0. \quad (5.66)$$

If we right-multiply by $-\gamma^0$, Eq. (5.66) becomes

$$-i\partial_\mu(\gamma^\mu\Psi)^\dagger - mc\Psi^\dagger = 0, \qquad (5.67)$$

which is the Hermitian conjugate of the Dirac equation.

5.3.2.2. *The Dirac Field Momentum*

From Eqs. (5.63) and (4.45), we obtain the field momentum

$$\wp = \frac{\delta\mathscr{L}}{\delta(\partial_0\Psi)} = i\bar{\Psi}\gamma_0 = \Psi^\dagger. \qquad (5.68)$$

5.3.2.3. *Constants of the Motion*

To find constants of the motion for a Dirac field, we apply Eq. (4.61) to the Lagrangian, Eq. (5.63). The energy-momentum tensor is

$$\mathrm{T}^{\mu}_{\mathrm{V}} = \frac{\delta\mathscr{L}}{\delta(\partial_\mu\Psi)}\partial_{\mathrm{V}}\Psi = i\hbar\bar{\Psi}\gamma^{\mu}\partial_{\mathrm{V}}\Psi, \tag{5.69}$$

where we have ignored the $\mathscr{L}\delta^{\mu}_{\mathrm{V}}$ term in Eq. (4.61) because for a Dirac field, Eqs. (5.62) and (5.64) show that $\mathscr{L} = 0$. The energy density is thus given by

$$\mathcal{H}/c = T^0_0 = i\hbar\bar{\Psi}\gamma^0\partial_0\Psi = \frac{1}{c}i\hbar\Psi^{\dagger}\partial_t\Psi = \frac{1}{c}\wp\partial_t\Psi \tag{5.70}$$

and the Hamiltonian is

$$H = \int d\boldsymbol{r}\mathcal{H} = i\hbar\int d\boldsymbol{r}\Psi^{\dagger}\partial_t\Psi. \tag{5.71}$$

A more useful form for the Hamiltonian is obtained by making use of Eq. (5.53) to write

$$i\hbar\partial_t = c(-i\hbar\boldsymbol{\alpha}\cdot\nabla + \beta mc), \tag{5.72}$$

whereupon Eq. (5.70) becomes in the γ^{μ} notation

$$H = c\int d\boldsymbol{r}\bar{\Psi}(-i\hbar\gamma\cdot\nabla + mc)\Psi. \tag{5.73}$$

Similarly, the total particle momentum is $\boldsymbol{P}_j = -\int d\boldsymbol{r}T^0_j$; that is,

$$\boldsymbol{P} = \int d\boldsymbol{r}\Psi^{\dagger}\frac{\hbar}{i}\nabla\Psi. \tag{5.74}$$

To find the expression for the charge-current density, we proceed as for complex scalar quantum fields, but now with the Lagrangian given in Eq. (5.63). First, we introduce a global phase shift α by making the replacements[6]

$$\Psi(x) \to \Psi(x)e^{i\alpha}, \quad \bar{\Psi}(x) \to \bar{\Psi}(x)e^{i\alpha}. \tag{5.75}$$

The Dirac Lagrangian now becomes

$$\mathscr{L} = \bar{\Psi}e^{-i\alpha}(i\hbar\gamma^\mu\partial_\mu - mc)\Psi e^{i\alpha}. \tag{5.76}$$

Taking the total derivative with respect to α gives

$$\begin{aligned}
\frac{d\mathscr{L}}{d\alpha} &= \frac{\partial\mathscr{L}}{\partial(\partial^\mu\bar{\psi})}\frac{\partial(\partial^\mu\bar{\psi})}{\partial\alpha} + \frac{\partial\mathscr{L}}{\partial\bar{\psi}}\frac{\partial\bar{\psi}}{\partial\alpha} + \frac{\partial\mathscr{L}}{\partial(\partial_\mu\psi)}\frac{\partial(\partial_\mu\psi)}{\partial\boldsymbol{\alpha}} + \frac{\partial\mathscr{L}}{\partial\psi}\frac{\partial\psi}{\partial\boldsymbol{\alpha}} \\
&= \cancel{\partial^\mu\left[\frac{\partial\mathscr{L}}{\partial(\partial^\mu\psi)}\frac{\partial\bar{\psi}}{\partial\boldsymbol{\alpha}}\right]} - \cancel{\partial^\mu\left[\frac{\partial\mathscr{L}}{\partial(\partial^\mu\bar{\psi})}\right]\frac{\partial\bar{\psi}}{\partial\boldsymbol{\alpha}} + \frac{\partial\mathscr{L}}{\partial\bar{\psi}}\frac{\partial\bar{\psi}}{\partial\boldsymbol{\alpha}}} \\
&\quad + \cancel{\partial^\mu\left[\frac{\partial\mathscr{L}}{\partial(\partial^\mu\psi)}\frac{\partial\psi}{\partial\boldsymbol{\alpha}}\right]} - \cancel{\partial_\mu\left[\frac{\partial\mathscr{L}}{\partial(\partial_\mu\bar{\psi})}\right]\frac{\partial\psi}{\partial\boldsymbol{\alpha}} + \frac{\partial\mathscr{L}}{\partial\psi}\frac{\partial\psi}{\partial\boldsymbol{\alpha}}}.
\end{aligned} \tag{5.77}$$

The first term in the second line of Eq. (5.77) is equal to zero because the Dirac Lagrangian does not depend upon $\partial^\mu\bar{\Psi}$. The last two terms in the second and third lines of Eq. (5.77) are equal to zero from Eq. (5.65) and its Hermitian conjugate. We are left with

$$\begin{aligned}
\frac{d\mathscr{L}}{d\alpha} &= \partial_\mu\left[\frac{\partial\mathscr{L}}{\partial(\partial_\mu\psi)}\frac{\partial\psi}{\partial\alpha}\right] = \partial_\mu[(i\hbar\bar{\Psi}e^{-i\alpha}\gamma^\mu)(i\Psi e^{i\alpha})] \\
&= -\hbar\partial_\mu(\bar{\Psi}\gamma^\mu\Psi).
\end{aligned} \tag{5.78}$$

Requiring $\mathscr{L}$ to be invariant with respect to α then gives the continuity equation

$$\partial_\mu(\bar{\Psi}\gamma^\mu\Psi) = 0. \tag{5.79}$$

We therefore identify the expression in parentheses in Eq. (5.79) as the 4-vector charge-current density

$$j^\mu = (\rho, \boldsymbol{j}) = \bar{\Psi}\gamma^\mu\Psi \tag{5.80}$$

[6]There should be no confusion with the Dirac operator α, since we are now using the γ^μ matrices in the Dirac theory.

and Eq. (5.79) takes the more familiar form

$$\frac{1}{c}\frac{\partial}{\partial t}\rho - \nabla \cdot \boldsymbol{j} = 0. \tag{5.81}$$

The charge density ρ is the $\mu = 0$ component of j^μ, namely

$$\rho = \bar{\Psi}\gamma^0\Psi = (\Psi^\dagger\gamma^0)\gamma^0\Psi = \Psi^\dagger\Psi, \tag{5.82}$$

and the current density is the 3-vector

$$\boldsymbol{j} = \Psi^\dagger\gamma^0\boldsymbol{\gamma}\Psi, \tag{5.83}$$

where the 3-vector $\boldsymbol{\gamma}$ is $(\gamma^1, \gamma^2, \gamma^3)$.

We have omitted the unit charge, which for electrons is $-e$. At this time this is a purely empirical constant, since no theory can yet predict its value. Without the factor $-e$, the density ρ is the probability density at space-time point x^μ and $\boldsymbol{j}$ is the probability current at that point.

5.3.2.4. *Coupling to the Electromagnetic Field*

Proceeding as in Section 5.2.3, we couple the Dirac Lagrangian to an electromagnetic field by considering the consequences of a *local* phase shift $\alpha(x)$:

$$\Psi(x) \to e^{i\alpha(x)}\Psi(x), \quad \Psi^\dagger(x) \to e^{-i\alpha(x)}\Psi^\dagger(x). \tag{5.84}$$

Then

$$\begin{aligned} \partial_\mu\Psi(x) &\to e^{i\alpha(x)}[\partial_\mu + ia^\mu(x)]\Psi(x), \\ \partial^\mu\Psi^\dagger(x) &\to e^{-i\alpha(x)}[\partial^\mu + ia_\mu(x)]\Psi^\dagger(x), \end{aligned} \tag{5.85}$$

where, as before,

$$a^\mu(x) = \partial_\mu\alpha(x). \tag{5.86}$$

The transformed Lagrangian density is

$$\begin{aligned} \mathscr{L} &= e^{-i\alpha(x)}\Psi^\dagger\gamma^0 e^{i\alpha(x)}\{i\hbar\gamma^\mu[\partial_\mu + ia^\mu(x] - mc1^{(4)}\}\Psi \\ &= \Psi^\dagger\gamma^0\{i\hbar\gamma^\mu[\partial_\mu + ia^\mu(x] - mc1^{(4)}\}\Psi. \end{aligned} \tag{5.87}$$

The canonical momenta are

$$\wp(x) = \frac{\partial \mathscr{L}}{\partial(\partial_0 \Psi)} = i\hbar\Psi^\dagger\gamma_0\gamma^0 = i\hbar\Psi^\dagger,$$
$$\wp^*(x) = \frac{\partial \mathscr{L}}{\partial(\partial^0 \Psi^\dagger)} = 0. \tag{5.88}$$

We have used the fact that $\mathscr{L}$ is independent of $\partial^0\Psi^\dagger$. We are now ready to find the transformed Dirac equation from the general property of $\mathscr{L}$ given in Eq. (4.41):

$$\gamma^0 \frac{\partial \mathscr{L}}{\partial \Psi^\dagger} = \{i\hbar\gamma^\mu[\partial_\mu - ia^\mu] - mc\}\Psi$$
$$\gamma^0 \partial_\mu \frac{\partial \mathscr{L}}{\partial(\partial_\mu \Psi^\dagger)} = 0, \tag{5.89}$$

since the Dirac Lagrangian does not depend upon $\partial_\mu\Psi^\dagger$. The Hamiltonian density can be found from Eqs. (5.70), (5.68), and (5.87) to be:

$$\mathcal{H} = c\bar{\Psi}[-i\hbar\gamma\cdot\nabla + mc]\Psi + \hbar c\bar{\Psi}\gamma^\mu\Psi a^\mu. \tag{5.90}$$

Comparison of Eqs. (5.90) and (5.73) shows that the Dirac Hamiltonian for a particle with a phase shift $\alpha(x)$ has the added term

$$\hbar c j^\mu a^\mu. \tag{5.91}$$

We have the same problem as we had in the discussion following Eq. (5.37), namely that the space component of a^μ being a gradient, cannot satisfy Maxwell's equations. The solution, as before, is to add in a 4-vector A with the property that under the phase shift of Eq. (5.83)

$$A_\mu(x) \to A_\mu(x) - \partial_\mu\alpha(x)$$
$$= A_\mu(x) - a_\mu(x). \tag{5.92}$$

Now the Dirac equation becomes

$$\{\hbar\gamma^\mu[\partial_\mu + i(A^\mu + a^\mu)] - ia^\mu - mc\}\Psi = 0, \tag{5.93}$$

so that the gradient a^μ is cancelled out and we are left with the 4-vector A which can satisfy Maxwell's equations and is gauge-invariant:

$$i\hbar\gamma^\mu[\partial_\mu + iA^\mu(x) - mc]\Psi(x) = 0. \tag{5.94}$$

Finally, to make complete the identification of the 4-vector A with the 4-potential of classical electrodynamics, we replace A with $(q/\hbar)A$. For electrons, $q = -e$, the empirically determined charge on the electron. The Dirac equation, Lagrangian, and Hamiltonian are therefore

$$i\hbar\gamma^\mu[\partial_\mu - mc]\Psi(x + e\mathrm{A}^\mu(x) = 0 \tag{5.95}$$

$$\Psi^\dagger(x)\gamma^0\{i\hbar\gamma^\mu\partial_\mu - mc\}\Psi(x) + j^\mu(x)\mathrm{A}_\mu(x) \tag{5.96}$$

$$\mathcal{H}c = \bar{\Psi}(x)(-i\hbar\gamma\cdot\nabla + mc)\Psi(x) + j^\mu(x)\cdot\mathrm{A}^\mu(x), \tag{5.97}$$

where now

$$j^\mu = e\bar{\Psi}\gamma^\mu\Psi. \tag{5.98}$$

According to Maxwell's equations, the term added to Eqs. (5.96) and (5.97) is

$$j^\mu(x)\cdot\mathrm{A}^\mu(x) = \rho V + \boldsymbol{j}\cdot\mathbf{A}, \tag{5.99}$$

where ρ is the charge density, V is the scalar potential of the electromagnetic field, $\boldsymbol{j}$ is the current density, and $\mathbf{A}$ is the vector potential of the electromagnetic field.

5.3.2.5. *Solutions for Free Dirac Particles*

In this section, we find solutions to the Dirac equation, Eq. (5.60), for the case of a free particle moving at constant momentum. We choose the direction of motion as that for which the corresponding Pauli matrix is diagonal, namely σ_3, and call that the z-axis in Cartesian 3-space so that the 4-momentum is

$$p = (E/c, 0, 0, p_z). \tag{5.100}$$

The solution we seek is therefore of the form

$$\Psi(x) = u(p)e^{-\frac{1}{\hbar}(Et - p_z Z)}, \tag{5.101}$$

where $\Psi(x)$ and $u(p)$ are 4-vector fields depending upon the 4-vectors x and p, respectively. Then

$$\gamma^{\mu}\partial_{\mu}\Psi = -\frac{i}{h}(\gamma^{0}E/c - \gamma^{z}p_{z})u(p)e^{-\frac{i}{h}(Et-p_{z}Z)}, \qquad (5.102)$$

so that the Dirac equation becomes

$$(\gamma^{0}E/c - \gamma^{Z}p_{z} - mc1^{(4)})u(p) = 0. \qquad (5.103)$$

Recall that the matrices are

$$\gamma^{\mu} = \begin{pmatrix} 0 & \sigma^{\mu} \\ \sigma_{\mu} & 0 \end{pmatrix}, \qquad (5.104)$$

where the zeros are 2×2 null matrices and we have defined σ^{μ} and σ_{μ} as the 4-vectors whose components are $1^{(2)}$ and the Pauli matrices $\boldsymbol{\sigma}$ given in Eq. (5.43):

$$\sigma^{\mu} = (1^{(2)}, \boldsymbol{\sigma}); \quad \sigma_{\mu} = (1^{(2)}, -\boldsymbol{\sigma}). \qquad (5.105)$$

Using Eqs. (5.104) and (5.105) to write out Eq. (5.103) in explicit matrix form gives

$$\begin{pmatrix} -mc1^{(2)} & (E/c)1^{(2)} - \sigma_{z}p_{z} \\ (E/c)1^{(2)} + \sigma_{z}p_{z} & -mc1^{(2)} \end{pmatrix} u(\boldsymbol{p}) = 0. \qquad (5.106)$$

In Eq. (5.106) the 4-vector field $u(p)$ has been written as two 2-component fields, $u = (u_a, u_b)$. Eq. (5.106) in the full 4×4 matrix form is

$$\begin{pmatrix} -mc & 0 & E/c - p_{z} & 0 \\ 0 & -mc & 0 & E/c + p_{z} \\ E/c + p_{z} & 0 & -mc & 0 \\ 0 & E/c - p_{z} & 0 & -mc \end{pmatrix} u(\boldsymbol{p}) = 0. \qquad (5.107)$$

With the help of the identities

$$\sqrt{E/c - p_{z}}\sqrt{E/c + p_{z}} = \sqrt{(E/c)^{2} - p_{z}^{2}} = mc \qquad (5.108)$$

it can be shown that Eq. (5.107) is satisfied by

$$u_{\uparrow}(\boldsymbol{p}) = \begin{pmatrix} \sqrt{E/c - p_z} \\ 0 \\ \sqrt{E/c + p_z} \\ 0 \end{pmatrix} \tag{5.109a}$$

and

$$u(\boldsymbol{p})_{\uparrow} = \begin{pmatrix} 0 \\ \sqrt{E/c + p_z} \\ 0 \\ \sqrt{E/c - p_z} \end{pmatrix}. \tag{5.109b}$$

What if we had chosen the form

$$\Psi(x) = v(p)e^{\frac{i}{\hbar}(Et - p_z z)} \tag{5.110}$$

instead of Eq. (5.101)? In this case, the matrix equation to be solved is

$$\begin{pmatrix} mc & 0 & E/c - p_z & 0 \\ 0 & mc & 0 & E/c + p_z \\ E/c + p_z & 0 & mc & 0 \\ 0 & E/c - p_z & 0 & mc \end{pmatrix} v(\boldsymbol{p}) = 0, \tag{5.111}$$

and the solutions are

$$v(\boldsymbol{p})_{\uparrow} = \begin{pmatrix} \sqrt{E/c - p_z} \\ 0 \\ -\sqrt{E/c + p_z} \\ 0 \end{pmatrix} \tag{5.112a}$$

and

$$v(\boldsymbol{p})_{\uparrow} = \begin{pmatrix} 0 \\ \sqrt{E/c + p_z} \\ 0 \\ -\sqrt{E/c - p_z} \end{pmatrix}. \tag{5.112b}$$

All four of the 4-vectors in Eqs. (5.109) and (5.112) are eigenstates of the spin operator S^2 with eigenvalue $3\hbar/2$. The arrow subscripts $\uparrow$ and $\downarrow$ indicate eigenvalues of S_z with eigenvalues $+\hbar/2$ and $-\hbar/2$, respectively.

5.3.2.6. *Quantization of the Dirac Field*

Using the quantization of the fields we have seen previously, we expect to use creation and annihilation operators in expressions like Eq. (5.20), except now we are dealing with four types of Dirac particles, corresponding to positive and negative frequencies in the Fourier expressions for the fields and their canonically conjugate momenta, namely $u(p)$ and $v(p)$ of Eqs. (5.109) and (5.112). These operators may be designated

$$\hat{a}^u_\uparrow(p)^\dagger, \quad \hat{a}^v_\uparrow(p)^\dagger, \quad \hat{a}^u_\downarrow(p)^\dagger, \quad \hat{a}^v_\downarrow(p)^\dagger, \quad \text{creators}$$
$$\hat{a}^u_\uparrow(p), \quad \hat{a}^v_\uparrow(p), \quad \hat{a}^u_\downarrow(p), \quad \hat{a}^v_\downarrow(p), \quad \text{annihilators}$$

There are many ways to group these operators into expressions for the fields $\Psi(x)$ and $\wp(x)$, but only one way that makes physical sense, and that is (for time $t = 0$),

$$\begin{aligned}\Psi(\boldsymbol{r}) = \int \frac{d\boldsymbol{p}}{2(\pi\hbar)^{3/2}} \frac{1}{\sqrt{2E(\boldsymbol{p})/c}} \\ \times \{[\hat{a}^u_\uparrow(\boldsymbol{p})u_\uparrow(\boldsymbol{p}) + \hat{a}^u_\downarrow(\boldsymbol{p})u_\downarrow(\boldsymbol{p})]e^{\frac{i}{\hbar}\boldsymbol{p}\cdot\boldsymbol{r}} \\ + [\hat{a}^v_\uparrow(\boldsymbol{p})^\dagger v_\uparrow(\boldsymbol{p}) + \hat{a}^v_\downarrow(\boldsymbol{p})^\dagger v_\downarrow(\boldsymbol{p})]e^{-\frac{i}{\hbar}\boldsymbol{p}\cdot\boldsymbol{r}}\}.\end{aligned} \tag{5.113}$$

The momentum field conjugate to $\Psi(\boldsymbol{r})$ was shown in Eq. (5.87) to be $\wp(\boldsymbol{r}) = i\hbar\Psi^\dagger(\boldsymbol{r})$. The Hermitian conjugate of Eqs. (5.113) is

$$\begin{aligned}\Psi^\dagger(\boldsymbol{r}) = \int \frac{d\boldsymbol{p}'}{2(\pi\hbar)^{3/2}} \frac{1}{\sqrt{2E(\boldsymbol{p}')/c}} \\ \times \{[\hat{a}^u_\uparrow(\boldsymbol{p}')^\dagger u_\uparrow(\boldsymbol{p}')^\dagger + \hat{a}^u_\downarrow(\boldsymbol{p}')^\dagger u_\downarrow(\boldsymbol{p}')^\dagger]e^{-\frac{i}{\hbar}\boldsymbol{p}'\cdot\boldsymbol{r}'} \\ + [\hat{a}^v_\uparrow(\boldsymbol{p}')v_\uparrow(\boldsymbol{p}')^\dagger + \hat{a}^v_\downarrow(\boldsymbol{p}')v_\downarrow(\boldsymbol{p}')^\dagger]e^{\frac{i}{\hbar}\boldsymbol{p}'\cdot\boldsymbol{r}'}\}.\end{aligned} \tag{5.114}$$

The choice of the expressions below the fraction bars in Eqs. (5.113) and (5.114) will become apparent later.

Our goal is to ensure that $[\Psi(\boldsymbol{r}), \Psi(\boldsymbol{r}')]_+ = \delta(\boldsymbol{r} - \boldsymbol{r}')$, where the *anti-commutator* $[a, b]_+$ is defined as $[a, b]_+ = ab + ba$. This is consistent with the Pauli exclusion principle, a property of fermions. Moreover, unless the anti-commutation rule is invoked, the Dirac field quantization leads to unbounded negative energy and the past being predicted by the future, both of which would be quite inconsistent with observation.

To calculate $\Psi(r)\Psi(r')^\dagger$ and $\Psi(r')^\dagger\Psi(r)$ we need explicit expressions for $u_\uparrow(\boldsymbol{p}), v_\uparrow(\boldsymbol{p}), u_\downarrow(\boldsymbol{p}), v_\downarrow(\boldsymbol{p})$; *i.e.*, Eqs. (5.109) and (5.112). The Hermitian conjugates of these real 4-vectors are just their transposes; *e.g.*,

$$u_\uparrow(\boldsymbol{p})^\dagger = (\sqrt{E/c - p_z}, 0, \sqrt{E/c + p_z}, 0). \tag{5.115}$$

The products such as $u_\uparrow(\boldsymbol{p})^\dagger u_\uparrow(\boldsymbol{p})$ are just numbers. All such products whose vectors have opposite spin will give zero, as will the combinations $u_\uparrow(\boldsymbol{p})^\dagger v_\uparrow(\boldsymbol{p}') + u_\downarrow(\boldsymbol{p})^\dagger v_\downarrow(\boldsymbol{p}')$ and $v_\uparrow(\boldsymbol{p})^\dagger u_\uparrow(\boldsymbol{p}') + v_\downarrow(\boldsymbol{p})^\dagger u_\downarrow(\boldsymbol{p}')$. This leaves only $u_\uparrow(\boldsymbol{p})^\dagger u_\uparrow(\boldsymbol{p}')$, $u_\downarrow(\boldsymbol{p})^\dagger u_\downarrow(\boldsymbol{p}')$, $v_\uparrow(\boldsymbol{p})^\dagger v_\uparrow(\boldsymbol{p}')$, and $v_\downarrow(\boldsymbol{p})^\dagger v_\downarrow(\boldsymbol{p}')$, which are four 4×4 matrices. We require the operators $\hat{a}$ to obey the anti-commutation rules

$$\begin{aligned}
[\hat{a}^u_\uparrow(\boldsymbol{p})^\dagger, \hat{a}^u_\uparrow(\boldsymbol{p}')]_+ &= \delta(\boldsymbol{p} - \boldsymbol{p}') \\
[\hat{a}^u_\downarrow(\boldsymbol{p})^\dagger, \hat{a}^u_\downarrow(\boldsymbol{p}')]_+ &= \delta(\boldsymbol{p} - \boldsymbol{p}') \\
[\hat{a}^u_\uparrow(\boldsymbol{p})^\dagger, \hat{a}^u_\uparrow(\boldsymbol{p}')]_+ &= \delta(\boldsymbol{p} - \boldsymbol{p}') \\
[\hat{a}^u_\downarrow(\boldsymbol{p})^\dagger, \hat{a}^u_\downarrow(\boldsymbol{p}')]_+ &= \delta(\boldsymbol{p} - \boldsymbol{p}')
\end{aligned} \tag{5.116}$$

(and all other anti-commutators zero); then after integration over $\boldsymbol{p}'$ the delta functions convert the $\boldsymbol{p}'$'s to $\boldsymbol{p}$'s. From Eqs. (5.109) and (5.112), the expressions and $u_\downarrow(\boldsymbol{p})^\dagger v_\uparrow(\boldsymbol{p}) + u_\downarrow(\boldsymbol{p})^\dagger v_\downarrow(\boldsymbol{p})$ and $v_\uparrow(\boldsymbol{p})^\dagger u_\uparrow(\boldsymbol{p}) + v_\downarrow(\boldsymbol{p})^\dagger u_\downarrow(\boldsymbol{p})$ are found to be

$$\begin{aligned}
&u_\uparrow(\boldsymbol{p})u_\uparrow(\boldsymbol{p})^\dagger + u_\downarrow(\boldsymbol{p})u_\downarrow(\boldsymbol{p})^\dagger \\
&\quad = \begin{pmatrix} E/c - p_z & 0 & mc & 0 \\ 0 & E/c + p_z & 0 & mc \\ mc & 0 & E/c + p_z & 0 \\ 0 & mc & 0 & E/c - p_z \end{pmatrix}
\end{aligned} \tag{5.117}$$

and

$$v_\uparrow(\boldsymbol{p})v_\uparrow(\boldsymbol{p})^\dagger + v_\downarrow(\boldsymbol{p})v_\downarrow(\boldsymbol{p})^\dagger$$
$$= \begin{pmatrix} E/c - p_z & 0 & -mc & 0 \\ 0 & E/c + p_z & 0 & -mc \\ -mc & 0 & E/c + p_z & 0 \\ 0 & -mc & 0 & E/c - p_z \end{pmatrix}. \qquad (5.118)$$

Since the integration over $\boldsymbol{p}_z$ is over the entire range, we may change the sign of $\boldsymbol{p}_z$ in the integral over the $\boldsymbol{v}$ fields so that the sum of Eqs. (5.117) and (5.118) is

$$u_\uparrow(\boldsymbol{p})u_\uparrow(\boldsymbol{p})^\dagger + u_\downarrow(\boldsymbol{p})u_\downarrow(\boldsymbol{p})^\dagger + v_\uparrow(-\boldsymbol{p})v_\uparrow(-\boldsymbol{p})^\dagger$$
$$+ v_\downarrow(-\boldsymbol{p})v_\downarrow(-\boldsymbol{p})^\dagger = (2E/c)1^{(4)}. \qquad (5.119)$$

Thus the product of the two $\sqrt{2E/c}$'s in the denominators of the expressions of the fields in Eqs. (5.113) and (5.114) is cancelled. Now all the momenta in the integral have been removed except in the $e^{\frac{i}{\hbar}\boldsymbol{p}\cdot(\boldsymbol{r}-\boldsymbol{r}')}$ factor. The remaining integral gives

$$\int d\boldsymbol{p} e^{\frac{i}{\hbar}\boldsymbol{p}\cdot(\boldsymbol{r}-\boldsymbol{r}')} = (2\pi\hbar)^3\delta(\boldsymbol{r}-\boldsymbol{r}'). \qquad (5.120)$$

The $(2\pi\hbar)^3$ from this result cancels the remaining factors in the denominators of Eqs. (5.113) and (5.114) so that the final result is

$$[\Psi(\boldsymbol{r}), \Psi^\dagger(\boldsymbol{r}')]_+ = \delta(\boldsymbol{r}-\boldsymbol{r}'), \qquad (5.121)$$

as required.

We now turn to finding the Hamiltonian in terms of the quantized Dirac field, a considerably simpler exercise than the one outlined in this section.

5.3.2.7. *The Quantized Dirac Hamiltonian*

Eq. (5.71) gives the Dirac Hamiltonian in its simplest and most direct form. To quantize it, we need $i\hbar\partial_t\Psi(x)$; using the time-dependent form for the quantized field given in Eq. (5.113):

$$\begin{aligned}\Psi(\boldsymbol{r},t) = \int \frac{d\boldsymbol{p}}{(2\pi\hbar)^{3/2}} \frac{1}{\sqrt{2E(\boldsymbol{p})/c}} \\ \times \{[\hat{a}^u_\uparrow(\boldsymbol{p})u_\uparrow(\boldsymbol{p}) + \hat{a}^u_\downarrow(\boldsymbol{p})u_\downarrow(\boldsymbol{p})]e^{\frac{i}{\hbar}(\boldsymbol{p}\cdot\boldsymbol{r}-Et/c)} \\ + [\hat{a}^v_\uparrow(\boldsymbol{p})^\dagger v_\uparrow(\boldsymbol{p}) + \hat{a}^v_\downarrow(\boldsymbol{p})^\dagger v_\downarrow(\boldsymbol{p})]e^{-\frac{i}{\hbar}(\boldsymbol{p}\cdot\boldsymbol{r}-Et/c)}\}.\end{aligned} \tag{5.122}$$

The result is

$$\begin{aligned}i\hbar\partial_t\Psi(\boldsymbol{r},t) = \int \frac{d\boldsymbol{p}}{(2\pi\hbar)^{3/2}} \frac{E(p)/c}{\sqrt{2E(\boldsymbol{p})/c}} \\ \times \{[\hat{a}^u_\uparrow(\boldsymbol{p})u_\uparrow(\boldsymbol{p}) + \hat{a}^u_\downarrow(\boldsymbol{p})u_\downarrow(\boldsymbol{p})]e^{\frac{i}{\hbar}(\boldsymbol{p}\cdot\boldsymbol{r}-Et)} \\ - [\hat{a}^v_\uparrow(\boldsymbol{p})^\dagger v_\uparrow(\boldsymbol{p}) + \hat{a}^v_\downarrow(\boldsymbol{p})^\dagger v_\downarrow(\boldsymbol{p})]e^{-\frac{i}{\hbar}(\boldsymbol{p}\cdot\boldsymbol{r}-Et)}\}.\end{aligned} \tag{5.123}$$

The Hamiltonian is then given by

$$\begin{aligned}\hat{H} &= i\hbar\int d\boldsymbol{r}\Psi(\boldsymbol{r})^\dagger[\partial_t\Psi(\boldsymbol{r},t)]_{t=0} \\ &= \int d\boldsymbol{r}\int\frac{d\boldsymbol{p}'}{(2\pi\hbar)^{3/2}}\int\frac{d\boldsymbol{p}}{(2\pi\hbar)^{3/2}}\frac{E(p)}{\sqrt{2E(p')/c}\sqrt{2E(p)/c}} \\ &\quad\times\{[\hat{a}^u_\uparrow(\boldsymbol{p})^\dagger u_\uparrow(\boldsymbol{p}')^\dagger + \hat{a}^u_\downarrow(\boldsymbol{p}')^\dagger u_\downarrow(\boldsymbol{p}')^\dagger]e^{-\frac{i}{\hbar}(\boldsymbol{p}'\cdot\boldsymbol{r})} \\ &\quad - [\hat{a}^v_\uparrow(-\boldsymbol{p}')v_\uparrow(-\boldsymbol{p}')^\dagger + \hat{a}^v_\downarrow(-\boldsymbol{p}')v_\downarrow(-\boldsymbol{p}')^\dagger]e^{-\frac{i}{\hbar}(\boldsymbol{p}'\cdot\boldsymbol{r})}\} \\ &\quad\times\{[\hat{a}^u_\uparrow(\boldsymbol{p})u_\uparrow(\boldsymbol{p}) + \hat{a}^u_\downarrow(\boldsymbol{p})u_\downarrow(\boldsymbol{p})]e^{\frac{i}{\hbar}(\boldsymbol{p}\cdot\boldsymbol{r})} \\ &\quad + [\hat{a}^v_\uparrow(-\boldsymbol{p})^\dagger v_\uparrow(-\boldsymbol{p}) + \hat{a}^v_\downarrow(-\boldsymbol{p})^\dagger v_\downarrow(-\boldsymbol{p})]e^{\frac{i}{\hbar}(\boldsymbol{p}\cdot\boldsymbol{r})}\},\end{aligned} \tag{5.124}$$

where we have reversed the signs of the integration variable in the v terms so that the integral over $\boldsymbol{p}'$ gives

$$\int d\boldsymbol{r} e^{\frac{i}{\hbar}\boldsymbol{r}\cdot(\boldsymbol{p}-\boldsymbol{p}')} = (2\pi\hbar)^3\delta(\boldsymbol{p}-\boldsymbol{p}').$$

As before, the u and v spinors pair only with the same spins, while pairings of any u with any v gives zero; *e.g.*,

$$u_\uparrow(\boldsymbol{p})^\dagger v_\uparrow(-\boldsymbol{p})$$

$$= (\sqrt{E/c - p_z} 0 \sqrt{E/c + p_z} 0) \begin{pmatrix} \sqrt{E/c + p_z} \\ 0 \\ -\sqrt{E/c - p_z} \\ 0 \end{pmatrix} = 0. \quad (5.125)$$

We need to evaluate the pairings $u_\uparrow(\boldsymbol{p})^\dagger u_\uparrow(\boldsymbol{p})$, $u_\downarrow(\boldsymbol{p})^\dagger u_\downarrow(\boldsymbol{p})$, $v_\downarrow(\boldsymbol{p})^\dagger v_\uparrow(\boldsymbol{p})$, and $v_\downarrow(\boldsymbol{p})^\dagger v_\downarrow(\boldsymbol{p})$, each of which equals $2E(p)/c$. For example,

$$u_\uparrow(\boldsymbol{p})^\dagger u_\uparrow(\boldsymbol{p}) = (\sqrt{E/c - p_z} 0 \sqrt{E/c + p_z} 0) \begin{pmatrix} \sqrt{E/c - p_z} \\ 0 \\ \sqrt{E/c + p_z} \\ 0 \end{pmatrix}$$

$$= 2E(\boldsymbol{p})/c. \quad (5.126)$$

The result is

$$\hat{H} = \int d\boldsymbol{p} E(\boldsymbol{p})\{[\hat{a}^u_\uparrow(\boldsymbol{p})^\dagger \hat{a}^u_\uparrow(\boldsymbol{p}) + \hat{a}^u_\downarrow(\boldsymbol{p})^\dagger \hat{a}^u_\downarrow(\boldsymbol{p})]$$

$$- [\hat{a}^v_\uparrow(-\boldsymbol{p})\hat{a}^v_\downarrow(-\boldsymbol{p})^\dagger + \hat{a}^v_\downarrow(-\boldsymbol{p})^\dagger \hat{a}^v_\downarrow(-\boldsymbol{p})]\}. \quad (5.127)$$

But when the anti-commutator for the last line is used, Eq. (5.127) becomes

$$\hat{H} = \int d\boldsymbol{p} E(\boldsymbol{p})\{[\hat{a}^u_\uparrow(\boldsymbol{p})^\dagger \hat{a}^u_\uparrow(\boldsymbol{p}) + \hat{a}^u_\downarrow(\boldsymbol{p})^\dagger \hat{a}^u_\downarrow(\boldsymbol{p})]$$

$$+ [\hat{a}^v_\uparrow(-\boldsymbol{p})^\dagger \hat{a}^v_\uparrow(-\boldsymbol{p}) + \hat{a}^v_\downarrow(-\boldsymbol{p})^\dagger \hat{a}^v_\downarrow(-\boldsymbol{p})]\delta(0)\}. \quad (5.128)$$

The first square bracket in Eq. (5.128) gives the number of electrons of both spins in a given field, while the second bracket gives the number of positrons; *e.g.*,

$$\begin{aligned}
&\hat{H}\hat{a}^u_\uparrow(\boldsymbol{p}_3)^\dagger\hat{a}^u_\downarrow(\boldsymbol{p}_2)^\dagger a^v_\uparrow(-\boldsymbol{p}_1)^\dagger|0\rangle \\
&\quad = [E(p_1)+E(p_2)+E(p_3)] \\
&\qquad \times \hat{a}^u_\uparrow(\boldsymbol{p}_3)^\dagger\hat{a}^u_\downarrow(\boldsymbol{p}_2)^\dagger a^v_\uparrow(-\boldsymbol{p}_1)^\dagger|0\rangle, \qquad (5.129)
\end{aligned}$$

where we have used the anti-commutator version of the contraction rule:

$$\overline{\hat{a}_\beta(\boldsymbol{p}_2)\hat{a}^\dagger_\alpha}(\boldsymbol{p}_1)|0\rangle = \delta_{\alpha\beta}\delta(\boldsymbol{p}_1,\boldsymbol{p}_2)|0\rangle$$

(see Chapters 3 and 6).

The infinity represented by $\delta(0)$ and redundant delta functions has been interpreted in several ways, but in calculations is just ignored! This is one of the reasons the quantum theory is questioned, even though it gives astonishingly accurate agreement with experiments.

5.3.2.8. *The Quantized Charge Density*

From Eq. (5.82), the charge density is just $\Psi^\dagger\Psi$; inserting the electron charge $-e$ and using Eqs. (5.113) and (5.114) give

$$\begin{aligned}
\hat{\rho} = \hat{\Psi}^\dagger\hat{\Psi}(\boldsymbol{r}) &= \int \frac{d\boldsymbol{p}'}{(2\pi\hbar)^{3/2}}\frac{1}{\sqrt{2E(p')/c}} \\
&\times \{[\hat{a}^u_\uparrow(\boldsymbol{p}')^\dagger u_\uparrow(\boldsymbol{p}')^\dagger + \hat{a}^u_\downarrow(\boldsymbol{p})^\dagger u_\downarrow(\boldsymbol{p}')^\dagger]e^{-\frac{i}{\hbar}\boldsymbol{p}'\cdot\boldsymbol{r}} \\
&+ [\hat{a}^v_\uparrow(\boldsymbol{p}')v_\uparrow(\boldsymbol{p}')^\dagger + \hat{a}^v_\downarrow(\boldsymbol{p}')v_\downarrow(\boldsymbol{p}')^\dagger]e^{\frac{i}{\hbar}\boldsymbol{p}'\cdot\boldsymbol{r}}\} \\
&\times \int \frac{d\boldsymbol{p}}{(2\pi\hbar)^{3/2}}\frac{1}{\sqrt{2E(p)/c}}\{[\hat{a}^u_\uparrow(\boldsymbol{p})u_\uparrow(\boldsymbol{p}) + \hat{a}^u_\downarrow(\boldsymbol{p})u_\downarrow(\boldsymbol{p})]e^{\frac{i}{\hbar}\boldsymbol{p}\cdot\boldsymbol{r}} \\
&+ [\hat{a}^v_\uparrow(\boldsymbol{p})^\dagger v_\uparrow(\boldsymbol{p}) + \hat{a}^v_\downarrow(\boldsymbol{p})^\dagger v_\downarrow(\boldsymbol{p})]e^{-\frac{i}{\hbar}\boldsymbol{p}\cdot\boldsymbol{r}}\}. \qquad (5.130)
\end{aligned}$$

We use the analogous procedures as for the Hamiltonian to obtain

$$\hat{\rho} = e \int d\boldsymbol{p}\{[\hat{a}^u_\uparrow(\boldsymbol{p})^\dagger \hat{a}^u_\uparrow(\boldsymbol{p}) + \hat{a}^u_\downarrow(\boldsymbol{p})^\dagger \hat{a}^u_\downarrow(\boldsymbol{p})]$$

$$-[\hat{a}^v_\uparrow(-\boldsymbol{p})^\dagger \hat{a}^v_\uparrow(-\boldsymbol{p}) + \hat{a}^v_\downarrow(-\boldsymbol{p})^\dagger \hat{a}^v_\downarrow(-\boldsymbol{p})] + \delta(0)\}. \quad (5.131)$$

The first bracket measures the total (negative) charge on all the electrons in the system, while the second bracket measures the total (positive) charge on all the positrons in the system. Again, we get that pesky $\delta(0)$, which we ignore!

5.4. Summary for Chapter 5

Using the basics of quantum field theory developed in Chapter 4 for spinless particles, we showed in this chapter how the theory needs to be modified for particles with both charge and spin.

Pauli's attempt at finding a wave equation for charged particles with spin led to a 2-component wave function and the formulation of spin in terms of the famous Pauli 2×2 matrices. Each component of this wave function obeys the Klein-Gordon equation. But Pauli's equation is not relativistic and his matrices seem to be mathematical constructs with little motivation from observation.

Dirac boldly tackled the problem of finding a relativistic wave equation for charged particles with spin $1/2$, notably electrons. Following Schrödinger, he sought an equation having a first-order time derivative. This led to a 4-component wave function and 4×4 matrices that incorporated the 2×2 Pauli matrices, which now appeared to be required for agreement with experiment. The requirement that electrons be fermions also is a requirement for the theory of positive electrons to be self-consistent.

Since the development of these ideas requires detailed manipulation of complicated equations containing symbols not to be trivially defined, we shall not repeat the developments in this brief summary, but refer the reader to relevant sections of the chapter.

5.5. Appendix A: Schwinger's Harmonic-Oscillator Model for Spin[7]

The properties of the Fock operators for harmonic oscillators allow the construction of operators that correspond to angular momentum. Consider two harmonic oscillators A and B with Fock operators and $\hat{a}$ and $\hat{b}$, respectively. Now let the vector operator $\hat{\boldsymbol{S}}$ be defined by

$$\hat{\boldsymbol{S}} = \frac{1}{2}(\hat{a}^\dagger \hat{b}^\dagger)\sigma \begin{pmatrix} \hat{a} \\ \hat{b} \end{pmatrix}, \tag{5A.1}$$

where $\boldsymbol{\sigma}$ is the 3-vector of Pauli matrices,

$$\boldsymbol{\sigma} = \begin{pmatrix} 0 & 1 \\ 1 & 0 \end{pmatrix} \hat{\mathbf{e}}_x + \begin{pmatrix} 0 & -i \\ i & 0 \end{pmatrix} \hat{\mathbf{e}}_y + \begin{pmatrix} 1 & 0 \\ 0 & -1 \end{pmatrix} \hat{\mathbf{e}}_z. \tag{5A.2}$$

The components of $\hat{\boldsymbol{S}}$ are therefore

$$\begin{aligned} \hat{S}_x &= \frac{1}{2}(\hat{a}^\dagger \hat{b} + \hat{b}^\dagger \hat{a}), \\ \hat{S}_y &= -\frac{i}{2}(\hat{a}^\dagger \hat{b} + \hat{b}^\dagger \hat{a}), \\ \hat{S}_z &= \frac{1}{2}(\hat{a}^\dagger \hat{b} + \hat{b}^\dagger \hat{b}), \end{aligned} \tag{5A.3}$$

Inspection of Eq. (4.3) shows $\hat{\boldsymbol{S}}$ to be Hermitian. From the determinantal form of the vector cross product, the definition of Eqs. (4.1)–(4.3), and the commutation relation given in Eq. (2.10), we obtain

$$\begin{aligned} \hat{\boldsymbol{S}} \times \hat{\boldsymbol{S}} &= \begin{vmatrix} e_x & e_y & e_z \\ \hat{S}_x & \hat{S}_y & \hat{S}_z \\ \hat{S}_x & \hat{S}_y & \hat{S}_z \end{vmatrix} = [S_y, S_z]_- \hat{\mathbf{e}}_x + [S_z, S_x]_- \hat{\mathbf{e}}_y + [S_x, S_y]_- \hat{\mathbf{e}}_z \\ &= iS_x \hat{\mathbf{e}}_x + iS_y \hat{\mathbf{e}}_y + iS_z \hat{\mathbf{e}}_z = i\hat{\boldsymbol{S}}, \end{aligned} \tag{5A.4}$$

the defining property of an angular momentum operator.

[7]Schwinger, J., in *Quantum Theory of Angular Momentum*, Academic Press (1965), p. 229.

5.5.1. *Spin Eigenvalues and Eigenvectors*

The eigenvalues of $\hat{S}_z$ are easy to find from the fact that $\hat{a}^\dagger\hat{a}$ and $\hat{b}^\dagger\hat{b}$ are the number operators $\hat{n}_a$ and $\hat{n}_b$, whose eigenvalues are n_a and n_b, respectively. Thus, from Eq. (4.3) we have

$$\hat{S}_z = \frac{1}{2}(\hat{n}_a - \hat{n}_b) \quad \text{with eigenvalues} \quad S_z = \frac{1}{2}(n_a - n_b).$$

5.5.2. *Rotation of the Spin Axes*

The axes of the spin vector whose operator is given by Eq. (4.1) may be rotated into an arbitrary orientation by successive application of rotation operators. For example, if we rotate the operator $\hat{S}_x$ counterclockwise about the z-axis by the angle φ, the rotated operator $\hat{S}_x(\varphi)$ is given by

$$\hat{S}_x(\varphi) = e^{i\hat{S}_z\varphi}\hat{S}_x e^{-i\hat{S}_z\varphi}. \tag{5A.5}$$

Chapter 6

The Perfect Molecular Gas*

6.1. Introduction

We have shown that creation and annihilation operators can be used for both elementary bosons (Chapter 4) and fermions (Chapter 5). In this chapter we apply quantum field theory to an ensemble of non-interacting molecules by an extension of the treatment of elementary particles in Chapters 4 and 5. Here, the field quanta are *molecules*, which may be either bosons or fermions, with momentum $\boldsymbol{p}$ and in internal state α. In the "classical" limit of high temperature and large volume and number of molecules, both types of molecules obey Boltzmann statistics. But even in this limit, quantum field theory remains a useful tool for deriving the properties of perfect gases and provides a basis for extension of the theory to imperfect gases and liquids, as we shall see in the next chapter. When the molecular phases in a coherent ensemble are randomized, the resulting ensemble operator becomes isodasic[1]; *i.e.*, it describes a gas with uniform density filling a volume V. When the entropy of the ensemble is maximized at constant N (average number of molecules) and E (average total energy), the ensemble operator describes a perfect gas at thermal equilibrium. In this case, the trace of the ensemble operator is the grand canonical partition function. Although the explicit

---

*Much of the material in this chapter has appeared in Porter, R. N., *AIP Conf. Proc.* **1102**, 219 (2009). It is reprinted here with the permission of AIP.

[1]From the Greek *iso* = same plus *dasys* = dense.

form of the isodasic ensemble operator is somewhat complicated, it obeys a very simple equation which facilitates the calculations. The thermodynamic properties of the gas derived directly from the isodasic ensemble operator by the application of quantum field theory are equivalent to those derived from the grand canonical partition function in the traditional way. The ensemble operator in the Fock operator form also provides a means of deriving statistical properties of Bose, Fermi, and Boltzmann gases.

It is intended that this chapter and the next on the quantum field theory of the perfect gas, imperfect gases, and liquids be self-contained. We therefore review briefly in Section 6.2 the molecular Fock operators introduced in earlier chapters and describe how the finite volume enters the present theory through a delta function with zero argument. Several of the operators that are used to calculate properties of the system of molecules are given there. A succinct form of Wick's theorem for bosons and fermions which we require for calculating the trace of the ensemble operator and average values is given in Section 6.3. In Section 6.4, we present the essential properties of a coherent ensemble of non-interacting molecules. We randomize the phases of the molecules in Section 6.5 and show that the resulting ensemble operator obeys a simple equation which is key to the calculation of properties of the perfect gas and to the extension of the theory to imperfect gases and liquids in Chapter 7. We determine the trace of the isodasic operator in Section 6.6 by a linked-cluster expansion and derive the distribution function for thermal equilibrium. In Section 6.7, we compare molecular statistics for coherent and isodasic ensembles.

We conclude with a summary and discussion in Section 6.8.

6.2. Molecular Fock Operators

A quantized molecular field is one in which the quanta are molecules. We define the Fock operator for the creation of a molecule in (internal) state α at point $\boldsymbol{x}$ to be $\hat{\psi}_\alpha(\boldsymbol{x})^\dagger$ and the corresponding annihilation operator to be its Hermitian conjugate, $\hat{\psi}_\alpha(\boldsymbol{x})$. These operators are related to the operators for creation and annihilation of a molecule in state α with linear momentum $\boldsymbol{p}$ (that is, $\hat{a}_{\alpha_\alpha}(\boldsymbol{p})^\dagger$ and $\hat{a}_\alpha(\boldsymbol{p})$, respectively) by the Fourier transform

$$\hat{\psi}_\alpha(\boldsymbol{x}) = h^{-\frac{3}{2}} \int d\boldsymbol{p} e^{\frac{i}{\hbar}\boldsymbol{p}\cdot\boldsymbol{x}} \hat{a}_\alpha(\boldsymbol{p}). \tag{6.1}$$

It should be noted that the Planck constant $\hbar = h/2\pi$ enters the Fourier transforms as the conversion factor between the wave length and the momentum, in accordance with de Broglie's rule, but disappears from the properties of macroscopic fluids in the Boltzmann limit. These molecular Fock operators have the usual Bose-Einstein commutation rules (upper sign) or Fermi-Dirac anti-commutation rules (lower sign):

$$[\hat{a}_\alpha(\boldsymbol{p}), \hat{a}_{\alpha'}(\boldsymbol{p}')]_\mp = [\hat{a}_\alpha(\boldsymbol{p})^\dagger, \hat{a}_{\alpha'}(\boldsymbol{p}')^\dagger]_\mp = 0 \tag{6.2a}$$

$$[\hat{a}_\alpha(\boldsymbol{p}), \hat{a}_{\alpha'}(\boldsymbol{p}')^\dagger]_\mp = \delta_{\alpha\alpha'}\delta(\boldsymbol{p} - \boldsymbol{p}'), \tag{6.2b}$$

where $[\hat{A}_1, \hat{A}_2]_\mp$ is the commutator $(-)$ or anti-commutator $(+)$ of the operators $\hat{A}_1$ and $\hat{A}_2$:

$$[\hat{A}_1, \hat{A}_2]_\mp = [\hat{A}_1\hat{A}_2 \mp \hat{A}_2\hat{A}_1]. \tag{6.2c}$$

In Eq. (6.2b), $\delta_{\alpha\alpha'}$ is the Kronecker delta defined as $\delta_{ij} = \begin{cases} 1 \text{ if } i = j \\ 0 \text{ if } i \neq j \end{cases}$, and $\delta(\boldsymbol{p} - \boldsymbol{p}')$ is the Dirac delta function with the Fourier integral representation

$$\delta(\boldsymbol{p} - \boldsymbol{p}') = h^{-3} \int d\boldsymbol{x} e^{\frac{i}{\hbar}(\boldsymbol{p}-\boldsymbol{p}')\cdot\boldsymbol{x}}. \tag{6.2d}$$

In the special case that $\boldsymbol{p}' = \boldsymbol{p}$, we have

$$[\hat{a}_\alpha(\boldsymbol{p}), \hat{a}_{\alpha'}(\boldsymbol{p})^\dagger]_\mp = \delta_{\alpha\alpha'}\delta(\boldsymbol{p} - \boldsymbol{p}), \tag{6.3}$$

where

$$\delta(\boldsymbol{p} - \boldsymbol{p}) = h^{-3} \int d\boldsymbol{x} e^{\frac{i}{\hbar}(\boldsymbol{p}-\boldsymbol{p})\cdot\boldsymbol{x}} = h^{-3} \int d\boldsymbol{x} = \frac{V}{h^3}, \tag{6.4}$$

with V the total volume of the system. This interpretation of $\delta(0)$, normally taken to be infinite, is nonetheless consistent with the fact that we assume a finite volume. The assumption of a continuous momentum variable is a reasonable approximation to the quantized values that result for a finite volume, since that volume is large enough that boundary conditions can be neglected in the calculation of bulk properties. Not only does the use of a continuum of values for both $\boldsymbol{x}$ and $\boldsymbol{p}$ allow for functional analysis of the equations we shall derive, but the use of Eq. (6.4) introduces the volume in the physically correct way, as we shall see in Section 6.7.

The state vector for a molecule in internal state α and with momentum $\boldsymbol{p}$ is $\hat{a}_\alpha(\boldsymbol{p})^\dagger|0\rangle$, where $|0\rangle$ is the vacuum *ket* defined by $\hat{a}_\alpha(\boldsymbol{p})|0\rangle = 0$ for all α and $\boldsymbol{p}$. For reasons that will become clear, we represent this state in density-matrix form as $\hat{a}_\alpha(\boldsymbol{p})^\dagger|0\rangle\langle 0|\hat{a}_\alpha(\boldsymbol{p})$, where $|0\rangle\langle 0|$ is the density matrix for the vacuum state, namely the ordered product of $|0\rangle$ and its Hermitian conjugate $\langle 0|$. A system of n molecules in specified states has the density matrix

$$|n\rangle\langle n| = \prod_{i=1}^{n} \hat{a}_{\alpha_{(n-i+1)}}(\boldsymbol{p}_{(n-i+1)})^\dagger|0\rangle\langle 0| \prod_{j=1}^{n} \hat{a}_{\alpha_j}(\boldsymbol{p}_j). \tag{6.5}$$

The indices ensure that the molecular states are paired as one moves out from the vacuum projection operator, beginning with state α_1, $\boldsymbol{p}_1$ closest to the vacuum and ending with state α_n, $\boldsymbol{p}_n$ farthest away.

To calculate the expectation value of observable property O of a multi-molecular state such as that of Eq. (6.5), we evaluate $\mathrm{Tr}[\hat{O}|n\rangle\langle n|/\mathrm{Tr}[|n\rangle\langle n|$, where $\hat{O}$ is the corresponding operator and Tr is the trace operator. Expressions for several such operators in the Fock formalism are:

Number density of species α in momentum space

$$\hat{N}_\alpha(\boldsymbol{p}) = \hat{a}_\alpha(\boldsymbol{p})^\dagger \hat{a}_\alpha(\boldsymbol{p}). \tag{6.6a}$$

Energy of species α

$$\hat{H}_\alpha = \int d\boldsymbol{p}\, \varepsilon_\alpha(p)\hat{a}_\alpha(\boldsymbol{p})^\dagger \hat{a}_\alpha(\boldsymbol{p}), \tag{6.6b}$$

where $\varepsilon_\alpha(p) = \varepsilon_\alpha + p^2/2m$.

Number density at point $\boldsymbol{x}$

$$\hat{D}_\alpha(\boldsymbol{x}) = \hat{\psi}_\alpha(\boldsymbol{x})^\dagger \hat{\psi}_\alpha(\boldsymbol{x}) = \frac{1}{h^3}\int d\boldsymbol{p} \int d\boldsymbol{p}' e^{\frac{i}{\hbar}(\boldsymbol{p}-\boldsymbol{p}')\cdot\boldsymbol{x}} \hat{a}_\alpha(\boldsymbol{p}')^\dagger \hat{a}_\alpha(\boldsymbol{p}). \tag{6.6c}$$

Two-point correlation function

$$\hat{D}_{\alpha\beta}(\boldsymbol{x_1}, \boldsymbol{x_2}) = \left(1 - \frac{1}{2}\delta_{\alpha\beta}\right) \hat{\psi}_\alpha(\boldsymbol{x}_1)^\dagger \hat{\psi}_\beta(\boldsymbol{x}_2)^\dagger \hat{\psi}_\beta(\boldsymbol{x}_2)\hat{\psi}_\alpha(\boldsymbol{x}_1). \quad (6.6\text{d})$$

Pair correlation function

$$\begin{aligned}\hat{D}_{\alpha\beta}(\boldsymbol{r}) &= \int d\boldsymbol{X}\hat{D}_{\alpha\beta}\left[\left(X + \frac{1}{2}\boldsymbol{r}\right), \left(X - \frac{1}{2}\boldsymbol{r}\right)\right] \\ &= \left(1 - \frac{1}{2}\delta_{\alpha\beta}\right)\frac{1}{h^3}\int d\boldsymbol{p}\int d\boldsymbol{p}'\int d\boldsymbol{q}e^{\frac{i}{\hbar}\boldsymbol{q}\cdot\boldsymbol{r}} \\ &\quad \times \hat{a}_\alpha(\boldsymbol{p} - \boldsymbol{q})^\dagger \hat{a}_\beta(\boldsymbol{p}' + \boldsymbol{q})^\dagger \hat{a}_\beta(\boldsymbol{p}')\hat{a}_\alpha(\boldsymbol{p}). \qquad (6.6\text{e})\end{aligned}$$

Normalization of the density matrix corresponding to a multi-molecular system requires that we calculate its trace. Using the cyclic property of the trace; that is, that the trace of a string of operators is invariant with respect to cyclic permutation of operators in the string, we obtain for the trace of Eq. (6.5)

$$\begin{aligned}&\mathrm{Tr}\left[\prod_{i=1}^{n}\hat{a}_{\alpha_{n-i+1}}(\boldsymbol{p}_{n-i+1})^\dagger|0\rangle\langle 0|\prod_{j=1}^{n}\hat{a}_{\alpha_j}(\boldsymbol{p}_j)\right] \\ &= \left\langle 0\left|\prod_{j=1}^{n}\hat{a}_{\alpha_j}(\boldsymbol{p}_j)\prod_{i=1}^{n}\hat{a}_{\alpha_{n-i+1}}(\boldsymbol{p}_{n-i+1})^\dagger\right|0\right\rangle. \qquad (6.7)\end{aligned}$$

The completion of the calculation of the trace indicated in Eq. (6.7) is greatly facilitated by the use of Wick's theorem to organize the application of the commutation rules of Eqs. (6.2).

6.3. Wick's Theorem[2]

The basic idea behind Wick's theorem is that any annihilation operator operating on the vacuum ket gives zero by definition, since the vacuum state contains no molecules to be annihilated. Thus we have

$$\hat{a}_\alpha(\boldsymbol{p})|0\rangle = 0 \qquad (6.8\text{a})$$

[2]See footnote on p. 56.

and its Hermitian conjugate

$$\langle 0|\hat{a}_\alpha(\boldsymbol{p})^\dagger = 0. \tag{6.8b}$$

This means that in a string of Fock operators averaged over the vacuum state, any annihilation operators must be placed to the left of any creation operators in order to get a non-zero result; for example,

$$\langle 0|\hat{a}_\alpha(\boldsymbol{p}_1)^\dagger \hat{a}_\beta(\boldsymbol{p}_2)|0\rangle = 0, \tag{6.9a}$$

but

$$\langle 0|\hat{a}_\beta(\boldsymbol{p}_2)\hat{a}_\alpha(\boldsymbol{p}_1)^\dagger|0\rangle \neq 0. \tag{6.9b}$$

If the commutation rule of Eq. (6.2b) is used, then Eq. (6.9b) can be written

$$\pm\langle 0|\hat{a}_\alpha(\boldsymbol{p}_1)^\dagger \hat{a}_\beta(\boldsymbol{p}_2)|0\rangle + \delta_{\alpha\beta}\delta(\boldsymbol{p}_1 - \boldsymbol{p}_2)\langle 0|0\rangle = \delta_{\alpha\beta}\delta(\boldsymbol{p}_1 - \boldsymbol{p}_2), \tag{6.10}$$

where we have used the convention that $\langle 0|0\rangle = 1$ to write the r.h.s. in Eq. (6.10). We see that we get the null result unless each $\hat{a}_\alpha(\boldsymbol{p})$ annihilates a creation operator to its right. This can be expressed by a line, called a *contraction*, connecting the two operators

$$\langle 0|\hat{a}_\beta(\boldsymbol{p}_2)\hat{a}_\alpha(\boldsymbol{p}_1)^\dagger|0\rangle = \delta_{\alpha\beta}\delta(\boldsymbol{p}_1 - \boldsymbol{p}_2)|0\rangle. \tag{6.11}$$

In a more extensive string, there are in general several ways of contracting all the annihilation operators; for example,

$$\begin{aligned}
&\langle 0|\hat{a}_{\alpha_1}(\boldsymbol{p}_1)\hat{a}_{\alpha_2}(\boldsymbol{p}_2)\hat{a}_{\alpha_3}(\boldsymbol{p}_3)^\dagger\hat{a}_{\alpha_4}(\boldsymbol{p}_4)^\dagger|0\rangle \\
&= \langle 0|\hat{a}_{\alpha_1}(\boldsymbol{p}_1)\hat{a}_{\alpha_2}(\boldsymbol{p}_2)\hat{a}_{\alpha_3}(\boldsymbol{p}_3)^\dagger\hat{a}_{\alpha_4}(\boldsymbol{p}_4)^\dagger|0\rangle \\
&\quad + \langle 0|\hat{a}_{\alpha_1}(\boldsymbol{p}_1)\hat{a}_{\alpha_2}(\boldsymbol{p}_2)\hat{a}_{\alpha_3}(\boldsymbol{p}_3)^\dagger\hat{a}_{\alpha_4}(\boldsymbol{p}_4)^\dagger|0\rangle \\
&= \delta_{\alpha_1\alpha_4}\delta(\boldsymbol{p}_1 - \boldsymbol{p}_4)\delta_{\alpha_2\alpha_3}\delta(\boldsymbol{p}_2 - \boldsymbol{p}_3) \\
&\quad \pm \delta_{\alpha_1\alpha_3}\delta(\boldsymbol{p}_1 - \boldsymbol{p}_3)\delta_{\alpha_2\alpha_4}\delta(\boldsymbol{p}_2 - \boldsymbol{p}_4).
\end{aligned} \tag{6.12}$$

All possible contractions must be added together and if there are an odd number of operators between the termini of a contraction, a $-$ sign must be assigned to that contraction if the two contracted molecules are fermions, in accordance with Eqs. (6.2). In our use of $\pm$ and $\mp$, the upper sign always refers to bosons and the lower sign to fermions. Each contracted operator is removed from the string; therefore, only one $\mp$ sign appears for the second contraction scheme in Eq. (6.12).

Since our calculations will always ultimately involve traces of operator strings operating on the vacuum state $|0\rangle\langle 0|$, Wick's theorem has the simple form: "All terms with any un-contracted Fock operators give a null result." It should be obvious that if the string contains an odd number of Fock operators, it will not be possible to fully contract them; the vacuum expectation of such a string will therefore necessarily be zero. In the applications of quantum field theory to ensembles of molecules, strings of Fock operators $\hat{O}$ are traced with the ensemble operator $\hat{\rho}$, which contains strings of $\hat{a}_\alpha(\boldsymbol{p})^\dagger$'s to the left and strings $\hat{a}_\alpha(\boldsymbol{p})$'s to the right. In this case, if $\hat{O} = \hat{A}\hat{B}$, the Fock operators in the operator product should be put into "normal order'; *i.e.*, with all the $\hat{a}_\alpha(\boldsymbol{p})^\dagger$'s to the left and all the $\hat{a}_\alpha(\boldsymbol{p})$'s to the right in the product $\hat{A}\hat{B}$. This means that a non-normal-ordered string is replaced by the sum of all possible contractions with the remaining operators in normal order, taking note of the appropriate signs for commuting Fock operators for bosons and fermions as expressed by Eqs. (6.2). This complication is seldom encountered, since we are generally seeking effective one- and two-molecule operators, allowing all possible contractions in the product $\hat{A}\hat{B}$ which preserve the number of annihilation operators initially in $\hat{B}$. This will become clearer as the calculations proceed.

At this point, we illustrate the organizational benefits of Wick's theorem in calculating with strings of Fock operators and the diagrammatic representations of the various contraction schemes by addressing the normalization of states comprised of several molecules. The methods are used in detail in order to get a "hands on" feel for them.

As an example of the use of Wick's theorem and the ensuing combinatorics to normalize a nine-molecule state, consider the state vector

$$\hat{a}_{\alpha_9}(\boldsymbol{p}_9)^\dagger \hat{a}_{\alpha_8}(\boldsymbol{p}_8)^\dagger \hat{a}_{\alpha_7}(\boldsymbol{p}_7)^\dagger \hat{a}_{\alpha_6}(\boldsymbol{p}_6)^\dagger \hat{a}_{\alpha_5}(\boldsymbol{p}_5)^\dagger$$
$$\times \hat{a}_{\alpha_4}(\boldsymbol{p}_4)^\dagger \hat{a}_{\alpha_3}(\boldsymbol{p}_3)^\dagger \hat{a}_{\alpha_2}(\boldsymbol{p}_2)^\dagger \hat{a}_{\alpha_1}(\boldsymbol{p}_1)^\dagger |0\rangle,$$

which we abbreviate as $\hat{a}_9^\dagger \hat{a}_8^\dagger \hat{a}_7^\dagger \hat{a}_6^\dagger \hat{a}_5^\dagger \hat{a}_4^\dagger \hat{a}_3^\dagger \hat{a}_2^\dagger \hat{a}_1^\dagger |0\rangle$. Its Hermitian conjugate is $\langle 0|\hat{a}_1\hat{a}_2\hat{a}_3\hat{a}_4\hat{a}_5\hat{a}_6\hat{a}_7\hat{a}_8\hat{a}_9$.

We now calculate the normalization bracket,

$$\langle 0|\hat{a}_1\hat{a}_2\hat{a}_3\hat{a}_4\hat{a}_5\hat{a}_6\hat{a}_7\hat{a}_8\hat{a}_9\hat{a}_9^\dagger \hat{a}_8^\dagger \hat{a}_7^\dagger \hat{a}_6^\dagger \hat{a}_5^\dagger \hat{a}_4^\dagger \hat{a}_3^\dagger \hat{a}_2^\dagger \hat{a}_1^\dagger |0\rangle.$$

According to Wick's theorem, each annihilator must be contracted with a creator. We may accomplish the contractions by taking all permutations of the string of creators and then, for each permutation, contract from the center outward. This will give a total of $9! = 362,880$ contraction schemes. But the contractions will fall into several different categories; for example, one contraction scheme can be diagramed as

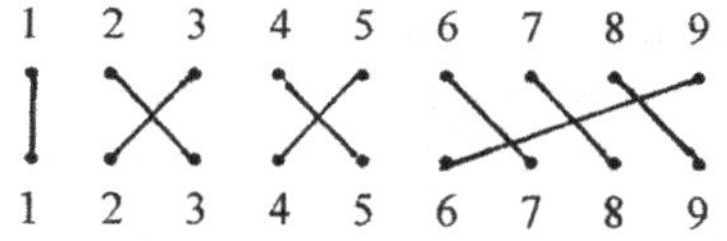

It is easy to see the contraction patterns in this diagram; in particular, one can see how many different contractions of a given type are among the total of 362,880 possible contractions. In this diagram we have: 1 cluster of 1 molecule, 2 clusters of 2 molecules, and 1 cluster of 4 molecules. The cluster of 1 can contain any of the 9 molecules; then the 2 clusters of 2 can be chosen in $8!/(2!\cdot 4!)$ ways, leaving the remaining 4 molecules for the cluster of 4. This gives $9\cdot 8!/(2!\cdot 2!\cdot 4!)$ ways of choosing the molecules for this contraction pattern. For the cluster of 4, there are 3 ways to chose the contraction partner of molecule 6 on the bottom; then there are 2 ways to contract molecule 7 and 1 remaining way each to contract molecules 8 and 9; this gives an additional factor of 3!. Furthermore, the 2 clusters of 2 are identical, so we must divide by 2 to get a total of $3!\cdot 9\cdot 8!/(8\cdot 4!) = 11,340$ contraction patterns of this type. All the molecules in each cluster are contracted into a single state $\alpha, \boldsymbol{p}$.

In addition, there is 1 "redundant" contraction in each cluster that leads to the equivalent of the factor $\delta(p - p) = V/h^3$. Thus, each of the 11,340 contraction patterns in this example contains the factor $(V/h^3)^4$. The crossing of lines in the diagram indicate odd permutations of operators that produce $\pm$ signs in accordance with the commutation rules of Eqs. (6.2). This diagram has 5 crossings and therefore acquires a $\pm$ sign.

On page 159 is a complete analysis of the contractions for the normalization of a 6-molecule state, where we have use the symbol γ for V/h^3. For $n = 6$, there are 11 types of diagrams, as shown. In the first column to the right of the diagrams is the number of diagrams of the type shown times γ raised to a power equal to the number of clusters in the diagram. A $\pm$ sign is affixed for those diagrams with an odd number of crossed lines. The total number of contraction schemes is the sum of the numbers in the first column, namely $6! = 720$. As a check on our count, we recalculate the normalization under the assumption that two of the molecules in the 6-molecule state vector are in the same single-molecule state $\alpha, \boldsymbol{p}$. The second column gives these results for each type of diagram. Note that the power of γ is raised by 1, since the additional redundancy yields an additional factor of V/h^3. Note that the signs do not change because the number of crossings remains the same. The third column gives the difference between the first two columns; that is, the results for the number of diagrams that remain unchanged by setting two single-molecule states equal. These entries will have the original powers of γ. The sum of the second and third columns thus gives the total result for the normalization of the 6-molecule vector in which 2 of the molecules are in the same single-molecule state. Note that for fermions (lower sign) this sum vanishes in each power of γ. This is required by the Pauli principle, that disallows multi-particle states in which two or more identical fermions are in the same single-particle state. The null result when 2 fermionic molecules are in the same state is a confirmation of our enumeration of the diagrams of the various types. In case of macroscopic volumes, V/h^3 is of course very large, so that the largest power of γ dominates. This means that in the classical limit only the single "direct" diagram on line 1 needs to be included.

But in the quantum regime of "degenerate" Bose and Fermi gases, all the diagrams must be included.

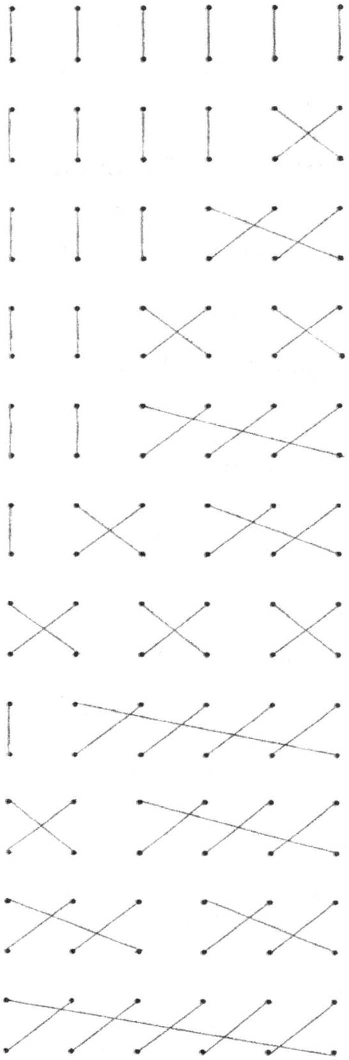

1	γ^6		γ^6
2	$\pm 15\gamma^5$	$\pm \gamma^6$	$\pm 14\gamma^5$
3	$40\gamma^4$	$8\gamma^5$	$32\gamma^4$
4	$45\gamma^4$	$6\gamma^5$	$39\gamma^4$
5	$\pm 90\gamma^3$	$\pm 36\gamma^4$	$\pm 54\gamma^3$
6	$\pm 120\gamma^3$	$\pm 32\gamma^4$	$\pm 88\gamma^3$
7	$\pm 15\gamma^3$	$\pm 3\gamma^4$	$\pm 12\gamma^3$
8	$144\gamma^2$	$96\gamma^3$	$48\gamma^2$
9	$90\gamma^2$	$42\gamma^3$	$48\gamma^2$
10	$40\gamma^3$	$16\gamma^4$	$24\gamma^3$
11	$\pm 120\gamma^3$	$\pm 120\gamma^4$	$0\gamma^3$
	720	360	360

The reader can decide for oneself whether this is feasible for a mol or so of molecules! Fortunately, there is an alternative approach to multi-molecule ensembles which allows normalization to be evaluated for any number of molecules and at the same time has relatively simple recipes for calculating expectation values for the measurable variables.

6.4. Coherent Ensembles of Independent Molecules

The ensemble operator for coherent bosonic molecules is obtained by a straightforward extension of the theory of the eigenstates of the annihilation operator:

$$\begin{aligned}\hat{\rho}_{\mathrm{c}} = \exp\left[\int d\boldsymbol{p}\sum_{\alpha'}\chi_{\alpha'}(\boldsymbol{p})\hat{a}_{\alpha'}(\boldsymbol{p}')^{\dagger}\right]|0\rangle\langle 0| \\ \times \exp\left[\int d\boldsymbol{p}\sum_{\alpha}\chi^{*}_{\alpha}(\boldsymbol{p})\hat{a}_{\alpha}(\boldsymbol{p})\right]. \end{aligned} \tag{6.13}$$

To show that

$$\hat{a}_{\alpha}(\boldsymbol{p})\hat{\rho}_{\mathrm{c}} = \chi_{\alpha}(\boldsymbol{p})\hat{\rho}_{\mathrm{c}}, \tag{6.14}$$

it is convenient to write the first exponential of Eq. (6.13) in the expanded form

$$\begin{aligned}\hat{\rho}_{\mathrm{c}} = |0\rangle + \sum_{n=1}^{\infty}\frac{1}{n!}\prod_{j=1}^{n}\Bigg[\int d\boldsymbol{p}_{(n-j+1)}\sum_{\alpha_{(n-j+1)}}\chi_{\alpha_{n-j+1}} \\ \times (\boldsymbol{p}_{(n-j+1)})\hat{a}_{\alpha_{(n-j+1)}}(\boldsymbol{p}_{(n-j+1)})^{\dagger}\Bigg]|0\rangle \\ \times \langle 0|\exp\left[\int d\boldsymbol{p}\sum_{\alpha}\chi_{\alpha}(\boldsymbol{p})^{*}\hat{a}_{\alpha}(\boldsymbol{p})\right]. \end{aligned} \tag{6.15}$$

Now for the $n = 0$ term

$$\hat{a}_{\alpha}(\boldsymbol{p})|0\rangle = 0. \tag{6.16}$$

For $n = 1$

$$\hat{a}_\alpha(\boldsymbol{p}) \left[\int d\boldsymbol{p}_1 \sum_{\alpha_1} \chi_{\alpha_1}(\boldsymbol{p}_1) \hat{a}_{\alpha_1}(\boldsymbol{p}_1)^\dagger \right] |0\rangle$$

$$= \left[\int d\boldsymbol{p}_1 \sum_{\alpha_1} \chi_{\alpha_1}(\boldsymbol{p}_1) \overbrace{\hat{a}_\alpha(\boldsymbol{p}) \hat{a}_{\alpha_1}}(\boldsymbol{p}_1)^\dagger \right] |0\rangle$$

$$\times \left[\int d\boldsymbol{p}_1 \sum_{\alpha_1} \chi_{\alpha_1}(\boldsymbol{p}_1) \delta_{\alpha\alpha_1} \delta(\boldsymbol{p} - \boldsymbol{p}_1) \right] |0\rangle$$

$$= \chi_{\alpha_1}(\boldsymbol{p}_1)|0\rangle. \tag{6.17}$$

For $n > 1$

$$\hat{a}_\alpha(\boldsymbol{p}) \sum_{n=2}^{\infty} \frac{1}{n!} \prod_{j=1}^{n} \left[\int d\boldsymbol{p}_{(n-j+1)} \sum_{\alpha_{(n-j+1)}} \chi_{\alpha_{(n-j+1)}} \right.$$

$$\left. \times (\boldsymbol{p}_{(n-j+1)}) \hat{a}_{\alpha_{(n-j+1)}} (\boldsymbol{p}_{(n-j+1)})^\dagger \right] |0\rangle$$

$$= \chi_\alpha(\boldsymbol{p}) \sum_{n=2}^{\infty} \frac{n}{n!} \prod_{j=1}^{n-1} \left[\int d\boldsymbol{p}_{(n-j)} \sum_{\alpha_{(n-j)}} \chi_{\alpha_{(n-j)}} \right.$$

$$\left. \times (\boldsymbol{p}_{(n-j)}) \hat{a}_{\alpha_{(n-j)}} (\boldsymbol{p}_{(n-j)})^\dagger \right] |0\rangle. \tag{6.18}$$

The factor n in Eq. (6.18) comes from the choice of n opportunities for contracting an $\hat{a}_\alpha(\boldsymbol{p})$ with an $\hat{a}_\alpha(\boldsymbol{p}')^\dagger$. Combining Eqs. (6.16)–(6.18) completes the derivation of Eq. (6.14). A similar treatment (or, more simply, taking the Hermitian conjugate) gives

$$\hat{\rho}_c \hat{a}_\alpha(\boldsymbol{p})^\dagger = \chi_\alpha(\boldsymbol{p})^\dagger \hat{\rho}_c. \tag{6.19}$$

The expectation for the number density of molecules in momentum space is now easily calculated for the coherent ensemble. From Eqs. (6.6a), (6.14) and (6.19) we have

$$\begin{aligned} \mathrm{Tr}[N_\alpha(\boldsymbol{p})\hat{\rho}_c]/\mathrm{Tr}[\hat{\rho}_c] &= \mathrm{Tr}[\hat{a}_\alpha(\boldsymbol{p})^\dagger \hat{a}_\alpha(\boldsymbol{p})\hat{\rho}_c]/\mathrm{Tr}[\hat{\rho}_c] \\ &= \mathrm{Tr}[\hat{a}_\alpha(\boldsymbol{p})\hat{\rho}_c \hat{a}_\alpha(\boldsymbol{p})^\dagger]/\mathrm{Tr}[\hat{\rho}_c] = \chi_\alpha(\boldsymbol{p})\chi_\alpha(\boldsymbol{p})^*. \end{aligned} \tag{6.20}$$

The expectation for the total number of molecules is therefore

$$N = \sum_{\alpha} \int d\boldsymbol{p} \chi_\alpha(\boldsymbol{p}) \chi_\alpha(\boldsymbol{p})^*. \tag{6.21}$$

Eqs. (6.20) and (6.21) give the physical interpretation of $\chi_\alpha(\boldsymbol{p})$ as the amplitude of the number density in momentum space of molecules in state α.

The expectation for the density in coordinate space obtained from Eqs. (6.6b), (6.14), and (6.19) is

$$\begin{aligned}
&\mathrm{Tr}[\hat{D}_\alpha(\boldsymbol{x})\hat{\rho}_c]/\mathrm{Tr}[\hat{\rho}_c] \\
&\quad = \frac{1}{h^3} \int d\boldsymbol{p} \int d\boldsymbol{p}' e^{\frac{i}{\hbar}(\boldsymbol{p}-\boldsymbol{p}')\cdot\boldsymbol{x}} \mathrm{Tr}[\hat{a}_\alpha(\boldsymbol{p}')^\dagger \hat{a}_\alpha(\boldsymbol{p})\hat{\rho}_c]/\mathrm{Tr}[\hat{\rho}_c] \\
&\quad = \frac{1}{h^3} \int d\boldsymbol{p} \int d\boldsymbol{p}' e^{\frac{i}{\hbar}(\boldsymbol{p}-\boldsymbol{p}')\cdot\boldsymbol{x}} \chi_\alpha(\boldsymbol{p}) \chi_\alpha(\boldsymbol{p}')^*.
\end{aligned} \tag{6.22}$$

As should be expected, the coherent ensemble does not represent a perfect gas, since the molecular density given by Eq. (6.22) is dependent upon the position $\boldsymbol{x}$ if $\chi_\alpha(\boldsymbol{p})$ is a distribution other than a delta function, as is consistent with the uncertainty principle.

Let us now consider an ensemble operator of the form of Eq. (6.13) for the case of fermionic molecules. Looking at the equivalent form shown in Eq. (6.15), we see that an interchange of a pair of adjacent annihilation or creation operators produces a change in sign, according to Eq. (6.2a) when the lower sign is used. But since the result of the interchange is only to exchange labels of the integration and summation variables to which $\hat{\rho}_c$ must be invariant, each term of $\hat{\rho}_c$ for which $n, m > 1$ must vanish. Thus, for fermions $\hat{\rho}_c$ has the following severely restrictive form with at most one fermion in a given state in the ensemble:

$$\begin{aligned}
\hat{\rho}_c^f = &\left[1 + \sum_{\alpha'} \int d\boldsymbol{p}' \chi_{\alpha'}(\boldsymbol{p}') \hat{a}_\alpha(\boldsymbol{p}')^\dagger\right] \\
&\times |0\rangle\langle 0| \left[1 + \sum_{\alpha'} \int d\boldsymbol{p} \chi_\alpha(\boldsymbol{p}) \hat{a}_\alpha(\boldsymbol{p})\right].
\end{aligned} \tag{6.23}$$

6.5. The Isodasic Molecular Ensemble

The coherent ensemble fails to represent a perfect gas for an even more profound reason than that the density of a coherent ensemble is density dependent. In a gas, inter-molecular collisions and collisions with the walls of the container produce phase shifts. A short time after the preparation of the ensemble, the phases of the molecules will be quite random. Suppose we start with a coherent ensemble, allow collisions to occur so that phase randomization occurs. Return to Eq. (6.13), but this time explicitly include the phases $\varphi_\alpha(\boldsymbol{p})$ of the amplitudes of the distribution functions $\chi_\alpha(\boldsymbol{p})$; that is, we write

$$\chi_\alpha(\boldsymbol{p}) = |\chi_\alpha(\boldsymbol{p})| e^{i\varphi_\alpha(\boldsymbol{p})}.$$

Expanding the exponential operators in Eq. (6.13) and omitting the terms with $n = 0$, we have

$$\hat{\rho} = \sum_{n=1}^{\infty}\sum_{m=1}^{\infty}\frac{1}{n!}\frac{1}{m!}\prod_{j=1}^{n}\left[\sum_{\alpha_j}\int d\boldsymbol{p}_j |\chi_{\alpha_j}(\boldsymbol{p}_j)| e^{i\varphi_{\alpha_j}(\boldsymbol{p})}\hat{a}_{\alpha_j}(\boldsymbol{p}_j)^\dagger\right]|0\rangle$$
$$\times \langle 0| \prod_{k=1}^{m}\left[\sum_{\alpha_k}\int d\boldsymbol{p}_k |\chi_{\alpha_k}(\boldsymbol{p}_k)| e^{i\varphi_{\alpha_k}(\boldsymbol{p})}\hat{a}_{\alpha_k}(\boldsymbol{p}_k)\right]. \qquad (6.24)$$

Taking the phases to be random, we treat $\varphi_\alpha(\boldsymbol{p})$ as a multi-valued function of $\alpha, \boldsymbol{p}$. Thus Eq. (6.24) is to be taken as a superposition over all possible phases; that is, we must sum over all possible values of the phases. Since $\hat{\rho}$ is not yet normalized we may average each phase factor separately:

$$\langle e^{i\varphi_\alpha(\boldsymbol{p})}\rangle = \frac{1}{2\pi}\int_0^{2\pi} d\varphi_\alpha(\boldsymbol{p}) e^{i\varphi_\alpha(\boldsymbol{p})} = 0, \qquad (6.25)$$

unless $\varphi_\alpha(\boldsymbol{p}) = 0$ (or, equivalently, unless $\varphi_\alpha(\boldsymbol{p})$ is fixed at some constant value). The existence of the coherent ensemble thus requires that all the molecular phases be independent of the α, $\boldsymbol{p}$ state of an individual molecule. On the other hand, we may pair each state α_j,

$\boldsymbol{p}_j$ on the left with a state α_k, $\boldsymbol{p}_k$ on the right. Then averaging over this pair of phase values gives

$$\begin{aligned}
&\langle e^{i[\varphi_{\alpha_j}(\boldsymbol{p}_j)-\varphi_{\alpha_k}(\boldsymbol{p}_k)]}\rangle \\
&= \frac{1}{(2\pi)^2}\int_0^{2\pi} d\varphi_{\alpha_j}(\boldsymbol{p}_j)\int_0^{2\pi} d\varphi_{\alpha_k}(\boldsymbol{p}_k)e^{i[\varphi_{\alpha_j}(\boldsymbol{p}_j)-\varphi_{\alpha_k}(\boldsymbol{p}_k)]} \\
&= 1; \quad \varphi_{\alpha_j}(\boldsymbol{p}_j) = \varphi_{\alpha_k}(\boldsymbol{p}_k) \\
&= 0; \quad \varphi_{\alpha_j}(\boldsymbol{p}_j) \neq \varphi_{\alpha_k}(\boldsymbol{p}_k). \qquad (6.26)
\end{aligned}$$

The randomization of phases thus forces the pairing of states that are created on the left of the vacuum with states of the annihilators on the right in the ensemble operator and requires these paired states to have the same phase; that is, their phases may not be varied separately. For those terms with $m \neq n$ there will be one or more states that cannot be paired; according to Eq. (6.25), this means that one or more zero factors will appear in such terms, thus eliminating these off-diagonal terms. There are $n!$ ways to pair states to have the same phase which implies that the states have identical histories; that is, that the states label the same molecule.

At this point, it is useful to take the momentum space to be divided into equal discrete infinitesimal cubes $\Delta\boldsymbol{p}_q$, the center of which is $\boldsymbol{p}_q$. The integrals in Eq. (6.24) are now discrete sums over the cubes $\Delta\boldsymbol{p}_j$ and $\Delta\boldsymbol{p}_k$. We therefore take the results of phase randomization as given by Eq. (6.26) to mean that δ_{ij} is to be inserted for each pair of states so that for each paired states we have the sums

$$\begin{aligned}
&\cdots\sum_j \Delta\boldsymbol{p}\chi_{\alpha_j}(\boldsymbol{p}_j)\hat{a}_{\alpha_j}(\boldsymbol{p}_j)^\dagger\cdots\sum_k \Delta\boldsymbol{p}\chi_{\alpha_k}(\boldsymbol{p}_k)^*\cdots|0\rangle\langle 0| \\
&\quad\cdots\hat{a}_{\alpha_k}(\boldsymbol{p}_k)\cdots\delta_{ij}\cdots \\
&= \cdots\sum_j(\Delta\boldsymbol{p})^2|\chi_{\alpha_j}(\boldsymbol{p}_j|^2\hat{a}_{\alpha_j}(\boldsymbol{p}_j)^\dagger\cdots|0\rangle\langle 0|\cdots\hat{a}_{\alpha_j}(\boldsymbol{p}_j)\cdots .
\end{aligned}$$

Reverting back to the integral notation, we have

$$\cdots \sum_{\alpha_j} \int d\boldsymbol{p}_j \Delta\boldsymbol{p} |\chi_{\alpha_j}(\boldsymbol{p}_j)|^2 \hat{a}_{\alpha_j}(\boldsymbol{p}_j)^\dagger \cdots |0\rangle\langle 0| \cdots \hat{a}_{\alpha_j}(\boldsymbol{p}_j) \cdots$$

$$= \cdots \sum_{\alpha_j} \int d\boldsymbol{p}_j \boldsymbol{\Phi}_{\alpha_j}(\boldsymbol{p}_j) \hat{a}_{\alpha_j}(\boldsymbol{p}_j)^\dagger \cdots |0\rangle\langle 0| \cdots \hat{a}_{\alpha_j}(\boldsymbol{p}_j) \cdots,$$

where we have written $\boldsymbol{\Phi}_\alpha(\boldsymbol{p})$ for $\Delta\boldsymbol{p}|\chi_\alpha(\boldsymbol{p})|^2$.

The result is the ensemble operator $\hat{\rho}_0$.

$$\hat{\rho}_0 = |0\rangle\langle 0| + \sum_{n=1}^{\infty} \frac{1}{n!} \sum_{\alpha_1} \int d\boldsymbol{p}_1 \boldsymbol{\Phi}_{\alpha_1}(\boldsymbol{p}_1) \cdots \sum_{\alpha_n} \int d\boldsymbol{p}_n \boldsymbol{\Phi}_{\alpha_n}(\boldsymbol{p}_n)$$
$$\times \prod_{j=1}^{n} [\hat{a}_{\alpha_{(n-j+1)}}(\boldsymbol{p}_{(n-j+1)})^\dagger] |0\rangle\langle 0| \prod_{k=1}^{n} [\hat{a}_{\alpha_k}(\boldsymbol{p}_k)], \tag{6.27}$$

where we have nested the paired operators as in Eq. (6.5). This nesting of operators is important; it ensures that fermions as well as bosons can be included in the formalism of the operator $\hat{\rho}_0$, for an interchange of a pair of summation and integration variables corresponds to the interchange of both a pair of annihilation operators and a pair of creation operators in mirrored positions with respect to the vacuum operator. Thus if a minus sign is incurred for one interchange it will be cancelled by a minus sign incurred by the other.

Although the form of Eq. (6.27) is somewhat complicated, this ensemble operator obeys a simple equation. Operating on the left with $\hat{a}_\alpha(\boldsymbol{p})$ gives

$$\hat{a}_\alpha(\boldsymbol{p})\hat{\rho}_0 = \boldsymbol{\Phi}_\alpha(\boldsymbol{p}) \Bigg\{ |0\rangle\langle 0| + \sum_{n=2}^{\infty} \frac{n}{n!} \sum_{\alpha_2} \int d\boldsymbol{p}_2 \boldsymbol{\Phi}_{\alpha_2}(\boldsymbol{p}_2)$$
$$\cdots \sum_{\alpha_n} \int d\boldsymbol{p}_n \boldsymbol{\Phi}_{\alpha_n}(\boldsymbol{p}_n)$$
$$\times \prod_{j=1}^{n-1} [\hat{a}_{\alpha_{(n-j)}}(\boldsymbol{p}_{(n-j)})^\dagger] |0\rangle\langle 0| \prod_{k=1}^{n-1} [\hat{a}_{\alpha_k}(\boldsymbol{p}_k)] \Bigg\} \hat{a}_\alpha(\boldsymbol{p})$$
$$= \boldsymbol{\Phi}_\alpha(\boldsymbol{p}) \hat{\rho}_0 \hat{a}_\alpha(\boldsymbol{p}). \tag{6.28}$$

The Hermitian conjugate of Eq. (6.28) is

$$\hat{\rho}_0\hat{a}_\alpha(\boldsymbol{p})^\dagger = \boldsymbol{\Phi}_\alpha\hat{a}_\alpha(\boldsymbol{p})^\dagger\hat{\rho}_0. \tag{6.29}$$

The number density in momentum space is now easily calculated. From the invariance of the trace of a string of operators under cyclic permutations of the string, we have

$$\begin{aligned}\mathrm{Tr}[\hat{N}_\alpha(\boldsymbol{p})\hat{\rho}_0] &= \mathrm{Tr}[\hat{a}_\alpha(\boldsymbol{p})\hat{a}_\alpha(\boldsymbol{p})^\dagger\hat{\rho}_0]\\ &= \boldsymbol{\Phi}_\alpha(\boldsymbol{p})\mathrm{Tr}[\hat{a}_\alpha(\boldsymbol{p})^\dagger\hat{\rho}_0\hat{a}_\alpha(\boldsymbol{p})].\end{aligned} \tag{6.30a}$$

Putting the operators in normal order and using Eq. (6.4) for the resulting $\delta(\boldsymbol{p}-\boldsymbol{p})$ gives

$$\begin{aligned}\mathrm{Tr}[\hat{N}_\alpha(\boldsymbol{p})\hat{\rho}_0] &= \boldsymbol{\Phi}_{\boldsymbol{\alpha}}(\boldsymbol{p})\frac{V}{h^3}\mathrm{Tr}\hat{\rho}_0 \pm \boldsymbol{\Phi}_{\boldsymbol{\alpha}}(\boldsymbol{p})\mathrm{Tr}[\hat{a}_\alpha(\boldsymbol{p})^\dagger\hat{a}_a(\boldsymbol{p})\hat{\rho}_0]\\ &= \boldsymbol{\Phi}_{\boldsymbol{\alpha}}(\boldsymbol{p})\frac{V}{h^3}\mathrm{Tr}\hat{\rho}_0 \pm \boldsymbol{\Phi}_{\boldsymbol{\alpha}}(\boldsymbol{p})\mathrm{Tr}[\hat{N}_\alpha(\boldsymbol{p})\hat{\rho}_0].\end{aligned} \tag{6.30b}$$

From the invariance of the trace of a string of operators under cyclic permutations of the string, we have

$$\begin{aligned}\mathrm{Tr}[\hat{N}_\alpha(\boldsymbol{p})\hat{\rho}_0] &= \boldsymbol{\Phi}_{\boldsymbol{\alpha}}(\boldsymbol{p})\frac{V}{h^3}\mathrm{Tr}\hat{\rho}_0 \pm \boldsymbol{\Phi}_{\boldsymbol{\alpha}}(\boldsymbol{p})\mathrm{Tr}[\hat{a}_\alpha^\dagger(\boldsymbol{p})\hat{a}_a(\boldsymbol{p})\hat{\rho}_0]\\ &= \boldsymbol{\Phi}_{\boldsymbol{\alpha}}(\boldsymbol{p})\frac{V}{h^3}\mathrm{Tr}\hat{\rho}_0 \pm \boldsymbol{\Phi}_{\boldsymbol{\alpha}}(\boldsymbol{p})\mathrm{Tr}[\hat{N}_\alpha(\boldsymbol{p})\hat{\rho}_0].\end{aligned} \tag{6.30c}$$

Solving Eq. (6.30c) for $\mathrm{Tr}[\hat{N}_\alpha(\boldsymbol{p})\hat{\rho}_0]$, we obtain

$$\mathrm{Tr}[\hat{N}_\alpha(\boldsymbol{p})\hat{\rho}_0] = \frac{V}{h^3}\frac{\boldsymbol{\Phi}_\alpha(\boldsymbol{p})}{[1 \mp \boldsymbol{\Phi}_\alpha(\boldsymbol{p})]}\mathrm{Tr}[\hat{\rho}_0]. \tag{6.30d}$$

In the same way, we have for the expectation of the energy

$$\mathrm{Tr}[\hat{H}_\alpha\hat{\boldsymbol{\rho}}_0] = \frac{V}{h^3}\int d\boldsymbol{p}\varepsilon_\alpha(\boldsymbol{p})\frac{\boldsymbol{\Phi}_\alpha(\boldsymbol{p})}{[1 \mp \boldsymbol{\Phi}_\alpha(\boldsymbol{p})]}\mathrm{Tr}[\hat{\boldsymbol{\rho}}_0]. \tag{6.31}$$

For the density at point $\boldsymbol{x}$, we obtain

$$\mathrm{Tr}[\hat{D}_\alpha(\boldsymbol{x})\hat{\boldsymbol{\rho}}_0] = \frac{1}{h^3}\int d\boldsymbol{p}\frac{\boldsymbol{\Phi}_\alpha(\boldsymbol{p})}{[1 \mp \boldsymbol{\Phi}_\alpha(\boldsymbol{p})]}\mathrm{Tr}\hat{\boldsymbol{\rho}}_0. \tag{6.32}$$

The density in configuration space is seen to be uniform. Furthermore, comparison of Eqs. (6.30d) and (6.32) shows that

$$\mathrm{Tr}[\hat{D}_\alpha(\boldsymbol{x})\hat{\rho}_0] = \frac{1}{V}\int d\boldsymbol{p}\mathrm{Tr}[\hat{N}_\alpha(\boldsymbol{p})\hat{\rho}_0] = \frac{N_\alpha}{V}\mathrm{Tr}\hat{\rho}_0, \tag{6.33}$$

where N_α is the total number of molecules in the state α and V is the total volume as in Eq. (6.4). Since an operator $\hat{\rho}_0$ that satisfies Eq. (6.28) describes an ensemble of non-interacting molecules with uniform density, $\hat{\rho}_0$ can properly be called an *isodasic* ensemble operator.

A word about the renormalization of the ensemble operator when the phases are randomized is now in order. The Boltzmann limit of Eq. (6.30d) gives $N_\alpha(\boldsymbol{p}) = (V/h^3)\boldsymbol{\Phi}_{\boldsymbol{\alpha}}(\boldsymbol{p})$. Comparison with Eq. (6.20) shows that $\boldsymbol{\Phi}_{\boldsymbol{\alpha}}(\boldsymbol{p})$ should be identified with $(h^3/V)|\chi_\alpha(\boldsymbol{p})|^2$, which is consistent with the discussion leading to Eq. (6.27), since according to the uncertainty principle, the smallest value of $\Delta\boldsymbol{p}$ with physical meaning is h^3/V.

Note that Eqs. (6.30c)–(6.32), which give the degenerate boson (−) and fermion (+) perfect gas results, follow directly from Eq. (6.29) without any additional *ad hoc* statistical reasoning.

Although we divide these results by $\mathrm{Tr}\,\hat{\rho}_0$ to obtain the average values of the respective observable variables, it is instructive to calculate the trace explicitly.

6.6. Trace of the Isodasic Operator

The trace of the isodasic operator $\hat{\rho}_0$ whose form as given in Eq. (6.27) is

$$\begin{aligned}\mathrm{Tr}[\hat{\rho}_0] = 1 + \sum_{n=1}^{\infty}\frac{1}{n!}\sum_{\alpha_1}\int d\boldsymbol{p}_1\boldsymbol{\Phi}_{\alpha_1}(\boldsymbol{p}_1)\cdots\sum_{\alpha_n}\int d\boldsymbol{p}_n\boldsymbol{\Phi}_{\alpha_n}(\boldsymbol{p}_n)\\ \times\langle 0|\prod_{k=1}^{n}[\hat{a}_{\alpha_k}(\boldsymbol{p}_k)]\prod_{j=1}^{n}[\hat{a}_{\alpha_{(n-j+1)}}(\boldsymbol{p}_{(n-j+1)})^\dagger]|0\rangle.\end{aligned} \tag{6.34}$$

Note that the vacuum expectation in the second line of Eq. (6.34) has the same form as the nomalization of an n-molecule state vector as shown in Eq. (6.7). We have already given some examples of the application of Wick's theorem to this problem. But now we need to derive an explicit expression for the general case and to sum over all values of n. The first step is to define "occupation numbers" for the clusters that appear in the contraction patterns. Let m be the *order* of a given cluster; *i.e.*, the number of molecular states in the cluster. For example, in row 6 on page 159 there are 1 cluster of order 1, 1 cluster of order 2, and 1 cluster of order 3. Let ℓ_m be the *occupation number* of the clusters in this pattern: $\ell_1 = 1$, $\ell_2 = 1$, $\ell_3 = 1$, $\ell_4 = 0$, *etc.* For the pattern on row 10, we have $\ell_1 = 0$, $\ell_2 = 0$, and $\ell_3 = 2$, $\ell_4 = 0$, *etc.* We may now fully describe any contraction pattern by giving the set of occupation numbers $\{\ell_m\}$. The total number of molecular states in the contraction pattern is $n = \sum_m m\ell_m$. With the use of the ℓ_m's to enumerate the occupied clusters in the pattern, the number of ways to choose the states for a given pattern is therefore

$$\frac{n!}{\prod_{m=1}^{\infty}(m\ell_m)!}. \tag{6.35}$$

Consider now the ℓ_m clusters of order m in the pattern. There are a total of $m\ell_m$ states in these clusters. The number of ways to choose the m states for the first of the clusters is

$$\frac{(m\ell_m)!}{m![m(\ell_m - 1)]!}. \tag{6.36}$$

The next cluster of m states can be chosen in

$$\frac{[m(\ell_m - 1)]!}{m![m(\ell_m - 2)]!}$$

ways, and so on. The number of ways of choosing operators for all ℓ_m of the clusters of order m is thus

$$\frac{(m\ell_m)!}{m![m(\ell_m - 1)]!}\frac{[m(\ell_m - 1)]!}{m![m(\ell_m - 2)]!}\cdots\frac{(2m)!}{(m!)^2} = \frac{(m\ell_m)!}{(m!)^{\ell_m}}.$$

But since the ℓ_m clusters are indistinguishable, we must divide by $\ell_m!$ to get the number of ways of choosing states for the ℓ_m clusters of order m in the given pattern:

$$\frac{(m\ell_m)!}{(m!)^{\ell_m}\ell_m!}.$$

Combining this result with the expression in Eq. (6.35) and taking into account that each cluster of order m can be formed in $(m-1)!$ ways as we pointed out in the discussion on page 158, we have for the number of ways of assigning the n states to the contraction pattern described by the set of occupation numbers $\{\ell\}$

$$\frac{n!}{\prod_{m=1}^{\infty}(m\ell_m)!}\prod_{m=1}^{\infty}\frac{(m\ell_m)!}{(m!)^{\ell_m}\ell_m!}[(m-1)!]^{\ell_m} = \frac{n!}{\prod_{m=1}^{\infty}(m)^{\ell_m}\ell_m!}. \tag{6.37}$$

This result may be checked by evaluating Eq. (6.37) for the contraction patterns on page 159. For the case of row 5, $\ell_1 = 2$, $\ell_2 = 0$, $\ell_3 = 0$, $\ell_4 = 1$, giving

$$6!\frac{1}{1^2 2!}\frac{(3!)^1}{(4!)^1 1!} = 90,$$

in agreement with the first column in row 5 on page 159.

Since the states $\alpha, \boldsymbol{p}$ are summed and integrated with integrand $\boldsymbol{\Phi}_{\alpha}(\boldsymbol{p})$ in Eq. (6.34), each cluster of order m gives a factor

$$(\pm 1)^{(m-1)}\frac{V}{h^3}\sum_{\alpha}\int d\boldsymbol{p}[\boldsymbol{\Phi}_{\alpha}(\boldsymbol{p})]^m, \tag{6.38}$$

where we have included the factor $(\pm 1)^{(m-1)}$ as discussed at the end of the first paragraph on page 158. We are now ready to sum the terms in $\mathrm{Tr}[\hat{\rho}_0]$. Combining Eqs. (6.37) and (6.38), we obtain the contribution to Eq. (6.34) from the contraction pattern $\{\ell\}$

$$\frac{1}{n!}\frac{n!}{\prod\limits_{m=1}^{\infty}(m)^{\ell_m}\ell_m}\prod_{m=1}^{\infty}\left[(\pm 1)^{(m-1)}\frac{V}{h^3}\sum_{\alpha}\int d\boldsymbol{p}[\boldsymbol{\Phi}_{\alpha}(\boldsymbol{p})^m\right]^{\ell_m}$$

$$= \prod_{m=1}^{\infty}\frac{1}{\ell_m!}\left[\frac{(\pm 1)^{(m-1)}}{m}\frac{V}{h^3}\sum_{\alpha}\int d\boldsymbol{p}[\boldsymbol{\Phi}_{\alpha}(\boldsymbol{p})]^m\right]^{\ell_m}. \tag{6.39}$$

To complete the calculation for $\mathrm{Tr}[\hat{\rho}_0]$, we must sum over all such patterns, with each ℓ_m ranging from zero to infinity (this sum will include the null pattern for $n = 0$):

$$\begin{aligned}\mathrm{Tr}\hat{\rho}_0 &= \prod_{m=1}^{\infty}\sum_{\ell_m=0}^{\infty}\frac{1}{\ell_m!}\left[\frac{(\pm 1)^{(m-1)}}{m}\frac{V}{h^3}\sum_{\alpha}\int d\boldsymbol{p}[\boldsymbol{\Phi}_\alpha(\boldsymbol{p})]^m\right]^{\ell_m} \\ &= \exp\left\{\frac{V}{h^3}\sum_{\alpha}\int d\boldsymbol{p}\sum_{m=1}^{\infty}\left[\frac{(\pm 1)^{(m-1)}}{m}[\boldsymbol{\Phi}_\alpha(\boldsymbol{p})]^m\right]\right\}. \qquad (6.40)\end{aligned}$$

The remaining sum over m gives the final result

$$\mathrm{Tr}\hat{\rho}_0 = \exp\left\{\mp\frac{V}{h^3}\sum_{\alpha}\int d\boldsymbol{p}\ln[1\mp\boldsymbol{\Phi}_\alpha(\boldsymbol{p})]\right\}. \qquad (6.41)$$

Eq. (6.41) is seen to be the generalized expression for the grand canonical partition function for non-interacting molecules. The result is remarkable, in that it has been derived only by randomization of the molecular phases and without any assumptions about equilibrium. The only requirement on $\boldsymbol{\Phi}_\alpha(\boldsymbol{p})$ is that it be any distribution function for which $\boldsymbol{\Phi}_\alpha(\boldsymbol{p}) < 0$ for all values of α, $\boldsymbol{p}$ so that the sums in $\hat{\rho}_0$ converge.

The Boltzmann limit for Eq. (6.41) may be obtained by including in the trace calculations only "direct" contractions (*i.e.*, clusters of order $m = 1$), each of which gives a factor V/h^3, or, equivalently from Eq. (6.41) with the limit

$$\ln[1\mp\boldsymbol{\Phi}_\alpha(\boldsymbol{p})] \xrightarrow[\boldsymbol{\Phi}_\alpha(\boldsymbol{p})\ll 1]{} \mp\boldsymbol{\Phi}_\alpha(\boldsymbol{p}). \qquad (6.42)$$

This gives

$$\mathrm{Tr}[\hat{\rho}_0]_{\mathrm{B}} = \exp\left\{\frac{V}{h^3}\sum_{\alpha}\int d\boldsymbol{p}\boldsymbol{\Phi}_\alpha(\boldsymbol{p})\right\} = e^{N_{\mathrm{B}}}, \qquad (6.43)$$

which is the grand canonical partition function for a Boltzmann gas.

We now find the explicit form for $\boldsymbol{\Phi}_\alpha(\boldsymbol{p})$ at thermal equilibrium by maximizing the entropy under the constraints of constant N

and E. The entropy operator which gives results that are consistent with the usual thermodynamic properties is

$$\hat{S}/k_{\rm B} = \ln \mathrm{Tr}\hat{\rho}_0 - \int d\boldsymbol{p} \sum_{\alpha} \ln \boldsymbol{\Phi}_\alpha(\boldsymbol{p}) \hat{a}^\dagger_\alpha(\boldsymbol{p}) \hat{a}(\boldsymbol{p}), \tag{6.44}$$

from which we obtain

$$\mathrm{Tr}[(\hat{S}/k_{\rm B})\hat{\rho}_0]/\mathrm{Tr}\,\hat{\rho}_0 = \ln \mathrm{Tr}\hat{\rho}_0 - \frac{V}{h^3} \int d\boldsymbol{p} \sum_{\alpha} \frac{\boldsymbol{\Phi}_\alpha(\boldsymbol{p}) \ln \boldsymbol{\Phi}_\alpha(\boldsymbol{p})}{1 \mp \boldsymbol{\Phi}_\alpha(\boldsymbol{p})}. \tag{6.45}$$

The Lagrange functional to be maximized is

$$\mathscr{L}[\boldsymbol{\Phi}_\alpha(\boldsymbol{p})] = S/k_{\rm B} - \beta(E - \mu N). \tag{6.46}$$

From Eqs. (6.44), (6.45), (6.30d), and (6.31), we have

$$\frac{\delta \mathscr{L}}{\delta \boldsymbol{\Phi}_\alpha(\boldsymbol{p})} = -\frac{V}{h^3}[1 \mp \boldsymbol{\Phi}_\alpha(\boldsymbol{p})]^{-2}\{\ln \boldsymbol{\Phi}_\alpha(\boldsymbol{p}) + \beta[\varepsilon_\alpha(p) - \mu]\} = 0, \tag{6.47}$$

from which we get

$$\boldsymbol{\Phi}_\alpha(\boldsymbol{p}) = e^{\beta[\mu - \varepsilon_\alpha(p)]}. \tag{6.48}$$

This is the expected result, after the identification $\beta = 1/k_{\rm B}$. We find the chemical potential μ to have the usual form by integrating and summing Eq. (6.30c) in the Boltzmann limit:

$$\mu = \frac{1}{\beta} \ln \left[\frac{N}{V} \frac{h^3}{Q} \left(\frac{\beta}{2\pi m} \right)^{\frac{3}{2}} \right], \tag{6.49}$$

where the "internal partition function" is

$$Q = \sum_{\alpha} e^{-\beta \varepsilon_\alpha}. \tag{6.50}$$

Thus, Eq. (6.47) can be written

$$\boldsymbol{\Phi}_\alpha(\boldsymbol{p}) = \frac{N}{V} \frac{h^3}{Q} \left(\frac{\beta}{2\pi m} \right)^{\frac{3}{2}} e^{-\beta \varepsilon_\alpha(\boldsymbol{p})}. \tag{6.51}$$

Insertion of Eq. (6.51) into Eq. (6.44) gives the Sackur-Tetrode entropy

$$S/k_{\mathrm{B}} = N \ln \left[\frac{V}{N} \left(\frac{2\pi m}{\beta} \right)^{\frac{3}{2}} h^3 e^{\frac{5}{2}} \right] - \sum_{\alpha} P_\alpha \ln P_\alpha, \tag{6.52}$$

where

$$P_\alpha = e^{-\beta\varepsilon_\alpha}/Q. \tag{6.53}$$

To complete our discussion of the derivation of properties of perfect gases from quantum field theory, we note that the mechanical pressure in a homogeneous system with translation energy E_{tr} is

$$P = \frac{2}{3}\frac{E_{\mathrm{tr}}}{V} = \frac{2}{3}\frac{1}{V}\frac{1}{2m}\frac{\int d\boldsymbol{p}p^2 e^{-\beta p^2/2m}}{\int d\boldsymbol{p}e^{-\beta p^2/2m}} = \frac{N}{\beta V}, \tag{6.54}$$

in accordance with the ideal gas law. The maximum value of $\mathscr{L}$ is found from Eqs. (6.46) and (6.47) to be

$$\mathscr{L}_{\mathrm{max}} = \ln \mathrm{Tr}\hat{\rho}_0 = S/k_{\mathrm{B}} - \beta(E - \mu N). \tag{6.55}$$

Using Eq. (6.43) for $\mathrm{Tr}[\rho_0]$ in the Boltzmann limit and Eq. (6.54) for N, we have the expression for the Gibbs free energy:

$$G \equiv \mu N = E + PV - TS. \tag{6.56}$$

We have now shown that an ensemble operator that obeys Eq. (6.28) gives the properties of a perfect gas at thermal equilibrium.

6.7. Molecular Statistics for Coherent and Isodasic Ensembles

Not only does $\hat{\rho}_0$ contain all the thermodynamic information about perfect gases, it incorporates the statistics of subsets of independent molecules. An example is the probability that n molecules are in state α. The probability of observing a given state in an ensemble $\hat{\rho}$ is given by the trace of the product of $\hat{\rho}$ with the projection operator for that state.

The projection operator for n molecules in the state α, regardless of the values of their momenta, is

$$\hat{P}_{\alpha}^{(n)} = \frac{1}{n!}\int d\boldsymbol{p}_1 \cdots \int d\boldsymbol{p}_n \prod_{j=1}^{n} [\hat{a}_{\alpha}(\boldsymbol{p}_{(n+1-j)})^{\dagger}]|0\rangle\langle 0| \prod_{k=1}^{n} [\hat{a}_{\alpha}(\boldsymbol{p}_k)]. \tag{6.57}$$

It is easy to verify that the projection property $\hat{P}^2 = \hat{P}$ is obeyed for this operator. The calculation of the corresponding probability P_{α}^{n} requires the trace

$$\mathrm{Tr}[\hat{P}_{\alpha}^{(n)}\hat{\rho}] = \frac{1}{n!}\int d\boldsymbol{p}_1 \cdots \int d\boldsymbol{p}_n \times \left\langle 0 \left| \prod_{k=1}^{n} [\hat{a}_{\alpha}(\boldsymbol{p}_k)]\hat{\rho} \prod_{j=1}^{n} [\hat{a}_{\alpha}(\boldsymbol{p}_{(n+1-j)})^{\dagger}] \right| 0 \right\rangle. \tag{6.58}$$

If the ensemble is coherent, we may use Eq. (6.14) to evaluate the bracket in Eq. (6.58) to obtain the probability

$$\begin{aligned}\mathrm{Tr}[\hat{P}_{\alpha}^{(n)}\hat{\rho}_{\mathrm{c}}]/\mathrm{Tr}[\hat{\rho}_{\mathrm{c}}] &= \frac{1}{n!}\left[\int d\boldsymbol{p}|\chi_{\alpha}(\boldsymbol{p})|^2\right]^n \langle 0|\hat{\rho}_{\mathrm{c}}|0\rangle e^{-N} \\ &= \frac{1}{n!}(N_{\alpha})^n e^{-N},\end{aligned} \tag{6.59}$$

where we have used Eq. (6.20), the result that $\mathrm{Tr}\,\hat{\rho}_{\mathrm{c}} = e^{N}$, which follows from the application of Wick's theorem and Eq. (6.20), and the fact that $\langle 0|\hat{\rho}_{\mathrm{c}}|0\rangle = 1$. If we ask for the probability that n_1 molecules are in state α_1 and n_2 molecules are in state α_2, regardless of their momentum values, we must use the projection operator

$$\begin{aligned}\hat{P}_{(\alpha_1,\alpha_2)}^{(n_1,n_2)} &= \frac{1}{n_1!}\frac{1}{n_2!}\int d\boldsymbol{p}_1 \cdots \int d\boldsymbol{p}_{n_1} \int d\boldsymbol{p}'_1 \cdots \int d\boldsymbol{p}'_{n_2} \\ &\quad \times \prod_{j'=1}^{n_2} [\hat{a}_{\alpha_2}(\boldsymbol{p}'_{(n_2+1-j')})^{\dagger}] \prod_{j=1}^{n_1} [\hat{a}_{\alpha_1}(\boldsymbol{p}_{(n_1+1-j)})^{\dagger}] \\ &\quad \times |0\rangle\langle 0| \prod_{k=1}^{n_1} [\hat{a}_{\alpha_1}(\boldsymbol{p}_k)] \prod_{k'=1}^{n_2} [\hat{a}_{\alpha_2}(\boldsymbol{p}_{k'})].\end{aligned} \tag{6.60}$$

The corresponding probability is

$$P_{(\alpha_1,\alpha_2)}^{(n_1,n_2)} = \frac{1}{n_1!}(N_{\alpha_1})^{n_1}\frac{1}{n_2!}(N_{\alpha_2})^{n_2}e^{-N}. \tag{6.61}$$

To find the probability that n molecules are in state α, regardless of the number of molecules in any other state, and regardless of any of the momentum values, we need to generalize Eqs. (6.60) and (6.61) to give the Poisson distribution:

$$\begin{aligned} P_\alpha^{(n)} &= \frac{1}{n!}(N_\alpha)^n \prod_{j(\alpha_j\neq\alpha)}^{\infty}\left[\sum_{n_j=0}^{\infty}\frac{1}{n_j!}(N_{\alpha_j})^{n_j}\right]e^{-N} \\ &= \frac{1}{n!}(N_\alpha)^n \prod_{j(\alpha_j\neq\alpha)}^{\infty}[e^{N_{\alpha_j}}]e^{-N} \\ &= \frac{1}{n!}(N_\alpha)^n\left[e^{\sum_{\alpha_j\neq\alpha}N_{\alpha_j}}\right]e^{-N} = \frac{1}{n!}(N_\alpha)^n e^{-N_\alpha}. \end{aligned} \tag{6.62}$$

If the ensemble is isodasic, we use Eq. (6.28) to write Eq. (6.58) in the form

$$\begin{aligned} \mathrm{Tr}[\hat{P}_\alpha^{(n)}\hat{\rho}_0] &= \frac{1}{n!}\int d\boldsymbol{p}_1\boldsymbol{\Phi}_\alpha(\boldsymbol{p}_1)\cdots\int d\boldsymbol{p}_n\boldsymbol{\Phi}_\alpha(\boldsymbol{p}_n) \\ &\quad\times\langle 0|\hat{\rho}_0\prod_{k=1}^{n}[\hat{a}_\alpha(\boldsymbol{p}_k)]\prod_{j=1}^{n}[(\hat{a}_\alpha(\boldsymbol{p}_{(n+1-j)}))^\dagger]|0\rangle. \end{aligned} \tag{6.63}$$

We may replace $\hat{\rho}_0$ with 1 in the second line of Eq. (6.63), since the vacuum *bra* to its left projects only its vacuum term. The vacuum expectation of the remaining operator string may be calculated from a linked-cluster analysis of the fully contracted terms as was done for the calculation of $\mathrm{Tr}\,\hat{\rho}_0$, except that here there is no sum over n.

The calculation in the Boltzmann limit is quite simple, since only the clusters of order unity need be considered. In that case, we again obtain the Poisson distribution as given in Eq. (6.62).

We now ask: what is the probability of finding n molecules in internal state α and in the volume $\Delta\boldsymbol{p}$ at $\boldsymbol{p}$ in momentum space.

The corresponding projection operator is

$$\hat{P}_{\alpha}^{(n)}(\boldsymbol{p}, \Delta\boldsymbol{p}) = \frac{1}{n!}\int_{\Delta\boldsymbol{p}}\int d\boldsymbol{p}_1 \cdots \int_{\Delta\boldsymbol{p}} d\boldsymbol{p}_n \times \prod_{j=1}^{n}[\hat{a}_{\alpha}(\boldsymbol{p}_{(n+1-j)})^{\dagger}]|0\rangle\langle 0|\prod_{k=1}^{n}[\hat{a}_{\alpha}(\boldsymbol{p}_k)]. \tag{6.64}$$

For a coherent ensemble, the calculation of $\boldsymbol{P}_{\alpha}^{(n)}$ proceeds in much the same way as before, leading to the Poisson distribution

$$P_{\alpha}^{(n)}(\boldsymbol{p}, \Delta\boldsymbol{p}) = \frac{1}{n!}\left[\int_{\Delta p} d\boldsymbol{p}'|\chi_{\alpha}(\boldsymbol{p})|^2\right]^n \exp\left[-\int_{\Delta p} d\boldsymbol{p}''|\chi_{\alpha}(\boldsymbol{p}'')|^2\right]. \tag{6.65}$$

As $\Delta\boldsymbol{p}$ becomes small, the limiting form for Eq. (6.65) is

$$P_{\alpha}^{(n)}(\boldsymbol{p}, \Delta\boldsymbol{p}) = \frac{1}{n!}[\Delta\boldsymbol{p}|\chi_{\alpha}(\boldsymbol{p})|^2]^n \exp[-\Delta\boldsymbol{p}|\chi_{\alpha}(\boldsymbol{p})|^2]. \tag{6.66}$$

When $\Delta\boldsymbol{p}$ is taken down to the smallest momentum-space volume with physical meaning, we replace it by $\boldsymbol{h}^3/V$ so that $\Delta(\boldsymbol{p})|\chi_{\alpha}(\boldsymbol{p})|^2$ becomes $\boldsymbol{\Phi}_{\alpha}(\boldsymbol{p})$ to give

$$P_{\alpha}^{(n)}(\boldsymbol{p}) = \frac{1}{n!}[\boldsymbol{\Phi}_{\alpha}(\boldsymbol{p})]^n \exp[-\boldsymbol{\Phi}_{\alpha}(\boldsymbol{p})] \tag{6.67}$$

for coherent ensembles of bosonic molecules.

For isodasic ensembles, the results are more interesting than those for coherent ensembles. In this case, the trace to be evaluated is

$$\begin{aligned}\mathrm{Tr}[\hat{P}_{\alpha}^{(n)}(\boldsymbol{p}, \Delta\boldsymbol{p})\hat{\rho}_0] &= \frac{1}{n!}\int_{\Delta\boldsymbol{p}} d\boldsymbol{p}_1 \cdots \int_{\Delta\boldsymbol{p}} d\boldsymbol{p}_n \langle 0|\prod_{k=1}^{n}[\hat{a}_{\alpha}(\boldsymbol{p}_k)]\hat{\rho}_0 \\ &\quad \times \prod_{j=1}^{n}[\hat{a}_{\alpha}(\boldsymbol{p}_{(n+1-j)})^{\dagger}]|0\rangle\end{aligned}$$

$$= \frac{1}{n!} \int_{\Delta p} d\boldsymbol{p}_1 \boldsymbol{\Phi}_\alpha(\boldsymbol{p}_1) \cdots \int_{\Delta p} d\boldsymbol{p}_n \boldsymbol{\Phi}_\alpha(\boldsymbol{p}_n) \langle 0|\hat{\rho}_0 \prod_{k=1}^{n} [\hat{a}_\alpha(\boldsymbol{p}_k)]$$

$$\times \prod_{j=1}^{n} [\hat{a}_\alpha(\boldsymbol{p}_{(n+1-j)})^\dagger]|0\rangle$$

$$= \frac{1}{n!} \int_{\Delta p} d\boldsymbol{p}_1 \boldsymbol{\Phi}_\alpha(\boldsymbol{p}_1) \cdots \int_{\Delta p} d\boldsymbol{p}_n \boldsymbol{\Phi}_\alpha(\boldsymbol{p}_n) \langle 0| \prod_{k=1}^{n} [\hat{a}_\alpha(\boldsymbol{p}_k)]$$

$$\times \prod_{j=1}^{n} [\hat{a}_\alpha(\boldsymbol{p}_{(n+1-j)})^\dagger]|0\rangle. \tag{6.68}$$

Eq. (6.68) is evaluated by summing all completely contracted terms that fall into the linked-cluster patterns such as those shown on page 159 for the case of $n = 6$. The calculations are greatly simplified in the limit of $\Delta \boldsymbol{p} \to h^3/V$; then for each linked cluster we have instead of the factor $\gamma = V/h^3$ generated by redundant contractions,

$$\int_{\Delta p} d\boldsymbol{p}\delta(\boldsymbol{p} - \boldsymbol{p})\boldsymbol{\Phi}_\alpha^m(\boldsymbol{p}) = \frac{V}{h^3}\frac{h^3}{V}\boldsymbol{\Phi}_\alpha^m(\boldsymbol{p}) = \boldsymbol{\Phi}_\alpha^m(\boldsymbol{p}), \tag{6.69}$$

and since $\ell_m m = n$, for the pattern as a whole we have a factor $\boldsymbol{\Phi}^n(\boldsymbol{p})$. Thus, the terms in the linked-cluster diagrams become identical, except for the sign and the numerical coefficient. The results for

$$\mathrm{Tr}[P_\alpha^{(n)}(\boldsymbol{p}, \Delta\boldsymbol{p})\hat{\rho}_0]$$

are:

$$n = 0 \qquad 1 \tag{6.70a}$$

$$n = 1 \qquad \boldsymbol{\Phi}_\alpha(\boldsymbol{p}) \tag{6.70b}$$

$$n > 1 \qquad \frac{1}{2}[\boldsymbol{\Phi}_\alpha^n(\boldsymbol{p}) \pm \boldsymbol{\Phi}_\alpha^n(\boldsymbol{p})]. \tag{6.70c}$$

For fermions, Eq. (6.70c) equals zero, so that only the terms with $n = 0, 1$ contribute:

$$\mathrm{Tr}[\hat{P}_\alpha^{(n)}(\boldsymbol{p}, \Delta\boldsymbol{p})\hat{\rho}_0^{\mathrm{F}}] = \delta_{n0} + \boldsymbol{\Phi}_\alpha(\boldsymbol{p})\delta_{n1}, \tag{6.71}$$

while for bosons we have for all n

$$\mathrm{Tr}[\hat{P}_\alpha^{(n)}(\boldsymbol{p}, \Delta\boldsymbol{p})\hat{\rho}_0^{\mathrm{B}}] = \boldsymbol{\Phi}_\alpha^n(\boldsymbol{p}). \tag{6.72}$$

These distributions may be normalized by summing over n to obtain, respectively,

$$\sum_{n=0}^{\infty}[\delta_{n0} + \boldsymbol{\Phi}_\alpha(\boldsymbol{p})\delta_{n1}] = 1 + \boldsymbol{\Phi}_\alpha(\boldsymbol{p}), \tag{6.73}$$

$$\sum_{n=0}^{\infty}\boldsymbol{\Phi}_\alpha^n = [1 - \boldsymbol{\Phi}_\alpha^n(\boldsymbol{p})]^{-1}. \tag{6.74}$$

Dividing Eq. (6.71) by Eq. (6.73) and Eq. (6.72) by Eq. (6.74) give the normalized distributions in momentum space for fermions and bosons, respectively,

$$P_\alpha^{(n)}(\boldsymbol{p})^{\mathrm{F}} = \frac{\delta_{n0} + \boldsymbol{\Phi}_\alpha(\boldsymbol{p})\delta_{n1}}{1 + \boldsymbol{\Phi}_\alpha(\boldsymbol{p})}, \tag{6.75}$$

$$P_\alpha^{(n)}(\boldsymbol{p})^{\mathrm{B}} = \boldsymbol{\Phi}_\alpha^n(\boldsymbol{p}) - \boldsymbol{\Phi}_\alpha^{n+1}(\boldsymbol{p}). \tag{6.76}$$

The average occupation number for bosons is easily obtained from these distributions:

$$\begin{aligned}\langle N_\alpha(\boldsymbol{p})\rangle^{\mathrm{B}}\Delta\boldsymbol{p} &= \sum_{n=0}^{\infty} nP_\alpha^{(n)}(\boldsymbol{p})^{\mathrm{B}} \\ &= \sum_{n=1}^{\infty} n[\boldsymbol{\Phi}_\alpha^n(\boldsymbol{p}) - \boldsymbol{\Phi}_\alpha^{(n+1)}(\boldsymbol{p})] = \frac{\boldsymbol{\Phi}_\alpha(\boldsymbol{p})}{1 - \boldsymbol{\Phi}_\alpha(\boldsymbol{p})}. \end{aligned} \tag{6.78}$$

Compare these results with the direct result obtained for the number density in an isodasic ensemble given in Eq. (6.30d). In this respect, it should again be noted that $\Delta\boldsymbol{p} \to h^3/V$ in the limit of small $\Delta\boldsymbol{p}$.

To calculate the probability that there are n molecules in internal state α in the sub-volume $\Delta\boldsymbol{p}$ located at the point $\boldsymbol{x}$, the relevant projection operator is

$$\begin{aligned}\hat{P}_{\alpha}^{(n)}(\boldsymbol{x}_0, \Delta V) \\ &= \frac{1}{n!}\int_{\Delta V} d\boldsymbol{x}_1 \cdots \int_{\Delta V} d\boldsymbol{x}_n \\ &\times \prod_{j=1}^{n}[\hat{\psi}_{\alpha}(\boldsymbol{x}_{(n-j+1)})^{\dagger}]|0\rangle\langle 0|\prod_{l=1}^{n}[\hat{\psi}_{\alpha}(\boldsymbol{x}_l)].\end{aligned} \tag{6.79}$$

The probability is obtained by tracing this projection operator over $\hat{\rho}_0$ and dividing by $\mathrm{Tr}\,\hat{\rho}_0$, then multiplying by the similarly obtained probabilities that there are any number of molecules in the complementary volume $V - \Delta V$ and that there are any number of molecules in ΔV which are in state $\beta \neq \alpha$. The calculations proceed in a similar way as for the momentum-space distributions. The result for bosons is

$$[P_{\alpha}^{(n)}(\boldsymbol{x}_0, \Delta V)]_{\mathrm{B}} = \left(N_{\alpha}^{\mathrm{B}}\frac{\Delta V}{V}\right)^{n} - \left(N_{\alpha}^{\mathrm{B}}\frac{\Delta V}{V}\right)^{(n-1)}, \tag{6.80}$$

whereas for fermions it is

$$[P_{\alpha}^{(n)}(\boldsymbol{x}_0, \Delta V)]_{\mathrm{F}} = \frac{\delta_{n0} + N_{\alpha}^{\mathrm{F}}\frac{\Delta V}{V}\delta_{n1}}{1 + N_{\alpha}^{\mathrm{F}}\frac{\Delta V}{V}}. \tag{6.81}$$

To get the Boltzmann limit, we ignore all of the "exchange" contractions in calculating the traces with the projection operators; that is, we include only contraction schemes that give the highest powers V/h^3. Then we get the Poisson distribution

$$[P_{\alpha}^{n}(\boldsymbol{x}_0, \Delta V)]_{\mathrm{B}} = \frac{1}{n!}\left(N_{\alpha}^{\mathrm{B}}\frac{\Delta V}{V}\right)^{n} e^{-(N_{\alpha}^{\mathrm{B}}\frac{\Delta V}{V})}. \tag{6.82a}$$

When the sub-volume ΔV is taken to be the total volume V, Eq. (6.82a) becomes identical with Eq. (6.62), showing that in *a coherent ensemble bosons obey Boltzmann and not Bose-Einstein statistics*!

An observable manifestation of the difference between the statistics of coherent and isodasic ensembles is their pair correlation functions. The simplest calculation is that for the coherent ensemble.

We use Eq. (6.6e) for the pair-correlation operator $\hat{D}_{\alpha\beta}(\boldsymbol{r})$ and Eq. (6.14) for the action of $\hat{a}_\alpha(\boldsymbol{p})$ upon $\hat{\rho}_c$ to get

$$\mathrm{Tr}[\hat{D}_{\alpha\beta}(\boldsymbol{r})\hat{\rho}_c]/\mathrm{Tr}\,\hat{\rho}_c = \left(1 - \frac{1}{2}\delta_{\alpha\beta}\right)\frac{1}{h^3}\int d\boldsymbol{p}\int d\boldsymbol{p}'\int d\boldsymbol{q} e^{\frac{1}{\hbar}\boldsymbol{q}\cdot\boldsymbol{r}} \times \chi^*_\alpha(\boldsymbol{p}-\boldsymbol{q})\chi^*_\beta(\boldsymbol{p}'+\boldsymbol{q})\chi_\beta(\boldsymbol{p}')\chi_\alpha(\boldsymbol{p}). \quad (6.82b)$$

The evaluation of Eq. (6.82b) requires knowledge of the momentum distribution amplitudes $\chi_\gamma(\boldsymbol{p})$; $\gamma = \alpha, \beta$. However, in most cases in which properties of coherent ensembles may be observed, the system is a Bose condensate in which the momenta are narrowly clustered near zero. This means that the range of $\boldsymbol{q}$ is very narrow; thus, there will be practically no observable dependence of the pair correlation function upon the distance $\boldsymbol{r}$ for condensed non-interacting bosons:

$$\mathrm{Tr}[D_{\alpha\beta}(\boldsymbol{r})\hat{\rho}_0]/\mathrm{Tr}\,\hat{\rho}_0 = \frac{N_\alpha N_\beta}{V}. \quad (6.83)$$

For isodasic ensembles, a calculation of $\mathrm{Tr}[\hat{D}_{\alpha\beta}(\boldsymbol{r})\hat{\rho}_0]$ is somewhat more complicated. But for $\alpha \neq \beta$, the allowed contractions lead to a $\delta(\boldsymbol{q})$ factor, so that the result is the same as for the density of α, β pairs: there are no quantum effects. For $\alpha = \beta$, if we simplify the calculation by ignoring the $\boldsymbol{q}$-dependence in denominators such as that in Eq. (6.32), the fully contracted direct term gives the pair density and the fully contracted exchange term gives the pair density times an exponential function of $\boldsymbol{r}^2$:

$$\mathrm{Tr}[\hat{D}_{\alpha\alpha}\hat{\rho}_0]/\mathrm{Tr}\,\hat{\rho}_0 = \frac{1}{2}\frac{N_\alpha^2}{V}(1 \pm e^{-\frac{m}{\beta\hbar^2}r^2}). \quad (6.84)$$

In the fermion case (lower sign), the "Fermi hole" is clearly seen: the probability is zero that two fermions occupy the same point in space. For bosons, the isodasic pair correlation function exhibits a maximum at $r = 0$, indicating "bunching" of non-interacting bosons.[3] In the presence of interaction, the strong short-range

[3]The bunching of bosonic atoms has been observed by Yasuda, M. and Shimizu, F., *Phys. Rev. Lett.* **77**, 3090 (1996); Fölling, S., *et al.*, *Nature* **434**, 481 (2005); Greiner, M., *et al.*, *Phys. Rev. Lett.* **94**, 110401 (2005); Schellekins, M., *et al.*, *Science* **310**, 648 (2005).

repulsion of a pair of atoms due to the behavior of the pair correlation of the fermionic *electrons* of which they are composed will require the expression in Eq. (6.84) to go to zero as $\boldsymbol{r}$ goes to zero even for bosonic atoms.

6.8. Summary for Chapter 6

In this chapter we have shown how quantum field theory can be applied to ensembles of non-interacting molecules by taking the field quanta to be the molecules in a specified internal state and linear momentum or, alternatively, position. The two types of ensembles we have treated are coherent and isodasic, the latter being derived from the former by randomizing the molecular phases in the coherent ensemble operator. The result is an ensemble with uniform density that describes a perfect gas of bosonic or fermionic molecules. The trace of this isodasic ensemble operator is calculated by the use of Wick's theorem together with a linked-cluster analysis of the contraction patterns; it turns out to be the grand canonical partition function for a perfect degenerate gas. The conditions of thermodynamic equilibrium are met by maximization of the entropy under the constraints of constant total energy and molecule number.

Molecules in the isodasic ensemble obey Bose-Einstein or Fermi-Dirac statistics. The bosonic molecules in the coherent ensemble obey Poisson statistics for distribution in sub-volumes ΔV, but when ΔV is the total volume V, coherent bosons are found to obey Boltzmann and not Bose-Einstein statistics! Calculation of the leading terms in the pair correlation function demonstrates the Fermi hole and boson bunching in isodasic ensembles, while the pair correlation function for coherent boson condensates would be expected to be very weakly dependent upon $\boldsymbol{r}$.

Perhaps the derivation of the statistical thermodynamics of perfect gases from quantum field theory applied to isodasic ensembles and the result that coherent ensembles do not obey Bose statistics are the most interesting results from this chapter.

Chapter 7

Real Gases and Phase Transitions*

7.1. Introduction

We have seen in Chapter 6 that quantum field theory provides a systematic investigative tool for ensembles of molecules. The isodasic grand-canonical ensemble operator for an ideal gas was presented there in terms of the Fock creation and annihilation operators. The ensemble operator $\hat{\rho}_0$ for a perfect gas was shown to obey a simple equation which is key to the calculation of quantum-statistical properties of bosonic and fermionic molecules; namely

$$\hat{a}_\alpha(\boldsymbol{p})\hat{\rho}_0 = \boldsymbol{\Phi}_\alpha(\boldsymbol{p})\hat{\rho}_0\hat{a}_\alpha(\boldsymbol{p}). \tag{7.1}$$

In this chapter we treat the effects of interaction of the molecules, first using a diagrammatic perturbation theory especially designed for this purpose, then by summing to infinite order the Feynman-like graphs corresponding to long- and short-range interactions so as to provide a model of the pair-correlation function for an ensemble of interacting molecules and a new avenue for the study of liquid-gas phase transitions. First-order perturbation theory in the case of ionic intermolecular forces yields the result obtained heuristically by Debye as a first approach to the study of ionic solutions. An outline of future work is included at the end.

---

*The material presented in this chapter has appeared in Porter, R. N., *AIP Conf. Proc.* **1102**, 219 (2009). It is reprinted here with the permission of AIP.

7.2. Ensembles of Interacting Molecules

The defining equation for isodasic ensemble operators, Eq. (7.1), facilitates the tractability of a perturbative treatment of molecular interactions. Let $\hat{\rho}_0(0)$ be the ensemble operator for which $\hat{\mathbf{\Phi}}(\beta)$ is replaced by $\mathrm{e}^{\beta\mu}$. (To simplify the notation, in this section we assume all molecules to be in the internal state α; the generalization to mixtures of different molecular species is straightforward.) It is clear that $\hat{\rho}_0(\beta)$ is recovered by the operation

$$\hat{\rho}_0(\beta) = e^{-\beta\hat{H}_0}\hat{\rho}_0(0) \tag{7.2}$$

since

$$e^{-\beta\hat{H}_0}\hat{a}^\dagger(\boldsymbol{p})e^{\beta\hat{H}_0} = e^{-\beta\varepsilon(p)}\hat{a}^\dagger(\boldsymbol{p}). \tag{7.3}$$

Then, for interacting molecules, the ensemble operator is given by

$$\begin{aligned}\hat{\rho}(\beta) &= e^{-\beta\hat{H}}\hat{\rho}_0(0) = e^{-\beta\hat{H}}e^{\beta\hat{H}_0}\hat{\rho}_0(\beta),\\ &= \hat{\boldsymbol{U}}(\beta)\hat{\rho}_0(\beta)\end{aligned} \tag{7.4}$$

where $\hat{\boldsymbol{U}}(\beta)$ is defined by Eq. (7.4) and

$$\hat{H} = \hat{H}_0 + \hat{V}. \tag{7.5}$$

For pair interactions, the potential-energy operator V is

$$\hat{V} = \frac{1}{2}\int d\boldsymbol{p}\int d\boldsymbol{p}'\int d\boldsymbol{q}\tilde{\boldsymbol{\nu}}(\boldsymbol{q})\hat{a}(\boldsymbol{p}-\boldsymbol{q})^\dagger\hat{a}(\boldsymbol{p}'+\boldsymbol{q})^\dagger\hat{a}(\boldsymbol{p})\hat{a}(\boldsymbol{p}), \tag{7.6}$$

where

$$\tilde{\boldsymbol{v}}(\boldsymbol{q}) = \frac{1}{h^3}\int d\boldsymbol{x}e^{-\frac{i}{\hbar}\boldsymbol{q}\cdot\boldsymbol{x}}v(\boldsymbol{x}). \tag{7.7}$$

Here, $\boldsymbol{q}$ is the momentum transferred upon collision, and $v(\boldsymbol{x})$ is the potential energy of a pair of molecules at the intermolecular distance $\boldsymbol{x}$.

The "temperature evolution operator" $\hat{\boldsymbol{U}}(\beta)$ satisfies the differential equation

$$\frac{\partial}{\partial\beta}\hat{\boldsymbol{U}}(\beta) = -e^{-\beta\hat{H}}(\hat{H}-\hat{H}_0)e^{-\beta\hat{H}_0} = -\hat{\boldsymbol{U}}(\beta)\hat{V}(\beta), \tag{7.8}$$

where

$$\hat{V}(\beta) = e^{-\beta \hat{H}_0} \hat{V} e^{\beta \hat{H}_0}.$$

The solution to Eq. (7.8) can be written as the expansion

$$\hat{\boldsymbol{U}}(\beta) = 1 + \sum_{n=1}^{\infty} \hat{\boldsymbol{U}}^{(n)}(\beta) = 1 + \sum_{n=1}^{\infty} (-1)^n \times \int_0^{\beta} d\boldsymbol{\kappa}_1 \cdots \int_0^{\boldsymbol{\kappa}_{n-1}} d\boldsymbol{\kappa}_n \hat{V}(\boldsymbol{\kappa}_n) \cdots \hat{V}(\boldsymbol{\kappa}_1). \quad (7.9)$$

The first-order contribution to $\hat{\boldsymbol{U}}(\beta)$ is

$$\hat{\boldsymbol{U}}^{(1)}(\beta) = -\int_0^{\beta} d\boldsymbol{k} \hat{V}(\boldsymbol{\kappa}) = -\frac{1}{2} \int_0^{\beta} d\boldsymbol{k} e^{-\frac{\boldsymbol{\kappa}}{m} \boldsymbol{q} \cdot (\boldsymbol{q} - \boldsymbol{p} + \boldsymbol{p}')} \times \tilde{v}(\boldsymbol{q}) \hat{a}^{\dagger}(\boldsymbol{p} - \boldsymbol{q}) \hat{a}^{\dagger}(\boldsymbol{p}' + \boldsymbol{q}) \hat{a}(\boldsymbol{p}') \hat{a}(\boldsymbol{p}). \quad (7.10)$$

We diagram this first-order interaction as

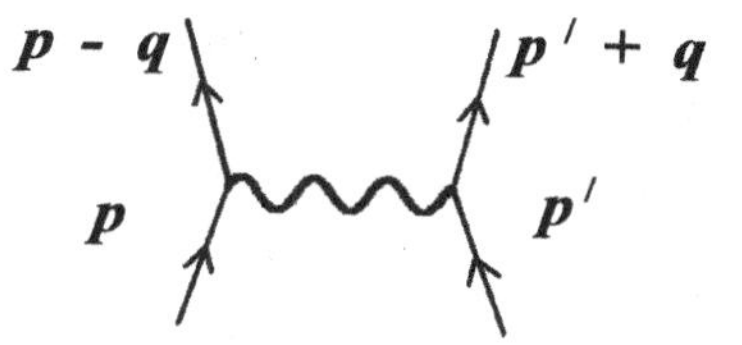

For second order, we have

$$\hat{\boldsymbol{U}}^{(2)}(\beta) = \int_0^{\beta} d\boldsymbol{\kappa}_1 \int_0^{\boldsymbol{\kappa}_2} d\boldsymbol{\kappa}_2 \hat{V}(\boldsymbol{\kappa}_2) \hat{V}(\boldsymbol{\kappa}_1). \quad (7.11)$$

7.2.1. *Short-range Correlations*

One contraction scheme for the two operators in Eq. (7.11) is

$$\hat{a}^{\dagger}(\boldsymbol{p}_2 - \boldsymbol{q}_2) \hat{a}^{\dagger}(\boldsymbol{p}_2' + \boldsymbol{q}_2) \hat{a}(\boldsymbol{p}_2') \hat{a}(\boldsymbol{p}_2) \hat{a}^{\dagger}(\boldsymbol{p}_1 - \boldsymbol{q}_1) \hat{a}^{\dagger}(\boldsymbol{p}_1' + \boldsymbol{q}_1) \hat{a}(\boldsymbol{p}_1') \hat{a}(\boldsymbol{p}_1) \quad (7.12)$$

which leads to the diagram

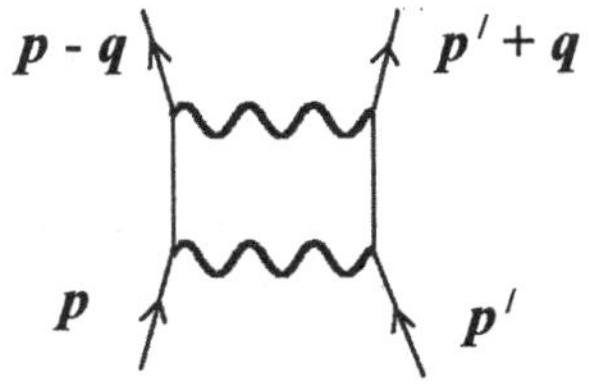

The initial momenta $\boldsymbol{p}_2$ and $\boldsymbol{p}_2'$ may be interchanged in $\hat{V}(\kappa_2)$ without changing the result; thus a factor of 2 needs to be inserted. The result (with some redefinition of integration variables) is

$$\begin{aligned}\hat{\boldsymbol{U}}_{\mathrm{L}}^{(2)}(\beta) &= \frac{1}{2}\int_0^{\beta} d\kappa \int_0^{\beta-\kappa} d\sigma \int d\boldsymbol{p}\int d\boldsymbol{p}'\int d\boldsymbol{q}\int d\boldsymbol{q}' \\ &\times \frac{1}{h^6}\int d\boldsymbol{x}_2\int d\boldsymbol{x}_1 e^{-\frac{i}{\hbar}\boldsymbol{q}'\cdot(\boldsymbol{x}_1-\boldsymbol{x}^2)} e^{-\frac{\sigma}{m}\boldsymbol{q}'\cdot(\boldsymbol{q}'-\boldsymbol{p}+\boldsymbol{p}')} \\ &\times e^{-\frac{i}{\hbar}\boldsymbol{q}\cdot\boldsymbol{x}_2} e^{-\frac{\kappa}{m}\boldsymbol{q}\cdot(\boldsymbol{q}-\boldsymbol{p}+\boldsymbol{p}')} \\ &\times v(\boldsymbol{x}_1)v(\boldsymbol{x}_2)\hat{a}(\boldsymbol{p}-\boldsymbol{q})^{\dagger}\hat{a}(\boldsymbol{p}'+\boldsymbol{q})^{\dagger}\hat{a}(\boldsymbol{p}')\hat{a}(\boldsymbol{p}). \end{aligned} \tag{7.13}$$

We use the subscript L to designate this contraction scheme, since the diagram has the form of a ladder.

In the high-temperature, low-impact limit, we may set

$$e^{-\frac{\sigma}{m}\boldsymbol{q}'\cdot(\boldsymbol{q}'-\boldsymbol{p}+\boldsymbol{p}')} \sim 1, \tag{7.14}$$

in which case Eq. (7.13) takes the simpler form

$$\begin{aligned}\hat{\boldsymbol{U}}_{\mathrm{L}}^{(2)}(\beta) &= \frac{1}{2}\int_0^{\beta} \kappa d\kappa \int d\boldsymbol{p}\int d\boldsymbol{p}'\int d\boldsymbol{q} e^{-\frac{\kappa}{m}\boldsymbol{q}\cdot(\boldsymbol{q}-\boldsymbol{p}+\boldsymbol{p}')} \\ &\times \frac{1}{h^3}\int d\boldsymbol{x} e^{-\frac{i}{\hbar}\boldsymbol{q}\cdot\boldsymbol{x}}[v(\boldsymbol{x})]^2 \hat{a}^{\dagger}(\boldsymbol{p}-\boldsymbol{q}) \\ &\times \hat{a}^{\dagger}(\boldsymbol{p}'+\boldsymbol{q})\hat{a}(\boldsymbol{p})\hat{a}(\boldsymbol{p}). \end{aligned} \tag{7.15}$$

Comparison of Eqs. (7.15) and (7.10) shows that the second-order ladder contribution is obtained from the first-order term by replacing $v(\boldsymbol{x})$ with $-\boldsymbol{\kappa}[v(\boldsymbol{x})]^2$.

For an n-rung ladder, the corresponding diagram is

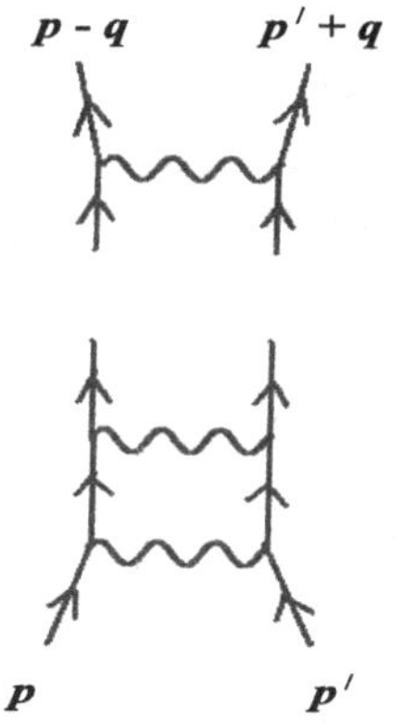

and the result for the potential-energy factor is

$$-\frac{1}{\boldsymbol{\kappa}}\frac{(-1)^n}{(n-1)!}\boldsymbol{\kappa}^n[v(\boldsymbol{x})]^n. \tag{7.16}$$

Summing the ladder contributions to infinite order gives

$$\boldsymbol{\nu}_{\mathrm{L}}(\boldsymbol{x};\boldsymbol{\kappa}) = -\frac{1}{\boldsymbol{\kappa}}\sum_{n-1}^{\infty}\frac{(-1)^n}{(n-1)!}\boldsymbol{\kappa}^n[\boldsymbol{v}(\boldsymbol{x})]^n = \boldsymbol{v}(\boldsymbol{x})e^{-\kappa\nu(\boldsymbol{x})} \tag{7.17}$$

where the $1/(n-1)!$ factor comes from the $\boldsymbol{\kappa}_2\cdots\boldsymbol{\kappa}_n$ integrations.

The total ladder contribution to $\hat{\boldsymbol{U}}(\beta)$ in the high-temperature, low-impact limit is thus

$$\hat{\boldsymbol{U}}_{\mathrm{L}}(\beta) = -\frac{1}{2}\int d\boldsymbol{p}\int d\boldsymbol{p}'\int d\boldsymbol{q}\frac{1}{h^3}\int d\boldsymbol{x}e^{-\frac{i}{\hbar}\boldsymbol{q}\cdot\boldsymbol{x}}[1-e^{-\beta\nu(\boldsymbol{x})}]$$
$$\times\,\hat{a}^{\dagger}(\boldsymbol{p}-\boldsymbol{q})\hat{a}^{\dagger}(\boldsymbol{p}'+\boldsymbol{q})\hat{a}(\boldsymbol{p}')\hat{a}(\boldsymbol{p}). \tag{7.18}$$

The ladders describe a pair of molecules that interact with one another n times; they represent *short-range correlations.*

7.2.2. *Long-range Correlations*

The other contraction scheme for is $V(\kappa_2)V(\kappa_1)$ is

$$\hat{a}^{\dagger}(\boldsymbol{p}_2-\boldsymbol{q}_2)\hat{a}^{\dagger}(\boldsymbol{p}_2'+\boldsymbol{q}_2)\hat{a}(\boldsymbol{p}_2')\hat{a}(\boldsymbol{p}_2)\hat{a}^{\dagger}(\boldsymbol{p}_1-\boldsymbol{q}_1)\hat{a}^{\dagger}(\boldsymbol{p}_1+\boldsymbol{q}_1)\hat{a}(\boldsymbol{p}_1')\hat{a}(\boldsymbol{p}_1). \tag{7.19}$$

This scheme seems to violate the rule that contractions must have $\hat{a}$ to the left of $\hat{a}^\dagger$. The answer is that the contraction is allowed by the cyclic property of the trace when $\hat{\boldsymbol{U}}(\beta)$ operates on $\hat{\rho}_0(\beta)$; in this way, $\hat{a}$ is moved to the left of $\hat{a}^\dagger$ and may therefore contract with it. The result of moving through $\rho_0(\beta)$, according to Eq. (7.1), is that a factor $\boldsymbol{\Phi}(\boldsymbol{p})$ is picked up. The contraction of Eq. (7.19) is diagrammed as

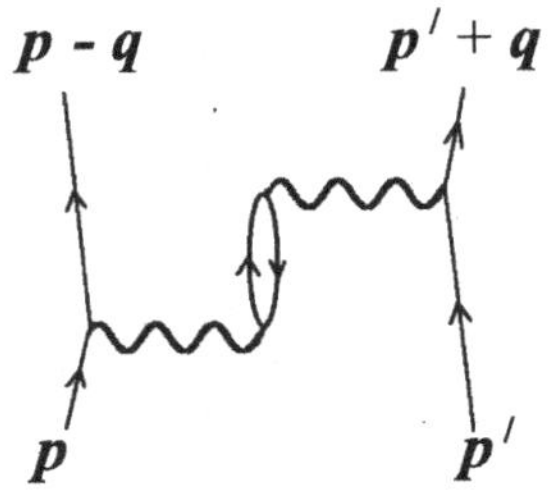

We call the above diagram a "ring" diagram and use the subscript R for its contribution to $\hat{U}(\beta)$. The nth order ring diagram is

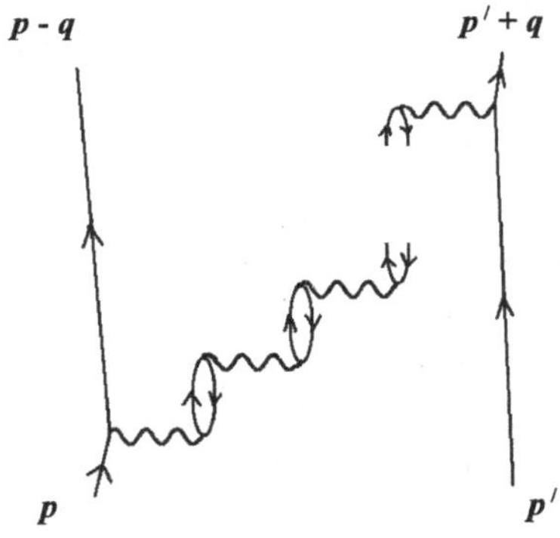

It is important to note that each ring produces a factor

$$\int d\boldsymbol{p}\, \boldsymbol{\Phi}(\boldsymbol{p}) = \frac{N}{V} h^3$$

and that holding $\boldsymbol{\kappa}_1$, which integrates to β in this case, above the other $\boldsymbol{\kappa}_2 \cdots \boldsymbol{\kappa}_n$, there are $(n-1)!$ ways to arrange the ladder without changing the result. Thus the $1/(n-1)!$ factor from the $\boldsymbol{\kappa}$ integrations is cancelled out. The result for the total ring contribution is

$$\hat{U}_{\mathrm{R}}(\beta) = \frac{1}{2}\int_0^\beta d\boldsymbol{\kappa} \int d\boldsymbol{p} \int d\boldsymbol{p}' \int d\boldsymbol{q}\, e^{-\frac{\kappa}{m}\boldsymbol{q}\cdot(\boldsymbol{q}-\boldsymbol{p}+\boldsymbol{p}')}$$
$$\times \frac{\tilde{v}}{1+\beta\frac{N}{V}h^3\tilde{v}(\boldsymbol{q})}\hat{a}^\dagger(\boldsymbol{p}-\boldsymbol{q})\hat{a}^\dagger(\boldsymbol{p}'+\boldsymbol{q})\hat{a}(\boldsymbol{p}')\hat{a}(\boldsymbol{p}). \quad (7.20)$$

Thus, the effective ring pair interaction is seen to be

$$\tilde{v}_{\mathrm{R}}(\boldsymbol{q};\beta) = \frac{\tilde{v}(\boldsymbol{q})}{1+\beta\frac{N}{V}h^3\tilde{v}(\boldsymbol{q})}. \quad (7.21)$$

The rings describe a collision of a molecule with another, collides with a third, and so on. They describe *long-range correlations.*

7.2.3. *Long-range Coulomb Correlations*

For the Coulomb repulsive interaction the pair potential takes the form

$$v(\boldsymbol{x}) = \frac{e^2}{x}. \quad (7.22)$$

This leads to the ring potential

$$\tilde{v}_{\mathrm{R}}(\boldsymbol{q};\beta) = \frac{e^2}{\pi h}\frac{1}{q^2+h^2k^2} \quad (7.23)$$

where

$$k = \left(4\pi e^2\beta\frac{N}{V}\right)^{\frac{1}{2}}. \quad (7.24)$$

We identify k with the Debye shielding radius.[1] Here it is obtained with the approximation being only first-order perturbation.

Going back to the $\boldsymbol{x}$ representation, we have

$$v_{\mathrm{R}}(\boldsymbol{x};\beta) = \frac{e^2}{x}e^{-kx}. \quad (7.25)$$

[1]Debye, P. and Hückel, E., *Phys. Zeits.* **24**, 185 (1928).

For an attractive Coulomb interaction, namely

$$v(\boldsymbol{x}) = -\frac{e^2}{x}, \tag{7.26}$$

we get the periodic effective potential

$$v_{\mathrm{R}}(\boldsymbol{x};\beta) = -\frac{e^2}{x}\cos kx. \tag{7.27}$$

These results suggest that a similar treatment might be useful for imperfect gases and liquids.

7.2.4. *Self-consistent Iterative Methods*

To describe all pair interactions as fully as possible, one may use the ring interaction in the calculation of the ladder contribution, then the resulting ladder interaction for the calculation of the ring contribution, and so on until separate converged results are obtained for rings and ladders. In what follows, we use these converged values to write

$$v_{\mathrm{L}}(\boldsymbol{x};\boldsymbol{\kappa}) = v_{\mathrm{R}}(\boldsymbol{x};\beta)e^{-\boldsymbol{\kappa} v_{\mathrm{R}}(\boldsymbol{x};\beta)} \tag{7.28}$$

and

$$\tilde{v}_{\mathrm{R}}(\boldsymbol{q};\beta) = \frac{\tilde{v}_{\mathrm{L}}(\boldsymbol{q};\beta)}{1+\beta\frac{N}{V}h^3\tilde{v}_{\mathrm{L}}(\boldsymbol{q};\beta)}. \tag{7.29}$$

7.3. The Pair-correlation Function for Interacting Molecules

From Eqs. (6.6e) and (7.18), we have for the pair correlation function in the ladder approximation to $\hat{\boldsymbol{U}}(\beta)$:

$$\begin{aligned}\mathrm{Tr}[\hat{D}(\boldsymbol{r})\hat{\boldsymbol{U}}(\beta)\hat{\rho}_0(\beta)] &= \mathrm{Tr}\{\hat{D}(\boldsymbol{r})[1+\hat{\boldsymbol{U}}_{\mathrm{L}}(\beta)]\hat{\rho}_0(\beta)\}\\ &= \frac{1}{2}\frac{N^2}{V}\{1-[1-e^{-\beta v_{\mathrm{R}}(\boldsymbol{r};\beta)}]\}\\ &= \frac{1}{2}\frac{N^2}{V}e^{-\beta v_{\mathrm{R}}(\boldsymbol{r};\beta)}.\end{aligned} \tag{7.30}$$

After dividing by the number of pairs per unit volume, $\mathrm{N}^2/2V$, we have

$$g(r) = e^{-\beta v_{\mathrm{R}}(\boldsymbol{r};\beta)} = e^{-\beta \int d\boldsymbol{q}\, e^{\frac{i}{h}\boldsymbol{q}\cdot\boldsymbol{r}} \left[\frac{\tilde{v}_{\mathrm{L}}(\boldsymbol{q};\beta)}{1+\beta\frac{N}{V}h^3\tilde{v}_{\mathrm{L}}(\boldsymbol{q};\beta)}\right]}. \tag{7.31}$$

We expect to have temperature- and density-dependent poles in the integrand in Eq. (7.31). Imaginary poles will give the behavior

$$g(r) \sim e^{\frac{A}{r}e^{-kr}} \tag{7.32}$$

which is characteristic of gases, while real poles will give the periodic behavior characteristic of liquids:

$$g(r) \sim e^{\frac{B}{r}\cos kr}. \tag{7.33}$$

Multiple poles will give sums of both types of contributions to the integral, which results in both liquid- and gas-type factors in $g(r)$ having two or more correlation lengths. At the value of β and N/V for which all the poles collapse to $q = 0$, we have the behavior

$$g(r) \sim 1, \tag{7.34}$$

which is characteristic of the critical point, where the correlation length is infinite.

7.4. Caveat

The results in the previous section cannot be expected to be numerically accurate, since there are perturbation diagrams representing correlations intermediate between short- and long-range. Besides, only pair interactions have been included, while multi-body potentials are expected to be important at short range. The important thing is that a qualitative model of phase transitions and the critical point has been found in which real and imaginary poles play a significant role.

7.5. Summary for Chapter 7

7.5.1. *What has been Accomplished so Far*

1. Quantum-field theory is a powerful tool for modeling real gases.
2. The defining equation for isodasic (uniform-density) ensemble operators facilitates evaluation of the properties of macroscopic ensembles of interacting molecules.
3. For repulsive interaction of ions, quantum field theory gives the Debye-Hückel theory of ionic solutions with the only approximation being the first-order perturbation.
4. The nesting of ladder and ring diagrams in infinite-order perturbation expansions gives a closed-form model for the effective pair interaction in an equilibrium ensemble.
5. The corresponding closed form for the pair correlation function allows gas and liquid phases to be interpreted in terms of poles in the integrand of a contour integral over the momentum transferred in molecular collisions.
6. The gas-liquid critical point is seen as the temperature and density at which all poles collapse to the point of zero momentum transfer.

7.5.2. *What Needs to be Done*

1. Exploration of the pole structure in the contour integral in a pair correlation function with realistic intermolecular potentials to confirm our expectations about relationship of these poles to the gas, liquid, and critical forms of the pair correlation function.
2. Numerical calculations of the self-consistent limit for the effective pair potential in which all diagrams are included besides the ladders-of-rings and rings-of-ladders scheme which we have outlined here.
3. Numerically precise calculations, based upon the theory of the isodasic ensemble presented here, of critical phenomena, including critical temperatures and densities and the critical exponents, using accurate experimental pair potentials.

4. Extension of the theory to tensor pair potentials so that complex molecular fluids can be studied.
5. Comparison of the utility of existing numerical techniques with this isodasic ensemble theory as an *ansatz* for numerical calculations of the properties of fluids.

Chapter 8

Photons

8.1. Introduction

In this chapter we describe the quantum-field ensemble operator for photons for both coherent and homogeneous light, the latter being analogous to the isodasic ensemble operator for molecules. This is accomplished by generalizing Glauber's theory of coherent light so that a focused light beam with a finite band width and a Gaussian profile can be described by an ensemble operator in the limits of both a coherent and an incoherent beam. The $\boldsymbol{k}$ vectors (analogous to the $\boldsymbol{p}$ vectors for material particles) are parameterized by characteristics of the prepared beam. Ensemble averaging is illustrated by calculating the photon density in $\boldsymbol{k}$-space (a Poisson distribution) and the spatial dependence of the light intensity in the beam. For homogeneous light (*e.g.*, incandescence), the ensemble operator gives the Planck distribution law in the limit of thermal equilibrium. The distinguishing mathematical and physical properties of these three ensemble operators are presented and the importance of phase in the representation of grand canonical ensembles of photons is illustrated.

In Section 8.2, we outline briefly the main features of quantum electrodynamics and in Section 8.3 we establish our notation by reviewing the relevant parts of the theory of coherence. In Section 8.4 we give the general form and some mathematical properties of the normalized ensemble operator for a coherent system of photons and derive the Poisson distribution for photons in a given sub-volume of $\boldsymbol{k}$-space. We obtain in Section 8.5 the explicit form for the photon number density in a focused coherent Gaussian pulse of light and calculate ensemble averages for the vector potential, energy density, and Pointing vector. We show how phase randomization leads to an ensemble operator for homogeneous radiation (*i.e.*, uniform energy density) and for a focused incoherent light pulse, and contrast the mathematical and physical properties of those ensembles with that for coherent light. A summary is given in Section 8.6.

8.2. Quantum Electrodynamics

First, we briefly review the classical theory of electrodynamics.

8.2.1. *The Classical Theory*

Consider an electromagnetic wave with frequency ω. The properties of the wave are described by two position- and time-dependent vector fields; namely, the electric field $\mathbf{E}(\boldsymbol{r}, t)$, and the magnetic field $\mathbf{B}(\boldsymbol{r}, t)$. Both of these fields can be derived from a scalar potential $\varphi(\boldsymbol{r}, t)$ and a vector potential $\mathbf{A}(\boldsymbol{r}, t)$; that is,

$$\mathbf{E}(\boldsymbol{r}, t) = -\nabla\varphi(\boldsymbol{r}, t) - \frac{\partial}{\partial t}\mathbf{A}(\boldsymbol{r}, t); \quad \mathbf{B}(\boldsymbol{r}, t) = \nabla \times \mathbf{A}(\boldsymbol{r}, t). \tag{8.1}$$

The $\mathbf{E}$ and $\mathbf{B}$ fields obey Maxwell's equations:

$$\begin{aligned}
\nabla \cdot \mathbf{E} &= \frac{\rho}{\varepsilon_0}, \\
\nabla \cdot \mathbf{B} &= 0, \\
\nabla \times \mathbf{E} &= -\frac{\partial \mathbf{B}}{\partial t}, \\
\nabla \times \mathbf{B} &= \frac{1}{c^2}\frac{\partial \mathbf{E}}{\partial t} + \mu_0 \mathbf{J},
\end{aligned} \tag{8.2}$$

where ρ is the charge density, $\mathbf{J}$ is the current density, ϵ_0 is the permittivity of the vacuum, and μ_0 is the vacuum permeability. The constants ϵ_0 and μ_0 are related to the speed of light c by $\epsilon_0\mu_0 = c^{-1}$. Photons are excitations of a quantized field; since the observable fields $\mathbf{E}$ and $\mathbf{B}$ are obtainable from the potentials φ and $\mathbf{A}$, we seek to quantize these potentials.

8.2.1.1. *The Electromagnetic 4-Vector Potential*

The form of the Maxwell equations suggests a simplification is possible if the scalar and vector potentials are expressed as components of a 4-vector in Minkowski space-time. We will use the Einstein summation convention, which requires that repeated indices in the same term are summed over all values of the index. Let

$$\partial_\alpha = \partial/\partial x^\alpha = \left(\frac{1}{c}\partial_t, \partial_x, \partial_y, \partial_z\right) \tag{8.3}$$

$$\mathrm{A}^\alpha = (\varphi/c, \mathrm{A}_x, \mathrm{A}_y, \mathrm{A}_z) \tag{8.4}$$

$$\mathrm{A}_\alpha = (\varphi/c, -\mathrm{A}_x, -\mathrm{A}_y, -\mathrm{A}_z). \tag{8.5}$$

The indices are lowered or raised by multiplying with the Minkowski metric tensor with signature (1, −1, −1, −1):

$$x^\alpha \eta^{\alpha\beta} x_\beta, \quad x_\alpha = \eta_{\alpha\beta} x^\beta. \tag{8.6}$$

By these rules we have the matrix

$$\eta^{\alpha\beta} = \eta_{\alpha\beta} = \begin{pmatrix} 1 & 0 & 0 & 0 \\ 0 & -1 & 0 & 0 \\ 0 & 0 & -1 & 0 \\ 0 & 0 & 0 & -1 \end{pmatrix}. \tag{8.7}$$

8.2.1.2. *Maxwell's Equations in Terms of the Potentials*

From Eqs. (8.1) and (8.2) we have

$$\nabla \cdot \mathbf{E} = -\nabla^2 \varphi - \frac{\partial}{\partial t} \nabla \cdot \mathbf{A} = \frac{\rho}{\epsilon_0} \tag{8.8}$$

and

$$\begin{aligned} \nabla \times \mathbf{B} &= \nabla \times (\nabla \times \mathbf{A}) = \nabla(\nabla \cdot \mathbf{A}) - \nabla^2 \mathbf{A} \\ &= \frac{1}{c^2} \nabla \frac{\partial \varphi}{\partial t} + \mu_0 \boldsymbol{j} - \frac{1}{c^2} \frac{\partial^2 \mathbf{A}}{\partial t^2}. \end{aligned} \tag{8.9}$$

The other two Maxwell equations are

$$\nabla \cdot \mathbf{B} = \nabla \cdot (\nabla \times \mathbf{A}) = 0,$$

which follows from the fact that the divergence of a curl is zero, and that

$$\nabla \times \mathbf{E} = -\nabla \times (\nabla \varphi) - \frac{\partial}{\partial t} \nabla \times \mathbf{A}.$$

These last equations reaffirm the definition of $\mathbf{E}$, since the curl of a gradient is zero.

We may rearrange Eqs. (8.8) and (8.9) for comparison:

$$\nabla^2 \varphi + \frac{\partial}{\partial t} \nabla \cdot \mathbf{A} = \frac{\rho}{\epsilon_0} \tag{8.10}$$

$$\frac{1}{c^2}\frac{\partial^2 \mathbf{A}}{\partial t^2} - \nabla^2 \mathbf{A} + \nabla \left(\nabla \cdot \mathbf{A} + \frac{1}{c^2}\frac{\partial \varphi}{\partial t} \right) = +\mu_0 \mathbf{J}. \tag{8.11}$$

Gauge invariance allows Eqs. (8.10) and (8.11) to be simplified.

8.2.1.3. *Gauge Invariance*

Since only the derivatives of φ and $\mathbf{A}$ are observable, the gradient of some scalar function $f(\boldsymbol{r}, \boldsymbol{t})$ can be added to $\mathbf{A}(\boldsymbol{r}, \boldsymbol{t})$ provided that the time derivative of f is subtracted from φ, for then

$$\mathbf{E} = -\frac{\partial}{\partial t}(\mathbf{A} + \nabla f) - \nabla \left(\varphi - \frac{\partial f}{\partial t} \right),$$

$$\mathbf{B} = \nabla \times (\mathbf{A} + \nabla f).$$

The terms involving f in the expression for $\mathbf{E}$ cancel as $(\partial/\partial t)\nabla = \nabla(\partial/\partial t)$ and the term involving f in the expression for $\mathbf{B}$ vanishes because $\nabla \cdot \nabla = 0$. A function f may be found for which

$$\nabla \cdot \mathbf{A} + \frac{1}{c^2}\frac{\partial \varphi}{\partial t} = 0; \tag{8.12}$$

this is the Lorenz gauge. In that choice of gauge Eqs. (8.10) and (8.11) become

$$\frac{1}{c^2}\frac{\partial \varphi}{\partial t^2} - \nabla^2 \varphi = \frac{\rho}{\epsilon_0} \tag{8.13}$$

$$\frac{1}{c^2} - \frac{\partial^2 \mathbf{A}}{\partial t^2} - \nabla^2 \mathbf{A} = \mu_0 \mathbf{J}, \tag{8.14}$$

so that both the scalar and vector potentials obey inhomogeneous wave equations. The Lorentz invariance of these wave equations is clear when we write them as

$$\Box \mathbf{A}^{\mu} = \mu_0 j^{\mu}, \tag{8.15}$$

where j^{μ} is the 4-vector current:

$$j^{\mu} = (c\rho, j_x, j_y, j_z) \tag{8.16}$$

and $\Box$ is the d'Alembertian operator:

$$\Box = \frac{1}{c^2}\frac{\partial^2}{\partial t^2} - \nabla^2. \tag{8.17}$$

8.2.1.4. *The Electromagnetic Tensor*

The fields **E** and **B** may be written in terms of the 4×4 *electromagnetic tensor* $F_{\alpha\beta}$, which is defined as

$$F_{\alpha\beta} = \partial_\alpha \mathrm{A}_\beta - \partial_\beta \mathrm{A}_\alpha = \frac{\partial}{\partial x^\alpha}\mathrm{A}_\beta - \frac{\partial}{\partial x^\beta}\mathrm{A}_\alpha. \tag{8.18}$$

Then using Eqs. (8.3) and (8.5) for x^α and A_β, respectively, we obtain

$$F_{\alpha\beta} = \begin{pmatrix} 0 & \mathrm{E}_x/c & \mathrm{E}_y/c & \mathrm{E}_z/c \\ -\mathrm{E}_x/c & 0 & -\mathrm{B}_z & \mathrm{B}_y \\ -\mathrm{E}_y/c & \mathrm{B}_z & 0 & -\mathrm{B}_x \\ -\mathrm{E}_z/c & -\mathrm{B}_y & \mathrm{B}_x & 0 \end{pmatrix}, \tag{8.19a}$$

$$F^{\alpha\beta} = \eta^{\alpha\mu} F_{\mu\mathrm{v}} \eta^{\mathrm{v}\beta} = \begin{pmatrix} 0 & -\mathrm{E}_x/c & -\mathrm{E}_y/c & -\mathrm{E}_z/c \\ \mathrm{E}_x/c & 0 & -\mathrm{B}_z & \mathrm{B}_y \\ \mathrm{E}_y/c & \mathrm{B}_z & 0 & -\mathrm{B}_x \\ \mathrm{E}_z/c & -\mathrm{B}_y & \mathrm{B}_x & 0 \end{pmatrix}. \tag{8.19b}$$

8.2.1.5. *Maxwell's Equations in Terms of the Electromagnetic Tensor*

These equations now take a very simple form in terms of $F^{\alpha\beta}$, for the covariant derivative is

$$\begin{aligned}\partial_\alpha F^{\alpha\beta} &= \frac{\partial}{\partial x^\alpha}(\partial^\alpha \mathrm{A}^\beta - \partial^\beta \mathrm{A}^\alpha)\\ &= \Box^2 \mathrm{A}^\beta - \partial^\beta\left(\frac{1}{c^2}\frac{\partial\varphi}{\partial t} + \nabla\cdot\mathbf{A}\right)\\ &= \Box^2\mathrm{A}^\beta, \end{aligned} \tag{8.20}$$

since the expression in large parentheses equals zero in the Lorenz gauge.

Thus, Eq. (8.15), which is equivalent to Maxwell's equations, takes the form

$$\partial_\alpha F^{\alpha\beta} = \mu_0 j^\beta. \tag{8.21}$$

Now the goal is to identify a Lagrangian density function that produces the Maxwell equations in the form of Eq. (8.21).

8.2.1.6. *The Lagrangian Density for Electromagnetic Fields*

Recall from Eq. (4.36) that the action is the integral over space and time of the Lagrangian density, $\mathscr{L}(\varphi, \partial\varphi)$, a function of all the relevant fields φ and their spatial derivatives $\partial\varphi$. Integration of $\mathscr{L}$ over space and time gives the action S:

$$S = \int dt\, d\boldsymbol{r} \mathscr{L}[\varphi(\boldsymbol{r},t),\, \partial\varphi(\boldsymbol{r},t)]. \tag{8.22}$$

The reasoning in Section 4.3.3 leading to Eq. (4.41) can be used to obtain the Euler-Lagrange equation for each component of the electromagnetic fields:

$$\partial_\mu \frac{\partial\mathscr{L}}{\partial(\partial_\mu \mathrm{A}_\alpha)} = \frac{\partial\mathscr{L}}{\partial \mathrm{A}_\alpha}. \tag{8.23}$$

The Lagrangian density we seek in its most compact form in the Lorenz gauge is

$$\mathscr{L} = -\frac{1}{4}F^{\gamma\delta}F_{\gamma\delta} - \mu_0 j^{\mathrm{v}}\mathrm{A}_{\mathrm{v}}. \tag{8.24}$$

Applying Eq. (8.23) to Eq. (8.24) requires that the tensor components $F^{\alpha\beta}$ and $F_{\alpha\beta}$ be written in terms of derivatives of the potential 4-vectors A_α and A^α; from Eq. (8.18), this gives for $\mathscr{L}$

$$\begin{aligned}\mathscr{L} &= -\frac{1}{4}(\partial^\gamma \mathrm{A}^\delta - \partial^\delta \mathrm{A}^\gamma)(\partial_\gamma \mathrm{A}_\delta - \partial_\delta \mathrm{A}_\gamma) - \mu_0 j^\mathrm{v} \mathrm{A}_\mathrm{v} \\ &- \frac{1}{4}(\partial^\gamma \mathrm{A}^\delta \partial_\gamma \mathrm{A}_\delta + \partial^\delta \mathrm{A}^\gamma \partial_\delta \mathrm{A}_\gamma \\ &- \partial^\delta \mathrm{A}^\gamma \partial_\gamma \mathrm{A}_\delta - \partial^\gamma \mathrm{A}^\delta \partial_\delta \mathrm{A}_\gamma) - \mu_0 j^\mathrm{v} \mathrm{A}_\mathrm{v}. \end{aligned} \tag{8.25}$$

Since we are summing over repeated indices, these "dummy" indices can be interchanged to give

$$\begin{aligned}\mathscr{L} &= -\frac{1}{4}(\partial^\gamma \mathrm{A}^\delta \partial_\gamma \mathrm{A}_\delta + \partial^\gamma \mathrm{A}^\delta \partial_\gamma \mathrm{A}_\delta - \partial^\delta \mathrm{A}^\gamma \partial_\gamma \mathrm{A}_\delta \\ &- \partial^\delta \mathrm{A}^\gamma \partial_\gamma \mathrm{A}_\delta - \mu_0 j^\mathrm{v} \mathrm{A}_\mathrm{v} \\ &- \frac{1}{2}(\partial^\gamma \mathrm{A}^\delta \partial_\gamma \mathrm{A}_\delta - \partial^\delta \mathrm{A}^\gamma \partial_\gamma \mathrm{A}_\delta) - \mu_0 j^\mathrm{v} \mathrm{A}_\mathrm{v}. \end{aligned} \tag{8.26}$$

Since we need $\partial\mathscr{L}/\partial(\partial_\mu \mathrm{A}_\alpha)$, we must lower the raised indices by substituting

$$\partial_\gamma \eta^{\gamma\sigma}\partial_\sigma, \quad \partial^\delta = \eta^{\delta\sigma}\partial_\sigma, \quad \mathrm{A}^\delta = \eta^{\delta\tau}\mathrm{A}_\tau, \quad \mathrm{A}^\gamma = \eta^{\gamma\tau}\mathrm{A}_\tau$$

to give

$$\mathscr{L} = \frac{1}{2}(\eta^{\gamma\sigma}\partial_\sigma \eta^{\delta\tau}\mathrm{A}_\tau \partial_\gamma \mathrm{A}_\delta - \eta^{\delta\sigma}\partial_\sigma \eta^{\gamma\tau}\mathrm{A}_\tau \partial_\gamma \mathrm{A}_\delta) - \mu_0 j^\mathrm{v} \mathrm{A}_\mathrm{v}. \tag{8.27}$$

Making use of the fact that the η's are constants and therefore $\partial\eta^{\gamma\sigma} = \eta^{\gamma\sigma}\partial$, we have for the l.h.s. of Eq. (8.23)

$$\begin{aligned}&\partial_\mu \frac{\partial \mathscr{L}}{\partial(\partial_\mu \mathrm{A}_\alpha)} \\ &= -\frac{1}{2}\partial_\mu(\partial^\mu \mathrm{A}^\alpha - \partial^\alpha \mathrm{A}^\mu + \eta^{\gamma\mu}\eta^{\delta\alpha}\partial_\gamma \mathrm{A}_\delta - \eta^{\delta\mu}\eta^{\gamma\alpha}\partial_\gamma \mathrm{A}_\delta \\ &-\partial_\mu(\partial^\mu \mathrm{A}^\alpha - \partial^\alpha \mathrm{A}^\mu) = -\partial_\mu F^{\mu\alpha}, \end{aligned} \tag{8.28}$$

while for the r.h.s. we have

$$\frac{\partial \mathscr{L}}{\partial \mathrm{A}_\alpha} = -\mu_0 j^\alpha. \tag{8.29}$$

Substituting Eqs. (8.28) and (8.29) into Eq. (8.23) gives Eq. (8.21), the required result.

8.2.1.7. *The Canonical Momentum and Hamiltonian*

Our next step is to find the canonical momentum density $\wp_\mu$ for each component of the potential 4-vector A_μ. From Noether's theorem and Eq. (8.28), $\wp_\mu$ is given by

$$\wp_\mu = \frac{\delta \mathscr{L}}{\delta(\partial_0 \mathrm{A}_\mu)} = -F^{0\mu} = (0, E_x/c, E_y/c, E_z/c). \tag{8.30}$$

The Hamiltonian density is then

$$\mathcal{H} = \wp^\mu \partial_0 \mathrm{A}_\mu - \mathscr{L} = \frac{1}{c}\mathbf{E} \cdot \left(-\frac{1}{c}\frac{\partial}{\partial t}\mathbf{A}\right) + F^{\alpha\beta}F_{\alpha\beta} + \mu_0 j^\alpha \mathrm{A}_\alpha. \tag{8.31}$$

From Eq. (8.1) we have

$$\frac{\partial}{\partial t}\mathbf{A} = -\mathbf{E} - \nabla\phi, \tag{8.32}$$

from Eqs. (8.19) we have

$$F^{\alpha\beta}F_{\alpha\beta} = 2(-\mathbf{E}^2/c^2 + \mathbf{B}^2). \tag{8.33}$$

Substituting Eqs. (8.32) and (8.33) into Eq. (8.31) gives

$$\mathcal{H}_{\text{em}} = \frac{1}{2}(\mathbf{E}^2/c^2 + \mathbf{B}^2) + \frac{1}{c^2}\mathbf{E} \cdot \nabla\varphi + \mu_0 j^\alpha \mathrm{A}_\alpha. \tag{8.34}$$

In "free space"; that is, in the absence of the sources φ and j^α, the Hamiltonian density is just

$$\mathcal{H}_{\text{vac}} = \frac{1}{2}(\mathbf{E}^2/c^2 + \mathbf{B}^2). \tag{8.35}$$

In the absence of φ and j^α, Eq. (8.15) becomes the homogeneous wave equation for $\mathbf{A}$

$$\frac{1}{c^2}\frac{\partial^2 \mathbf{A}}{\partial t^2} - \nabla^2 \mathbf{A} = 0, \tag{8.36}$$

the plane-wave solution to which is

$$\mathbf{A}(\boldsymbol{r}, t) = \int d\boldsymbol{k} K_{\text{norm}}(\boldsymbol{k}) \times [C(\boldsymbol{k}) e^{i(\boldsymbol{k}\cdot\boldsymbol{r} - \omega t)} + C^*(\boldsymbol{k}) e^{-i(\boldsymbol{k}\cdot\boldsymbol{r} - \omega t)}] \boldsymbol{e}_{\text{A}}. \tag{8.37}$$

In Eq. (8.37) K_{norm} is a normalization constant which we will determine later, the wave vector $\boldsymbol{k} = (\omega/c)\boldsymbol{e}_k$, where ω is the *angular frequency* and $\boldsymbol{e}_{\text{A}}$ and $\boldsymbol{e}_k$ are unit vectors in the direction of $\mathbf{A}$ and $\boldsymbol{k}$, respectively. Since $\nabla \cdot \mathbf{A} = 0$, this means that

$$\nabla \cdot \mathbf{A} = i \int d\boldsymbol{k} K_{\text{norm}}(\boldsymbol{k}) \times [C(\boldsymbol{k}) e^{i(\boldsymbol{k}\cdot\boldsymbol{r} - \omega t)} - C^*(\boldsymbol{k}) e^{-i(\boldsymbol{k}\cdot\boldsymbol{r} - \omega t)})] \boldsymbol{k} \cdot \boldsymbol{e}_{\text{A}} = 0. \tag{8.38}$$

Thus $\boldsymbol{k} \cdot \boldsymbol{e}_{\text{A}} = 0$, indicating that $\boldsymbol{e}_k$ and $\boldsymbol{e}_{\text{A}}$ are orthogonal; that is, the vector $\mathbf{A}$ is a wave whose amplitude is in a direction ($\boldsymbol{e}_{\text{A}}$) transverse (orthogonal) to its direction of propagation ($\boldsymbol{e}_k$). If we designate the two unit vectors that are orthogonal to $\boldsymbol{e}_k$ as $\boldsymbol{e}_1$ and $\boldsymbol{e}_2$ such that $\boldsymbol{e}_k$, $\boldsymbol{e}_1$, $\boldsymbol{e}_2$ form a right-handed system, then $\boldsymbol{e}_{\text{A}}$ can be either $\boldsymbol{e}_1$ or $\boldsymbol{e}_2$. Both of these possibilities must be included by adding subscripts to the coefficients $C(\boldsymbol{k})$ and $C^*(\boldsymbol{k})$ and the unit vector $\boldsymbol{e}_{\text{A}}$ and summing them:

$$\mathbf{A}(\boldsymbol{r}, t) = \int d\boldsymbol{k} K_{\text{norm}}(\boldsymbol{k}) \times \sum_{\sigma=1,2} [C_\sigma(\boldsymbol{k}) e^{i(\boldsymbol{k}\cdot\boldsymbol{r} - \omega t)} + C_\sigma^*(\boldsymbol{k}) e^{-i(\boldsymbol{k}\cdot\boldsymbol{r} - \omega t)}] \boldsymbol{e}_\sigma. \tag{8.39}$$

Our next step is to quantize the electromagnetic wave. This means replacing $C_\sigma(\boldsymbol{k})$ and $C_\sigma(\boldsymbol{k})^*$ by Fock operators.

8.2.2. *Quantization of the Fields*

The Fock operators we use in quantizing the electromagnetic fields are $\hat{C}_\sigma(\boldsymbol{k})$ and $\hat{C}_\sigma(\boldsymbol{k})^\dagger$, whose commutators are

$$[\hat{C}_\sigma(\boldsymbol{k}), \hat{C}_{\sigma'}(\boldsymbol{k}')^\dagger]_- = \delta_{\sigma\sigma'}\delta(\boldsymbol{k}-\boldsymbol{k}')$$
$$[\hat{C}_\sigma(\boldsymbol{k}), \hat{C}_{\sigma'}(\boldsymbol{k}')]_- = 0$$
$$[\hat{C}_\sigma(\boldsymbol{k})^\dagger, \hat{C}_{\sigma'(\boldsymbol{k}')^\dagger}]_- = 0. \quad (8.40)$$

8.2.2.1. *The Vector Potential*

From Eq. (8.39), the vector potential is given by the Hermitian operator

$$\hat{\mathbf{A}}(\boldsymbol{r}, t) = \int d\boldsymbol{k} K_{\text{norm}}(\boldsymbol{k})$$
$$\times \sum_{\sigma=1}^{2} [\hat{C}_\sigma(\boldsymbol{k}) e^{i(\boldsymbol{k}\cdot\boldsymbol{r}-\omega t)} + \hat{C}_\sigma(\boldsymbol{k})^\dagger e^{-i(\boldsymbol{k}\cdot\boldsymbol{r}-\omega t)}] \boldsymbol{e}_\sigma. \quad (8.41)$$

8.2.2.2. *The Electric Field*

Using the first of Eqs. (8.1) with $\varphi = 0$, we have

$$\hat{\mathbf{E}}(\boldsymbol{r}, t) = i \int d\boldsymbol{k} K_{\text{norm}}(\boldsymbol{k})\omega$$
$$\times \sum_{\sigma=1}^{2} [\hat{C}_\sigma(\boldsymbol{k}) e^{-i(\boldsymbol{k}\cdot\boldsymbol{r}-\omega t)} - \hat{C}_\sigma^\dagger(\boldsymbol{k})\, e^{-i(\boldsymbol{k}\cdot\boldsymbol{r}-\omega t)}] \boldsymbol{e}_\sigma. \quad (8.42)$$

The electric field is seen to be in the same direction as the vector potential.

8.2.2.3. *The Magnetic Field*

From the second of Eqs. (8.1), we have

$$\hat{\mathbf{B}}(\boldsymbol{r}, t) = i \int d\boldsymbol{k} k K_{\text{norm}}(\boldsymbol{k})$$
$$\times \sum_{\sigma=1}^{2} [\hat{C}_\sigma(\boldsymbol{k}) e^{i(\boldsymbol{k}\cdot\boldsymbol{r}-\omega t)} - \hat{C}_\sigma^\dagger(\boldsymbol{k}) e^{-i(\boldsymbol{k}\cdot\boldsymbol{r}-\omega t)}] \boldsymbol{e}_k \times \boldsymbol{e}_\sigma. \quad (8.43)$$

The magnetic field is seen to be orthogonal to both the electric field and the vector potential.

8.2.2.4. *The Hamiltonian*

We seek here the Hamiltonian itself; that is, the integral of the Hamiltonian density $\mathcal{H}$ over all space. Our procedure will be to find the operators for $|\mathbf{E}(\boldsymbol{r},\, t)|^2$ and $|\mathbf{B}(\boldsymbol{r},\, t)|^2$ then integrate them over $\boldsymbol{r}$, and insert the results into Eq. (8.35). From Eq. (8.42), this gives:

$$\int d\boldsymbol{r}|\hat{\mathbf{E}}(\boldsymbol{r},\, t)|^2 = \int d\boldsymbol{r}\left\{ i\int d\boldsymbol{k}K_{\text{norm}}(\boldsymbol{k}\omega)\sum_{\sigma=1}^{2}\left[\hat{C}_\sigma(\boldsymbol{k})e^{-i(\boldsymbol{k}\cdot\boldsymbol{r}-\omega t)}\right.\right.$$
$$\left. - \hat{C}_\sigma(\boldsymbol{k})^\dagger e^{-i(\boldsymbol{k}\cdot\boldsymbol{r}-\omega t)}\right] \boldsymbol{e}_\sigma \cdot i\int d\boldsymbol{k}'K_{\text{norm}}(\boldsymbol{k}\omega')\sum_{\sigma'=1}^{2}$$
$$\left.\times\left[\hat{C}_{\sigma'}(\boldsymbol{k}')e^{i(\boldsymbol{k}'\cdot\boldsymbol{r}-\omega' t)} - \hat{C}_{\sigma'}(\boldsymbol{k}')^\dagger e^{-i(\boldsymbol{k}'\cdot\boldsymbol{r}-\omega' t)}\right]\boldsymbol{e}_{\sigma'}\right\}. \tag{8.44}$$

In evaluating Eq. (8.44), note that $\boldsymbol{e}_\sigma \cdot \boldsymbol{e}_{\sigma^\dagger} = \delta_{\sigma\sigma'}$. The result is

$$\int d\boldsymbol{r}|\hat{\mathbf{E}}(\boldsymbol{r},\, t)|^2 = \int d\boldsymbol{k}K_{\text{norm}}(\boldsymbol{k})\int d\boldsymbol{k}'K_{\text{norm}}(\boldsymbol{k}')\omega'\omega$$
$$\times\sum_{\sigma=1}^{2}[\hat{C}_\sigma(\boldsymbol{k})^\dagger e^{i(\boldsymbol{k}'\cdot\boldsymbol{r}-\omega' t)}\hat{C}_\sigma(\boldsymbol{k}')e^{i(\boldsymbol{k}'\cdot\boldsymbol{r}-\omega' t)}$$
$$+\hat{C}_\sigma(\boldsymbol{k}')e^{i(\boldsymbol{k}'\cdot\boldsymbol{r}-\omega' t)}\hat{C}_\sigma(\boldsymbol{k})^\dagger e^{-i(\boldsymbol{k}\cdot\boldsymbol{r}-\omega t)}$$
$$-\hat{C}_\sigma(\boldsymbol{k})e^{i(\boldsymbol{k}\cdot\boldsymbol{r}-\omega t)}\hat{C}_\sigma(\boldsymbol{k}')e^{i(\boldsymbol{k}'\cdot\boldsymbol{r}-\omega' t)}$$
$$-\hat{C}_\sigma(\boldsymbol{k})^\dagger e^{-i(\boldsymbol{k}\cdot\boldsymbol{r}-\omega t)}\hat{C}_\sigma(\boldsymbol{k}')^\dagger e^{-i(\boldsymbol{k}'\cdot\boldsymbol{r}-\omega' t)}]. \tag{8.45}$$

The first and second terms in the brackets in Eq. (8.45) can be combined by using the commutation relation

$$\hat{C}_\sigma(\boldsymbol{k}')\hat{C}_\sigma(\boldsymbol{k})^\dagger = \hat{C}_\sigma(\boldsymbol{k}') + \delta(\boldsymbol{k}-\boldsymbol{k}'). \tag{8.46}$$

After applying Eq. (8.46) and some regrouping, Eq. (8.45) becomes

$$\int d\boldsymbol{r}|\hat{\mathbf{E}}(\boldsymbol{r},t)|^2 = \int d\boldsymbol{k}K_{\text{norm}}(\boldsymbol{k})\int d\boldsymbol{k}'K_{\text{norm}}(\boldsymbol{k}')\omega'\omega$$
$$\times\sum_{\sigma=1}^{2}\Bigg\{[2\hat{C}_\sigma(\boldsymbol{k})^\dagger\hat{C}_\sigma(\boldsymbol{k}')+\delta(\boldsymbol{k}-\boldsymbol{k}')]$$
$$\times\int d\boldsymbol{r}\,e^{-i(\boldsymbol{k}\cdot\boldsymbol{r}-\omega t)}\,e^{i(\boldsymbol{k}'\cdot\boldsymbol{r}-\omega' t)}$$
$$-\hat{C}_\sigma(\boldsymbol{k})\hat{C}_\sigma(\boldsymbol{k}')\int d\boldsymbol{r}\,e^{i(\boldsymbol{k}\cdot\boldsymbol{r}-\omega t)}\,e^{i(\boldsymbol{k}'\cdot\boldsymbol{r}-\omega' t)}$$
$$-\hat{C}_\sigma(\boldsymbol{k}')^\dagger\hat{C}_\sigma(\boldsymbol{k}')^\dagger\int d\boldsymbol{r}\,e^{-i(\boldsymbol{k}\cdot\boldsymbol{r}-\omega t)}\,e^{-i(\boldsymbol{k}'\cdot\boldsymbol{r}-\omega' t)}\Bigg\}. \tag{8.47}$$

The integrals in Eq. (8.47) over $\boldsymbol{r}$ will yield the delta functions $\int d\boldsymbol{r}\,e^{i(\boldsymbol{k}-\boldsymbol{k}')\cdot\boldsymbol{r}} = (2\pi)^3\delta(\boldsymbol{k}-\boldsymbol{k}')$ and since $\omega = ck$, the result is

$$\int d\boldsymbol{r}|\hat{\mathbf{E}}(\boldsymbol{r},t)|^2 = (2\pi)^3\int d\boldsymbol{k}K_{\text{norm}}(\boldsymbol{k})\int d\boldsymbol{k}'K_{\text{norm}}(\boldsymbol{k}')c^2kk'$$
$$\times\sum_{\sigma=1}^{2}\{[2\hat{C}_\sigma(\boldsymbol{k})^\dagger\hat{C}_\sigma(\boldsymbol{k}')+\delta(\boldsymbol{k}-\boldsymbol{k}')]\delta(\boldsymbol{k}-\boldsymbol{k}')$$
$$-\hat{C}_\sigma(\boldsymbol{k})\hat{C}_\sigma(\boldsymbol{k}')\delta(\boldsymbol{k}+\boldsymbol{k}')e^{-i(\omega+\omega')t}$$
$$-\hat{C}_\sigma(\boldsymbol{k})^\dagger\hat{C}_\sigma(\boldsymbol{k}')^\dagger\delta(\boldsymbol{k}+\boldsymbol{k}')e^{i(\omega+\omega')t}\}. \tag{8.48}$$

Finally, the integral over $\boldsymbol{k}'$ gives

$$\int d\boldsymbol{r}|\hat{\mathbf{E}}(\boldsymbol{r},t)|^2 = (2\pi)^3\int d\boldsymbol{k}[K_{\text{norm}}(\boldsymbol{k})]^2c^2k^2$$
$$\times\sum_{\sigma=1}^{2}\Bigg\{\left[2\hat{C}_\sigma(\boldsymbol{k})^\dagger\hat{C}_\sigma(\boldsymbol{k})+\frac{V}{(2\pi)^3}\right]$$
$$-\hat{C}_\sigma(\boldsymbol{k})\hat{C}_\sigma(-\boldsymbol{k})e^{-2i\omega t}-\hat{C}_\sigma(\boldsymbol{k})^\dagger\hat{C}_\sigma(-\boldsymbol{k})^\dagger e^{2i\omega t}\Bigg\}, \tag{8.49}$$

where we have used $\delta(0) = V/(2\pi)^3$; this term in V, lacking annihilation and creation operators, will give the vacuum a cosmic constant energy, which we will leave to the cosmologists to deal with and ignore here. The last two terms in Eq. (8.49), having two annihilation and two creation operators, respectively, will have zero contribution to eigenstates of the number operator, since they will cause annihilation of either the *bra* or the *ket* of the vacuum when expectation values of $|\hat{\mathbf{E}}(\boldsymbol{r}, t)|^2$ are taken. These terms will, however, contribute to coherent photon states under certain conditions (see Section 8.2.4). For the present purpose, to make $K_{\text{norm}}(\boldsymbol{k})$ consistent with observation, we will assume single-photon states and ignore the last line in Eq. (8.49), writing

$$\int d\boldsymbol{r}|\hat{\mathbf{E}}(\boldsymbol{r}, t)|^2 = 2(2\pi)^3 \int d\boldsymbol{k}[K_{\text{norm}}(\boldsymbol{k})]^2 c^2 k^2 \sum_{\sigma=1}^{2} \hat{C}_\sigma(\boldsymbol{k})^\dagger \hat{C}_\sigma(\boldsymbol{k}). \tag{8.50}$$

Similarly, the calculation for $\int d\boldsymbol{r}|\hat{\mathbf{B}}(\boldsymbol{r}, t)^2$ gives

$$\int d\boldsymbol{r}|\hat{\mathbf{B}}(\boldsymbol{r}, t)|^2 = 2(2\pi)^3 \int d\boldsymbol{k}[K_{\text{norm}}(\boldsymbol{k})]^2 k^2 \sum_{\sigma=1}^{2} \hat{C}_\sigma(\boldsymbol{k})^\dagger \hat{C}_\sigma(\boldsymbol{k}). \tag{8.51}$$

Thus, the **E** and **B** fields contribute equally to the electromagnetic energy. With Eqs. (8.50) and (8.51) inserted into Eq. (8.35), the Hamiltonian for radiation in the absence of sources is

$$\hat{\mathbf{H}} = 2(2\pi)^3 \int d\boldsymbol{k}[K_{\text{norm}}(\boldsymbol{k})]^2 k^2 \sum_{\sigma=1}^{2} \hat{C}_\sigma(\boldsymbol{k})^\dagger \hat{C}_\sigma(\boldsymbol{k}). \tag{8.52}$$

In order for the energy of a single photon to be $\hbar\omega(= \hbar ck)$, we must choose the normalization constant to be

$$K_{\text{norm}} = \left(\frac{\hbar c}{16\pi^3 k}\right)^{\frac{1}{2}}. \tag{8.53}$$

With Eq. (8.53) inserted into Eq. (8.52), the second-quantized form for the free-field radiation Hamiltonian operator is

$$\hat{\mathbf{H}}^0 = \int d\boldsymbol{k}\ \hbar ck \sum_{\sigma=1}^{2} \hat{C}_\sigma^\dagger(\boldsymbol{k})\hat{C}_\sigma(\boldsymbol{k}). \tag{8.54}$$

and the vector potential operator is

$$\hat{\mathbf{A}}(\mathbf{r}, t) = \left(\frac{\hbar c}{16\pi^3}\right)^{\frac{1}{2}} \int \frac{d\boldsymbol{k}}{\sqrt{k}} \times \sum_{\sigma=1}^{2} [\hat{C}_\sigma(\boldsymbol{k})e^{i(\boldsymbol{k}\cdot\boldsymbol{r}-\omega t)} + \hat{C}_\sigma(\boldsymbol{k})^\dagger e^{-i(\boldsymbol{k}\cdot\boldsymbol{r}-\omega t)}]\boldsymbol{e}_\sigma. \tag{8.55}$$

8.2.2.5. *Photon States*

The number operator for photons whose wave vectors are $\boldsymbol{k}$ and have polarization σ is

$$\hat{\mathbf{N}}_\sigma(\boldsymbol{k}) = \hat{C}_\sigma(\boldsymbol{k})^\dagger \hat{C}_\sigma(\boldsymbol{k}), \tag{8.56a}$$

and for the total number of photons,

$$\hat{\mathbf{N}} = \int d\boldsymbol{k} \sum_\sigma \hat{\mathbf{N}}_\sigma(\boldsymbol{k}) = \int d\boldsymbol{k} \sum_\sigma \hat{C}_\sigma(\boldsymbol{k})^\dagger\, \hat{C}_\sigma(\boldsymbol{k}). \tag{8.56b}$$

Suppose we have one photon with wave vector $\boldsymbol{k}_1$ and polarization σ_1. This state is represented by the *ket* $\hat{C}_{\sigma_1}(\boldsymbol{k}_1)^\dagger|\,0\rangle$. Operation on this *ket* with $\hat{\mathbf{N}}$ gives a term ending in $\hat{C}_\sigma(\boldsymbol{k})C_{\sigma_1}(\boldsymbol{k}_1)^\dagger$. From the boson commutation rules, this will give

$$\hat{C}_\sigma(\boldsymbol{k})C_{\sigma_1}(\boldsymbol{k}_1)^\dagger = C_{\sigma_1}(\boldsymbol{k}_1)^\dagger \hat{C}_\sigma(\boldsymbol{k}) + \delta_{\sigma\sigma_1}\delta(\boldsymbol{k} - \boldsymbol{k}_1);$$

thus we will get

$$\begin{aligned}\hat{\mathbf{N}}\hat{C}_{\sigma_1}(\boldsymbol{k}_1)^\dagger|\,0\rangle &= \int d\boldsymbol{r} \sum_\sigma \hat{C}_\sigma(\boldsymbol{k})^\dagger \hat{C}_\sigma(\boldsymbol{k})\hat{C}_{\sigma_1}(\boldsymbol{k}_1)^\dagger|\,0\rangle \\ &= \int d\boldsymbol{k} \sum_\sigma \hat{C}_\sigma(\boldsymbol{k})^\dagger[\hat{C}_\sigma(\boldsymbol{k})\hat{C}_{\sigma_1}(\boldsymbol{k}_1)^\dagger]|\,0\rangle\end{aligned}$$

$$= \int d\boldsymbol{k} \sum_{\sigma} \hat{C}_{\sigma}(\boldsymbol{k})^{\dagger}[\hat{C}_{\sigma 1}(\boldsymbol{k}_1)^{\dagger}\hat{C}_{\sigma}(\boldsymbol{k}) + \delta_{\sigma\sigma_1}\delta(\boldsymbol{k} - \boldsymbol{k}_1)]|\,0\rangle$$

$$= 1\hat{C}_{\sigma_1}(\boldsymbol{k}_1)^{\dagger}|\,0\rangle, \tag{8.57}$$

since the vacuum is annihilated in the first term in the third line; that is, $\hat{C}_{\sigma}(\boldsymbol{k})|\,0\rangle = 0$. The 1 in the last line emphasizes that this 1-photon state is an eigenstate of $\hat{\mathbf{N}}$ with eigenvalue 1. For a 3-photon state, the eigenvalue is 3, *etc.*:

$$\hat{\mathbf{N}}C_{\sigma 3}(\boldsymbol{k}_3)C_{\sigma 2}(\boldsymbol{k}_2)^{\dagger}C_{\sigma 1}(\boldsymbol{k}_1)^{\dagger}|\,0\rangle$$

$$= \int d\boldsymbol{k} \sum_{\sigma} \hat{C}_{\sigma}(\boldsymbol{k})^{\dagger}\hat{C}_{\sigma}(\boldsymbol{k})C_{\sigma 3}(\boldsymbol{k}_3)^{\dagger}C_{\sigma_2}(\boldsymbol{k}_2)^{\dagger}C_{\sigma_1}(\boldsymbol{k}_1)^{\dagger}|\,0\rangle$$

$$= 3C_{\sigma_3}(\boldsymbol{k}_3)^{\dagger}C_{\sigma_2}(\boldsymbol{k}_2)^{\dagger}C_{\sigma_1}(\boldsymbol{k}_1)^{\dagger}|\,0\rangle, \tag{8.58}$$

since there are three ways to pair $\hat{C}_{\sigma}(\boldsymbol{k})$ with a $C_{\sigma_j}(\boldsymbol{k}_j)^{\dagger}$. For any number of photons n, we have

$$\hat{\mathbf{N}}\prod_{j}^{n} \hat{C}_{\sigma_j}(\boldsymbol{k}_j)^{\dagger}|\,0\rangle = n\prod_{j}^{n} \hat{C}_{\sigma_j}(\boldsymbol{k}_j)^{\dagger}|\,0. \tag{8.59}$$

In most situations there will be several photons with the same $\boldsymbol{k}$ and σ, in which case we write for the state

$$\hat{\mathbf{N}}\prod_{j}^{n}[\hat{C}_{\sigma_j}(\boldsymbol{k}_j)^{\dagger}]^{n_j}|\,0\rangle = \sum_{j=1}^{n} n_j \prod_{j}^{n}[\hat{C}_{\sigma_j}(\boldsymbol{k}_j)^{\dagger}]^{n_j}|\,0\rangle. \tag{8.60}$$

The normalization integral for a 1-photon state is

$$\langle 0|\hat{C}_{\sigma}(\boldsymbol{k})\hat{C}_{\sigma}(\boldsymbol{k})^{\dagger}|\,0\rangle = \frac{V}{(2\pi)^3} \tag{8.61a}$$

so the normalized 1-photon *ket* is

$$\left[\frac{(2\pi)^3}{V}\right]^{-\frac{1}{2}} \hat{C}_{\sigma}(\boldsymbol{k})^{\dagger}|\,0\rangle;$$

the normalization bracket for a state of n distinct photons is thus

$$\langle 0|\prod_j^n \hat{C}_{\sigma_j}(\boldsymbol{k}_j)\prod_j^n \hat{C}_{\sigma_j}(\boldsymbol{k}_j)^\dagger|\,0\rangle = n!\left[\frac{V}{(2\pi)^3}\right]^n, \qquad (8.61\text{b})$$

since there are $n!$ ways of pairing n annihilation operators with n creation operators. The normalized *ket* is therefore

$$\frac{1}{\sqrt{n!}}\left[\frac{(2\pi)^3}{V}\right]^{\frac{n}{2}}\prod_j^n[\hat{C}_{\sigma_j}(\boldsymbol{k}_j)^\dagger]|\,0\rangle. \qquad (8.62)$$

If groups of photons have the same $\boldsymbol{k}$ and σ, the normalization bracket is

$$\langle 0|\prod_j^n[\hat{C}_{\sigma_j}(\boldsymbol{k}_j)]^{n_j}\prod_j^n[\hat{C}_{\sigma_j}(\boldsymbol{k}_j)^\dagger]^{n_j}|\,0\rangle = \left(\sum_{j=1}^n n_j\right)!\left[\frac{V}{(2\pi)^3}\right]^{(\sum_j^n n_j)} \qquad (8.63)$$

and the normalized *ket* is

$$\frac{1}{\sqrt{\left(\sum\limits_{j=1}^n n_j\right)!\left[\frac{V}{(2\pi)^3}\right]^{(\sum_j^n n_j)}}}\prod_j^n[\hat{C}_{\sigma_j}(\boldsymbol{k}_j)^\dagger]^{n_j}|\,0\rangle. \qquad (8.64)$$

When a system is a linear combination of eigenstates of the number operator, it is often desirable to project out a particular state in the mixture. For this we use *projection operators*. If an operator $\hat{\mathbf{P}}$ is a projection operator, then $\mathbf{P}^2 = \mathbf{P}$. To illustrate, if we wish to project a particular 1-photon state from a mixture of states, the projection operator is

$$\hat{\mathbf{P}}_\sigma(\boldsymbol{k}) = \hat{C}_\sigma(\boldsymbol{k})^\dagger|\,0\rangle\frac{(2\pi)^3}{V}\langle 0|\hat{C}_\sigma(\boldsymbol{k}). \qquad (8.65)$$

The square of $\hat{\mathbf{P}}_\sigma(\boldsymbol{k})$ is

$$\begin{aligned}
[\hat{\mathbf{P}}_\sigma(\boldsymbol{k})]^2 &= \hat{C}_\sigma(\boldsymbol{k})^\dagger|\,0\rangle\frac{(2\pi)^3}{V}\langle 0|\hat{C}_\sigma(\boldsymbol{k})\hat{C}_\sigma(\boldsymbol{k})^\dagger|\,0\rangle\frac{(2\pi)^3}{V}\langle 0|\hat{C}_\sigma(\boldsymbol{k}) \\
&= \hat{C}_\sigma(\boldsymbol{k})^\dagger|\,0\rangle\frac{(2\pi)^3}{V}\left\langle 0\left|\frac{V}{(2\pi)^3}\right|0\right\rangle\frac{(2\pi)^3}{V}\langle 0|\hat{C}_\sigma(\boldsymbol{k}) \\
&= \hat{C}_\sigma(\boldsymbol{k})^\dagger|\,0\rangle\frac{(2\pi)^3}{V}\langle 0|\hat{C}_\sigma(\boldsymbol{k}) = \hat{\mathbf{P}}_\sigma(\boldsymbol{k}).
\end{aligned} \tag{8.66}$$

Let $\hat{\mathbf{P}}_\sigma(\boldsymbol{k})$ operate on a mixture of a 1-photon state and a 2-photon state:

$$\begin{aligned}
&\hat{\mathbf{P}}_\sigma(\boldsymbol{k})[\hat{C}_{\sigma_1}(\boldsymbol{k}_1)^\dagger + \hat{C}_{\sigma_2}(\boldsymbol{k}_2)^\dagger\hat{C}_{\sigma_3}(\boldsymbol{k}_3)^\dagger]|\,0\rangle \\
&= \hat{C}_\sigma(\boldsymbol{k})^\dagger|\,0\rangle\frac{(2\pi)^3}{V}\langle 0|\hat{C}_\sigma(\boldsymbol{k})[\hat{C}_{\sigma_1}(\boldsymbol{k}_1)^\dagger + \hat{C}_{\sigma_2}(\boldsymbol{k}_2)^\dagger\hat{C}_{\sigma_3}(\boldsymbol{k}_3)^\dagger]|\,0\rangle \\
&= \hat{C}_\sigma(\boldsymbol{k})^\dagger|\,0\rangle\frac{(2\pi)^3}{V}\langle 0|\delta_{\sigma\sigma_1}\delta(\boldsymbol{k}-\boldsymbol{k}_1|\,0\rangle.
\end{aligned} \tag{8.67}$$

If $\sigma \neq \sigma_1$ and $\boldsymbol{k} \neq \boldsymbol{k}_1$ the last line of Eq. (8.67) is zero, meaning that the state to be projected is not in the mixture. But if $\sigma_1 = \sigma$ and $\boldsymbol{k}_1 = \boldsymbol{k}$, we have $\delta(\boldsymbol{k}-\boldsymbol{k}) = V/(2\pi)^3$, giving

$$\hat{\mathbf{P}}_\sigma(\boldsymbol{k})[\hat{C}_{\sigma_1}(\boldsymbol{k}_1)^\dagger + \hat{C}_{\sigma_2}(\boldsymbol{k}_2)^\dagger\hat{C}_{\sigma_3}(\boldsymbol{k}_3)^\dagger]|\,0\rangle = \hat{C}_\sigma(\boldsymbol{k})^\dagger|\,0\rangle. \tag{8.68}$$

To see *how many* 1-photon states are in the mixture, we use

$$\hat{\mathbf{P}}_1 = \int d\boldsymbol{r}\sum_\sigma \hat{C}_\sigma(\boldsymbol{k})^\dagger|\,0\rangle\langle 0\,|\hat{C}_\sigma(\boldsymbol{k}) \tag{8.69}$$

to get:

$$\begin{aligned}
&\hat{\mathbf{P}}_1\left[\sum_{j=1}^n \hat{C}_{\sigma_j}(\boldsymbol{k}_j)^\dagger + \hat{C}_{\sigma_2}(\boldsymbol{k})_2^\dagger\hat{C}_{\sigma_3}(\boldsymbol{k}_3)^\dagger\right]\left|0\right\rangle \\
&= n\left[\sum_{j=1}^n \hat{C}_{\sigma_j}(\boldsymbol{k}_j)^\dagger + \hat{C}_{\sigma_2}(\boldsymbol{k}_2)^\dagger\hat{C}_{\sigma_3}(\boldsymbol{k}_3)^\dagger\right]\left|0\right\rangle.
\end{aligned} \tag{8.70}$$

To proceed to the photon states that describe intense light beams, we need to use Wick's theorem.

8.2.3. *Wick's Theorem*[1,2]

An important concept for application of Wick's theorem is that of normal ordering.

8.2.3.1. *Normal Order*

For a string of Fock operators for photon creation and annihilation, the commutation rules must be used to put the string into *normal order*; for example,

$$\hat{C}_j\hat{C}_k^\dagger\hat{C}_l\hat{C}_m^\dagger = \hat{C}_k^\dagger\hat{C}_m^\dagger\hat{C}_j\hat{C}_l + \delta_{jk}\hat{C}_m^\dagger\hat{C}_l + \delta_{lm}\hat{C}_k^\dagger\hat{C}_j + \delta_{jk}\delta_m, \quad (8.71)$$

where we have written $\hat{C}_j$ for $\hat{C}_{\sigma_j}(\boldsymbol{k}_j)$ and δ_{jl} for $\delta_{\sigma_j\sigma_l}\,\delta(\boldsymbol{k_j}-\boldsymbol{k_l})$. In the following we use $\hat{\Gamma}_j$ for either $\hat{C}_j$ or $\hat{C}_j^\dagger$, and $\mathbb{N}[\Gamma_1\cdots\Gamma_n]$ for the normal-ordered product of the operators $\hat{\Gamma}_1\cdots\hat{\Gamma}_n$. The commutators are represented as *contractions* $\hat{\Gamma}_j\cdots\hat{\Gamma}_l$, which are zero unless $\hat{\Gamma}_j$ is the annihilation operator $\hat{C}_j$ and $\hat{\Gamma}_l$ is the creation operator $\hat{C}_l^\dagger$, in which case the contraction gives the factor δ_{jl} outside the normal-ordered product ($\mathbb{N}$-product) of the un-contracted operators.

The reason normal ordering is necessary is that the vacuum will be annihilated if a string in a bracket has an annihilation operator to the right of a creation operator. In the next section, Wick's theorem is stated and illustrated by application to simple examples.

8.2.3.2. *Illustrations of the use of Wick's Theorem*

In this notation of the last section, Wick's theorem, which can be proved by induction, is expressed in general as follows:

---

[1]See footnote on page 56.

[2]See Sections 3.3 and 3.4 for applications of Wick's theorem to the harmonic oscillator.

$$\hat{\Gamma}_i \cdots \hat{\Gamma}_n = N[\hat{\Gamma}_1 \cdots \hat{\Gamma}_n] + \sum_{j \neq l} N[\hat{\Gamma}_1 \cdots \hat{\Gamma}_j \cdots \hat{\Gamma}_l \cdots \hat{\Gamma}_n]$$

$$+ \sum_{j \neq m} \sum_{l \neq p} N[\hat{\Gamma}_1 \cdots \hat{\Gamma}_j \cdots \hat{\Gamma}_l \cdots \hat{\Gamma}_m \cdots \hat{\Gamma}_p \cdots \hat{\Gamma}_q]$$

$$+ \sum_{\text{all triple contractions}} N[\hat{\Gamma}_1 \cdots \hat{\Gamma}_j \cdots \hat{\Gamma}_l \cdots \hat{\Gamma}_m \cdots \hat{\Gamma}_p \cdots \hat{\Gamma}_q \cdots \hat{\Gamma}_r \cdots \hat{\Gamma}_n]$$

$$+ \cdots \sum_{\text{maximally contracted terms}} N[\hat{\Gamma}_1 \cdots \hat{\Gamma}_n]. \tag{8.72}$$

The first term is the $\mathbb{N}$-product of the original string of operators without any contractions; the second term is the $\mathbb{N}$-product of the string with one pair of operators removed by a contraction, summed over all such contractions; the third term is the sum of all $\mathbb{N}$-products after removal of four operators by two contractions, *etc.*; the last term is the sum of all $\mathbb{N}$-ordered products after removal of the maximum number of operators by contraction of annihilation-creation pairs. Each of the terms includes the factors δ_{jl} as indicated by the contraction lines.

As an illustration, Eq. (8.71) has the form

$$\hat{C}_j \hat{C}_k^\dagger \hat{C}_l \hat{C}_m^\dagger = \mathbb{N}[\hat{C}_j \hat{C}_k^\dagger \hat{C}_l \hat{C}_m] + \mathbb{N}[\hat{C}_j \hat{C}_k^\dagger \hat{C}_l \hat{C}_m] + \mathbb{N}[\hat{C}_j \hat{C}_k^\dagger \hat{C}_l \hat{C}_m^\dagger]$$
$$+ \mathbb{N}[\hat{C}_j \hat{C}_k^\dagger \hat{C}_l \hat{C}_m^\dagger]. \tag{8.73}$$

Because each term in Eq. (8.73) is an $\mathbb{N}$-product, if any annihilation operators are present in the string, only the terms in which all of them are removed will give non-zero results when operating on the vacuum *ket* $|0\rangle$. Similarly, if any creation operators are present, only terms in which all of them are removed will give non-zero results when operating on the vacuum *bra* $\langle 0|$. Therefore, the vacuum expectation of photon operators, $\langle 0|\hat{\Gamma}_1 \cdots \hat{\Gamma}_n|0\rangle$, is the sum of the fully contracted terms in Eq. (8.72) in which no operators are left uncontracted. Vacuum expectations of a string containing an excess of either creation or annihilation operators is equal to zero. This result was used in deriving the expression for the Hamiltonian operator [see Eq. (8.54) and the discussion leading to it].

8.3. Coherent Photons

In an intense light source there is a very large number of photons. Since photons are quantum numbers of the electromagnetic field, an intense light source approaches the Bohr Correspondence Limit and therefore begins to exhibit classical behavior. We exploit the analogy between the electromagnetic field and a system of harmonic oscillators. On this point, the reader should look at Chapter 3 Sections 3.4 and 3.5 for a more thorough treatment than we will give here.

Glauber[3] used the set of wave packets that have minimal-uncertainty products for the oscillator coordinates and their conjugate momenta as a complete, orthogonal basis for representing the states of a quantized electromagnetic field. While any state of the field can be expanded on such a basis, as the correspondence limit is approached, the states are dominated by fewer of these basis members, just as wave packets accurately describe behavior of harmonic oscillators in the correspondence limit. Glauber called members of this wave-packet basis coherent states, since a field in such a state has factorable correlation functions and therefore produces coherent optical effects, such as interference fringes. For a single mode, the *ket* for an un-normalized coherent state has the form[4]

$$|\chi\rangle = e^{\chi \hat{C}^\dagger}|0\rangle,$$

where χ is a complex number and and $\hat{C}^\dagger$ is the creation operator for the mode. The *ket* $|\chi\rangle$ is an *eigenket* of the annihilation operator

[3]Glauber, R. J., *Phys. Rev.* **130**, 2529 (1963); Cahill, K. E. and Glauber, R. J., *Phys. Rev.* **177**, 1857 (1969).

[4]Louisell, W. H., *Quantum Statistical Properties of Radiation*, Wiley (1973).

for the mode, since

$$\hat{C}|\chi\rangle = \sum_{n=0}^{\infty}\frac{\chi^n}{n!}\hat{C}(\hat{C}^\dagger)^n|0\rangle = \sum_{n=0}^{\infty}\frac{\chi^n}{n!}[(\hat{C}^\dagger)^n\hat{C} + n(\hat{C}^\dagger)^{n-1}|0\rangle$$
$$= \chi\sum_{n=1}^{\infty}\frac{\chi^{n-1}}{(n-1)!}(\hat{C}^\dagger)^{n-1}|0\rangle = \chi|\chi\rangle, \tag{8.74}$$

where we have used the form of Wick's theorem applicable to a single mode. For s discrete modes, a coherent-state *ket* takes the form

$$|\chi^{(s)}\rangle = \prod_{j=1}^{s} e^{[\chi_j\hat{C}_{\sigma_j}(\boldsymbol{k}_j)^\dagger]}|0\rangle = \exp\left[\sum_{j=1}^{s}\chi_j\hat{C}_{\sigma_j}(\boldsymbol{k}_j)^\dagger\right]\Big|0\Big\rangle. \tag{8.75}$$

The normalization bracket for this multi-mode state is

$$\langle\chi^{(s)}|\chi^{(s)}\rangle = \langle 0|\prod_{j=1}^{s} e^{[\chi_j^*\hat{C}_{\sigma_j}(\boldsymbol{k}_j)]}\prod_{l=1}^{s} e^{[\chi_l\hat{C}_{\sigma_l}(\boldsymbol{k}_l)^\dagger]}|0\rangle$$
$$= \langle 0|\prod_{j=1}^{s}\sum_{n_j=0}^{\infty}\frac{1}{n_j!}[\chi_j^*\hat{C}_{\sigma_j}(\boldsymbol{k}_j)]^{n_j}\prod_{l=1}^{s}\sum_{n_l=0}^{\infty}\frac{1}{n_l!}$$
$$\times [\chi_l\hat{C}_{\sigma_l}(\boldsymbol{k}_l)^\dagger]^{n_l}|0\rangle. \tag{8.76}$$

Full contraction of the operator products in Eq. (8.76) is possible with terms for which $n_l = n_j$; Eq. (8.76) thus reduces to

$$\langle 0|\prod_{j=1}^{s}\sum_{n_j=0}^{\infty}\frac{1}{(n_j!)^2}|\chi_j|^{2n_j}[\hat{C}_{\sigma_j}(\boldsymbol{k}_j)]^{n_j}$$
$$\times [\hat{C}_{\sigma_j}(\boldsymbol{k}_j)^\dagger]^{n_j}|0\rangle, \tag{8.77}$$

where we have used the fact that $\hat{C}_{\sigma_j}(\boldsymbol{k}_j)$ and $\hat{C}_{\sigma_l}(\boldsymbol{k}_l)^\dagger$ commute for $l \neq j$. There are $n_j!$ ways to contract all of the $\hat{C}_{\sigma_j}(\boldsymbol{k}_j)$'s with the $\hat{C}_{\sigma_j}(\boldsymbol{k}_j)^\dagger$'s, each contraction contributing the factor unity; the result is therefore

$$\prod_{j=1}^{s}\sum_{n_j=0}^{\infty}\frac{1}{n_j!}|\chi_j|^{2n_j} = \exp\left(\sum_{j=1}^{s}|\chi_j|^2\right). \tag{8.78}$$

From Eqs. (8.75) and (8.78), the normalized multi-mode coherent-state *ket* is

$$|\chi^{(s)}\rangle = \prod_{j=1}^{s}\exp\left[\chi_j\hat{C}_{\sigma_j}(\boldsymbol{k}_j)^\dagger - \frac{1}{2}|\chi_j|^2\right]|0\rangle. \tag{8.79}$$

To find the physical interpretation of the eigenvalues χ_j of the operators $\hat{C}^{\sigma_j}(\boldsymbol{k}_j)^\dagger$, we calculate the average number of photons in the mode j in a state represented by the *ket* of Eq. (8.79), using the fact that a coherent density operator for a system of photons in which N_j of them are in mode j, $\hat{\rho}_c(N_j)$, is an eigenstate of $\hat{C}_{\sigma_j}(\boldsymbol{k}_j)$ and $\hat{C}_{\sigma_j}(\boldsymbol{k}_j)^\dagger$ in the sense that

$$\begin{aligned}\hat{C}_{\sigma_j}(\boldsymbol{k}_j)\hat{\rho}_c[N_\sigma(\boldsymbol{k})] &= \hat{\rho}_c[N_\sigma(\boldsymbol{k})]\hat{C}_{\sigma_j}(\boldsymbol{k}_j)^\dagger \\ &= [N_{\sigma_j}(\boldsymbol{k}_j)]^{\frac{1}{2}}\hat{\rho}_c[N_\sigma(\boldsymbol{k})].\end{aligned} \tag{8.80}$$

The result is

$$N_j = \langle\chi_j|\hat{C}_{\sigma_l}(\boldsymbol{k}_l)^\dagger\hat{C}_{\sigma_j}(\boldsymbol{k}_j)|\chi_j\rangle = \langle\chi_j|\chi_j\chi_j^*|\chi_j\rangle = |\chi_j|^2. \tag{8.81}$$

Since χ_j is complex, we can write it as the product of an amplitude and a phase factor:

$$\chi_{\sigma_j}(\boldsymbol{k}_j) = [N_{\sigma_j}(\boldsymbol{k}_j)]^{\frac{1}{2}}e^{i\varphi_{\sigma_j\boldsymbol{k}_j})}, \tag{8.82}$$

where we have written out the mode characteristics explicitly. If the phases $\varphi_{\sigma_j}(\boldsymbol{k}_j)$ are single-valued functions of σ_j, $\boldsymbol{k}_j$, the phase factor in Eq. (8.82) may be incorporated into the operators $\hat{C}_{\sigma_j}(\boldsymbol{k}_j)^\dagger$ and $\hat{C}_{\sigma_j}(\boldsymbol{k}_j)$ without any compromise of physical completeness, since we evaluate brackets by contracting annihilation-creation operator pairs for the same $\sigma_j, \boldsymbol{k}_j$ in which case the phase factors cancel. It is only when $\varphi_{\sigma_j}(\boldsymbol{k}_j)$ is not uniquely determined by the mode label (for example, in a thermal light source) that the phases play a physical role. For coherent light sources (such as lasers), we take photons of the same mode to have the same phases and drop the phase factors.

The total average number of photons is given by

$$N_p = \sum_{j=1}^{s} N_{\sigma_j}(\boldsymbol{k}_j) = \sum_{j=1}^{s} |\chi_j(\boldsymbol{k}_j)|^2. \tag{8.83}$$

The normalized density operator for a coherent state of the electromagnetic field having photons in s distinct modes is seen from Eqs. (8.81), (8.82), and (8.83) to be

$$\hat{\rho}_{\mathrm{c}} = \exp\left\{\sum_{j=1}^{s}[N_{\sigma_j}(\boldsymbol{k}_j)]^{\frac{1}{2}}\hat{C}_{\sigma_j}(\boldsymbol{k}_j)^{\dagger}\right\}|\,0\rangle e^{-N_p}\langle 0|$$
$$\times \exp\left\{\sum_{l=1}^{s}[N_{\sigma_l}(\boldsymbol{k}_l)]^{\frac{1}{2}}\,\hat{C}_{\sigma_l}(\boldsymbol{k}_l)\right\}\Bigg|, \tag{8.84}$$

where we have replaced $\chi_j(\boldsymbol{k}_j)$ by the real number $[N_{\sigma_j}(\boldsymbol{k}_j)]^{\frac{1}{2}}$, namely the square root of the average number of photons in mode j. It is easily confirmed that $\mathrm{Tr}\,\hat{\rho}_{\mathrm{c}} = 1$.

For a coherent state, the number of photons is not fixed, but has a Poisson distribution. The probability that there are n_j photons in mode j is given by the trace of the product of the density operator $\hat{\rho}_{\mathrm{c}}(N_j)$ with the projection operator

$$\hat{\mathbf{P}}_j^{(n_j)} = \frac{1}{n_j!}[\hat{C}_{\sigma_j}(\boldsymbol{k}_j)^{\dagger}]^{n_j}|0\rangle\langle 0|[\hat{C}_{\sigma_j}(\boldsymbol{k}_j)]^{n_j}. \tag{8.85}$$

From Eqs. (8.84) and (8.85) the result is

$$\mathrm{Tr}[\hat{\mathbf{P}}_j^{(n_j)}\hat{\rho}_{\mathrm{c}}] = \frac{1}{n_j!}\langle 0|[\hat{C}_{\sigma_j}(\boldsymbol{k}_j)]^{n_j}\hat{\rho}_{\mathrm{c}}[\hat{C}_{\sigma_j}(\boldsymbol{k}_j)^{\dagger}]^{n_j}|0, \tag{8.86}$$

where we have used the invariance of the trace to cyclic permutations of operators. From Eqs. (8.80) and (8.86) we obtain

$$\mathrm{Tr}[\hat{\mathbf{P}}_j^{(n_j)}\hat{\rho}_{\mathrm{c}}] = \frac{1}{n_j!}[N_{\sigma_j}(\boldsymbol{k}_j)]^{n_j}e^{-N_p}. \tag{8.87}$$

In deriving Eq. (8.87) we have used the fact that only the leading terms in the expansion of the exponentials in Eq. (8.84) contribute to the bracket in Eq. (8.86) after Eq. (8.80) is applied. If we ask for the probability of finding n_j photons in mode j and n_l photons in mode l, we use the projection operator

$$\begin{aligned}\hat{\mathbf{P}}_{jl}^{(n_j,n_l)} &= \frac{1}{n_j!n_l!}[\hat{C}_{\sigma_j}(\boldsymbol{k}_j)^\dagger]^{n_j}[\hat{C}_{\sigma_l}(\boldsymbol{k}_l)^\dagger]^{n_l} \\ &\quad \times |0\rangle\langle 0|[\hat{C}_{\sigma_j}(\boldsymbol{k}_j)]^{n_j}[\hat{C}_{\sigma_l}(\boldsymbol{k}_l)]^{n_l} \end{aligned} \tag{8.88}$$

and obtain

$$\mathrm{Tr}[\hat{\mathbf{P}}_{jl}^{(n_j n_l)}\hat{\rho}_{\mathrm{c}}] = \frac{1}{n_j!}[N_{\sigma_j}(\boldsymbol{k}_j)]^{n_j}\frac{1}{n_l!}[N_{\sigma l}(\boldsymbol{k}_l)]^{n_l}e^{-N_p}. \tag{8.89}$$

To find the probability $\mathcal{P}(n_j)$ that there are n_j photons in mode j, regardless of the number of photons in the other modes, we must take the product of the r.h.s. of Eq. (8.88) over all $l \neq j$ and sum over all n_l to get

$$\begin{aligned}\mathcal{P}(n_j) &= \frac{1}{n_j!}[N_{\sigma_j}(\boldsymbol{k}_j)]^{n_j}\prod_{l\neq j}\sum_{n_l=0}^{\infty}\frac{1}{n_l!}[N_{\sigma_l}(\boldsymbol{k}_l)]^{n_l}e^{-N_p} \\ &= \frac{1}{n_j!}[N_{\sigma_j}(\boldsymbol{k}_j)]^{n_j}\prod_{lnej}e^{[N_{\sigma_l}(\boldsymbol{k}_l)]}e^{-N_p} \\ &= \frac{1}{n_j!}[N_{\sigma_j}(\boldsymbol{k}_j)]^{n_j}e^{[\sum_{l\neq j}N_{\sigma_l}(\boldsymbol{k}_l)-N_p]} = \frac{1}{n_j!}[N_{\sigma_j}(\boldsymbol{k}_j)]^{n_j}e^{-n_j}.\end{aligned} \tag{8.90}$$

This is the Poisson distribution for the number of photons in mode j.

We also note that $\hat{\rho}_{\mathrm{c}}$ is a projection operator. To see this, we evaluate

$$\begin{aligned}\hat{\rho}_{\mathrm{c}}^2 = \exp&\left\{\sum_{j=1}^{s}[N_{\sigma_j}(\boldsymbol{k}_j)]^{\frac{1}{2}}\hat{C}_{\sigma_j}(\boldsymbol{k}_j)^\dagger\right\} \\ &\times |0\rangle\langle 0|\exp\left\{\sum_{l=1}^{s}[N_{\sigma_l}(\boldsymbol{k}_l)]^{\frac{1}{2}}\hat{C}_{\sigma_l}(\boldsymbol{k})\right\}\hat{\rho}_{\mathrm{c}}.\end{aligned} \tag{8.91}$$

From Eq. (8.80), we obtain

$$\exp\left\{\sum_{l=1}^{s}[N_{\sigma_l}(\boldsymbol{k}_l)]^{\frac{1}{2}}\hat{C}_{\sigma_l}(\boldsymbol{k}_l)\right\}\hat{\rho}_{\rm c}$$

$$= \exp\left\{\sum_{l=1}^{s}[N_{\sigma_l}(\boldsymbol{k}_l)]^{\frac{1}{2}}\,[N_{\sigma_l}(\boldsymbol{k}_l)]^{\frac{1}{2}}\right\}\hat{\rho}_{\rm c} = e^{-N_p}\hat{\rho}_{\rm c}. \quad (8.92)$$

Inserting Eq. (8.92) into Eq. (8.91) gives

$$\hat{\rho}_{\rm c}^2 = \exp\left\{\sum_{j=1}^{s}[N_{\sigma_j}(\boldsymbol{k}_j)]^{\frac{1}{2}}\hat{C}_{\sigma_j}(\boldsymbol{k}_j)^{\dagger}\right\}|\,0\rangle\langle 0\,|\,\hat{\rho}_{\rm c} = \hat{\rho}_{\rm c}. \quad (8.93)$$

In reaching Eq. (8.93), we have used the fact that

$$\langle 0\,|\hat{\rho}_c = e^{-N_p}\langle 0\,|\exp\left\{\sum_{l=1}^{s}[N_{\sigma_l}(\boldsymbol{k}_l)]^{\frac{1}{2}}\hat{C}_{\sigma_l}(\boldsymbol{k}_l)\right\}. \quad (8.94)$$

Note also that Eq. (8.94) and its Hermitian conjugate lead to

$$\mathrm{Tr}\hat{\rho}_c = \langle 0\,|\,\hat{\rho}_{\rm c}|\,0\rangle = \langle 0|0\rangle = 1. \quad (8.95)$$

Thus, coherent photons in a single mode $\boldsymbol{k}\sigma$ are seen to be in pure states, which are defined by

$$\mathrm{Tr}\,\hat{\rho}^2 = \mathrm{Tr}\,\hat{\rho} = 1. \quad (8.96)$$

8.4. Ensemble Operator and Fields for a Coherent Laser Pulse

The intensity of a laser beam is concentrated about the beam axis, but the beam has a finite width and its frequency range is also narrow, but finite. We treat the system of photons in a laser pulse in the idealized limit of stability of the phase and amplitude of the source during the emission of the pulses. To obtain an ensemble operator that describes a pulse in such a beam, we need to allow the wave vector to be a continuous variable.

8.4.1. *Generalization to Continuous Values of k*

We generalize the treatment of coherent photons in Section 8.3 to a field with continuously variable wave vector $\boldsymbol{k}$. In particular, we replace the sums over j in Eq. (8.75), and following, with an integral over $\boldsymbol{k}$ and a sum over the two values of σ and replace $N_\sigma(\boldsymbol{k})$ with $\rho_\sigma{}^{\frac{1}{2}}(\boldsymbol{k})$; thus the ensemble operator in Eq. (8.84) becomes

$$\hat{\rho}_{\mathrm{c}} = \exp\left\{\int d\boldsymbol{k} \sum_{\sigma} \rho_\sigma{}^{\frac{1}{2}}(\boldsymbol{k})\hat{C}_\sigma(\boldsymbol{k})^\dagger\right\}$$

$$\times |0\rangle e^{-N_p}\langle 0| \exp\left\{\int d\boldsymbol{k}' \sum_{\sigma'} \rho_{\sigma'}{}^{\frac{1}{2}}(\boldsymbol{k}')\hat{C}_{\sigma'}(\boldsymbol{k}')\right\}. \qquad (8.97)$$

In writing this form for the ensemble operator for a laser pulse, we have assumed that the phase $\varphi_\sigma(\boldsymbol{k})$ associated with a given mode $\boldsymbol{k}\sigma$ is a single-valued function of $\boldsymbol{k}$ and σ and have dropped the phase factors in accordance with the discussion following Eq. (8.81). The distribution function $\rho_\sigma{}^{\frac{1}{2}}(\boldsymbol{k})$ is therefore taken to be real-valued. To obtain its physical interpretation, we calculate the trace of the product of $\hat{\rho}_{\mathrm{c}}$ with the number operator $\hat{\mathbf{N}}_\sigma(\boldsymbol{k})$ defined in Eq. (8.56a):

$$\mathrm{Tr}[\hat{\mathbf{N}}_\sigma(\boldsymbol{k})\hat{\rho}_{\mathrm{c}}] = \langle 0| \exp\left[\int d\boldsymbol{k} \sum_{\sigma} \rho_\sigma{}^{\frac{1}{2}}(\boldsymbol{k}')\hat{C}_\sigma(\boldsymbol{k})\right] \hat{C}_\sigma(\boldsymbol{k})^\dagger \hat{C}_\sigma(\boldsymbol{k})$$

$$\times \exp\left[\int d\boldsymbol{k}' \sum_{\sigma'} \rho_{\sigma'}{}^{\frac{1}{2}}(\boldsymbol{k}')\hat{C}_{\sigma'}(\boldsymbol{k}')^\dagger\right] |0\rangle e^{-N_p}. \qquad (8.98)$$

Eq. (8.98) is evaluated by expanding the right-most exponential; *i.e.*,

$$\hat{C}_\sigma(\boldsymbol{k}) e^{\int d\boldsymbol{k}' \sum_{\sigma'} \rho_{\sigma'}{}^{\frac{1}{2}}(\boldsymbol{k}')\hat{C}_{\sigma'}(\boldsymbol{k}')^\dagger} |0\rangle$$

$$= \hat{C}_\sigma(\boldsymbol{k}) \sum_{n=1}^{\infty} \frac{1}{n!} \left[\int d\boldsymbol{k}' \sum_{\sigma'} \rho_{\sigma'}{}^{\frac{1}{2}}(\boldsymbol{k}')\hat{C}_{\sigma'}(\boldsymbol{k}')^\dagger\right]^n |0\rangle$$

$$= \rho_\sigma{}^{\frac{1}{2}}(\boldsymbol{k}) \sum_{n=1}^{\infty} \frac{n}{n!} \left[\int d\boldsymbol{k}' \sum_{\sigma'} \rho_{\sigma'}{}^{\frac{1}{2}}(\boldsymbol{k}') \hat{C}_{\sigma'}(\boldsymbol{k}')^\dagger \right]^{n-1} |\,0\rangle$$

$$= \rho_\sigma{}^{\frac{1}{2}}(\boldsymbol{k}) \sum_{n=0}^{\infty} \frac{1}{n!} \left[\int d\boldsymbol{k}' \sum_{\sigma'} \rho_{\sigma'}{}^{\frac{1}{2}}(\boldsymbol{k}') \hat{C}_{\sigma'}(\boldsymbol{k}')^\dagger \right]^{n} |\,0\rangle$$

$$= \rho_\sigma{}^{\frac{1}{2}}(\boldsymbol{k}) e^{\int d\boldsymbol{k}' \sum_{\sigma'} \rho_\sigma{}^{\frac{1}{2}}(\boldsymbol{k}') \hat{C}_{\sigma'}(\boldsymbol{k})^\dagger} |\,0\rangle. \tag{8.99}$$

We have omitted the $n = 0$ term in line one of Eq. (8.99) for the reason $\hat{C}_\sigma(\boldsymbol{k})|\,0\rangle = 0$ and in line two we have used the fact that there are n ways of paring with $\hat{C}_\sigma(\boldsymbol{k})$ with a $\hat{C}_{\sigma'}(\boldsymbol{k}')^\dagger$.

Left-multiplying Eq. (8.99) by another annihilation operator gives

$$\hat{C}_\sigma(\boldsymbol{k}) \hat{C}_{\sigma'}(\boldsymbol{k}') e^{\int d\boldsymbol{k}'' \sum_{\sigma''} \rho_{\sigma''}{}^{\frac{1}{2}}(\boldsymbol{k}'') \hat{C}_{\sigma''}(\boldsymbol{k}'')^\dagger} |\,0\rangle$$
$$= \rho_\sigma(\boldsymbol{k})^{\frac{1}{2}} \rho_{\sigma'}(\boldsymbol{k}')^{\frac{1}{2}} e^{\int d\boldsymbol{k}'' \sum_{\sigma''} \rho_{\sigma''}{}^{\frac{1}{2}}(\boldsymbol{k}'') \hat{C}_{\sigma''}(\boldsymbol{k}'')^\dagger} |\,0\rangle \tag{8.100}$$

The Hermitian conjugates of Eqs. (8.99) and (8.100) are

$$\langle 0\,|\, e^{\int d\boldsymbol{k}' \sum_{\sigma'} \rho_{\sigma'}{}^{\frac{1}{2}}(\boldsymbol{k}') \hat{C}_{\sigma'}(\boldsymbol{k}')} \hat{C}_\sigma(\boldsymbol{k})^\dagger$$
$$= \langle 0\,|\, e^{\int d\boldsymbol{k}' \sum_{\sigma'} \rho_{\sigma'}{}^{\frac{1}{2}}(\boldsymbol{k}') \hat{C}_{\sigma'}(\boldsymbol{k}')} \rho_\sigma{}^{\frac{1}{2}}(\boldsymbol{k}) \tag{8.101}$$

and

$$\langle 0\,|\, e^{\int d\boldsymbol{k}'' \sum_{\sigma''} \rho_{\sigma''}{}^{\frac{1}{2}}(\boldsymbol{k}'') \hat{C}_{\sigma''}(\boldsymbol{k}'')} \hat{C}_\sigma(\boldsymbol{k}) \hat{C}_{\sigma'}(\boldsymbol{k}')^\dagger$$
$$= \langle 0\,|e^{\int d\boldsymbol{k}'' \sum_{\sigma''} \rho_{\sigma''}{}^{\frac{1}{2}}(\boldsymbol{k}'') \hat{C}_{\sigma''}(\boldsymbol{k}'')} \rho_\sigma{}^{\frac{1}{2}}(\boldsymbol{k}) \rho_\sigma{}^{\frac{1}{2}}(\boldsymbol{k}'). \tag{8.102}$$

As in the case with discrete values of $\boldsymbol{k}$, the coherent photon ensemble operator $\hat{\rho}_c$ is shown to be an eigenstate of $\hat{C}_\sigma(\boldsymbol{k})$ and $\hat{C}_\sigma(\boldsymbol{k})^\dagger$ in the sense of Eq. (8.80).

The average number density for photons in a given mode $\boldsymbol{k}$, σ becomes

$$\mathrm{Tr}[\hat{C}_\sigma(\boldsymbol{k}) \hat{C}_\sigma(\boldsymbol{k})^\dagger \hat{\rho}_c]/\mathrm{Tr}\hat{\rho}_c = \mathrm{Tr}[\hat{C}_\sigma(\boldsymbol{k}) \hat{\rho}_c \hat{C}_\sigma(\boldsymbol{k})^\dagger] = \rho_\sigma(\boldsymbol{k}). \tag{8.103}$$

with the aid of Eqs. (8.99) and (8.102) and the cyclic property of the trace. The total average number of photons is thus

$$N_p = \int dk \sum_\sigma \rho_\sigma(k). \qquad (8.104)$$

The normalization of $\hat{\rho}_c$ is verified by the use of Eqs. (8.99)–(8.103):

$$\begin{aligned}
\mathrm{Tr}\,\hat{\rho}_c &= \langle 0\,| e^{\int dk \sum_\sigma \rho_\sigma^{\frac{1}{2}}(k)\hat{C}_\sigma(k)}\, e^{\int dk' \sum_{\sigma'} \rho_{\sigma'}^{\frac{1}{2}}(k')\hat{C}_{\sigma'}(k')^\dagger} |0\rangle e^{-N_p} \\
&= \langle 0\,| e^{\int dk \sum_\sigma \rho_\sigma^{\frac{1}{2}}(k)\rho_\sigma^{\frac{1}{2}}(k)}\, e^{\int dk' \sum_{\sigma'} \rho_{\sigma'}^{\frac{1}{2}}(k')\hat{C}_{\sigma'}(k')^\dagger} \,|\,0\rangle e^{-N_p} \\
&= e^{\int dk \sum_\sigma \rho_\sigma^{\frac{1}{2}}(k)\rho_\sigma^{\frac{1}{2}}(k)} \langle 0 \,|\, e^{\int dk' \sum_{\sigma'} \rho_{\sigma'}^{\frac{1}{2}}(k')\hat{C}'_\sigma(k')^\dagger} |0\rangle e^{-N_p} \\
&= e^{\int dk \sum_\sigma \rho_\sigma(k)} e^{-N_p} = e^{N_p}\, e^{-N_p} = 1. \qquad (8.105)
\end{aligned}$$

The distribution $\rho_\sigma^{\frac{1}{2}}(k)$ is seen from Eq. (8.104) to be the square root of the number density in k-space for coherent photons.

8.4.2. *Poisson Distribution for Photons in a Continuum of Wave Vectors*

The extension of Eq. (8.90) to the case in which k is continuously variable requires that we calculate the probability for observing n photons in the finite cell Δk centered at k and with polarization σ, regardless of the number of photons in other modes. The projection operator in Eq. (8.85) becomes

$$\begin{aligned}
\hat{P}_\sigma^{(n)}(k, \Delta k) = \frac{1}{n!} \int_{\Delta k} dk_1 \cdots \int_{\Delta k} dk_n \\
\times \prod_{j=1}^{n} [\hat{C}_\sigma(k_{(n+1-j)}))^\dagger \,|\,0\rangle\langle 0\,| \prod_{l=1}^{n} [\hat{C}_\sigma(k_l). \qquad (8.106)
\end{aligned}$$

Here we have been careful to nest the operators so that $\hat{C}_{\sigma_1}(k_1)^\dagger$ and $\hat{C}_{\sigma_1}(k_1)$ are next to the vacuum projection, $\hat{C}_{\sigma_2}(k)_2)^\dagger$ and $\hat{C}_{\sigma_2}(k_2)$ are next to them, and so on. We now show that $\hat{P}_\sigma^{(n)}$ is a projection operator; *i.e.*, that it is equal to its square. The square of Eq. (8.106) is

$$[\hat{P}_\sigma^{(n)}(\boldsymbol{k},\Delta\boldsymbol{k})]^2 = \frac{1}{(n!)^2}\int_{\Delta\boldsymbol{k}} d\boldsymbol{k}_1 \cdots \int_{\Delta\boldsymbol{k}} d\boldsymbol{k}_n \prod_{j=1}^{n}[\hat{C}_\sigma(\boldsymbol{k}_{n+1-j})^\dagger]|\,0\rangle$$

$$\times \langle 0\,| \prod_{l'=1}^{n}[\hat{C}_\sigma(\boldsymbol{k}_l)] \times \int_{\Delta\boldsymbol{k}} d\boldsymbol{k}'_1 \cdots \int_{\Delta\boldsymbol{k}} d\boldsymbol{k}'_n \prod_{j'=1}^{n}$$

$$\times [\hat{C}_\sigma(\boldsymbol{k}'_{n+1-j'})^\dagger\,|\,0\rangle\langle 0\,| \prod_{l'=1}^{n}[\hat{C}_\sigma(\boldsymbol{k}'_{l'})]. \qquad (8.107)$$

There are $n!$ ways to fully contract the operators appearing in the bracket in Eq. (8.107), giving a string of functions that become unity when the integrals are carried out. The result is

$$[\hat{P}_\sigma^{(n)}(\boldsymbol{k},\Delta\boldsymbol{k})]^2 = \frac{1}{n!}\int_{\Delta\boldsymbol{k}} d\boldsymbol{k}_1 \cdots \int_{\Delta\boldsymbol{k}} d\boldsymbol{k}_n$$

$$\times \prod_{j=1}^{n}[\hat{C}_\sigma(\boldsymbol{k}^\dagger_{(n+1-j)}]|\,0\rangle\langle 0| \prod_{l=1}^{n}[\hat{C}_\sigma(\boldsymbol{k}_l)] = \hat{P}_\sigma^{(n)}. \qquad (8.108)$$

The reason for nesting the operators in the order prescribed is not apparent in the boson case, but for fermions, the nested form assures that the negative signs attached to some of the fully contracted terms will be removed when the product of annihilation operators to the right of the right-most vacuum *bra* is unscrambled into nested order.[5] Thus this nested form for the projection operators can be used for both fermions and bosons.

The probability for finding n photons in $\Delta\boldsymbol{k}$ at $\boldsymbol{k}$ and polarization σ but no photons in other modes in $\Delta\boldsymbol{k}$ at $\boldsymbol{k}$ is found by evaluating

$$\mathrm{Tr}[P_\sigma^{(n)}(\boldsymbol{k},\Delta\boldsymbol{k})\hat{\rho}_c] = \frac{1}{n!}\int_{\Delta\boldsymbol{k}} d\boldsymbol{k}_1 \cdots \int_{\Delta\boldsymbol{k}} d\boldsymbol{k}_n$$

$$\times \mathrm{Tr}\left\{\prod_{j=1}^{n}[\hat{C}_\sigma(\boldsymbol{k}_{(n+1-j)})^\dagger]|\,0\rangle\langle 0\,| \prod_{l=1}^{n}[\hat{C}_\sigma(\boldsymbol{k}_l)]\hat{\rho}_c\right\}$$

[5]See Chapter 6, for example.

$$= \frac{1}{n!}\int_{\Delta \boldsymbol{k}} d\boldsymbol{k}_1 \cdots \int_{\Delta \boldsymbol{k}} d\boldsymbol{k}_n \mathrm{Tr}\left\{\langle 0| \prod_{l=1}^{n}[\hat{C}_\sigma(\boldsymbol{k}_l)]\hat{\rho}_c \prod_{j=1}^{n}[\hat{C}_\sigma(\boldsymbol{k}_{n+1-j})^\dagger]|0\rangle\right\}$$

$$= \frac{1}{n!}\left[\int_{\Delta \boldsymbol{k}} d\boldsymbol{k}\rho_\sigma(\boldsymbol{k})\right]^n e^{-N_p}. \tag{8.109}$$

A projection operator for no photons in the $\boldsymbol{k}$-space complementary to the small volume $\Delta\boldsymbol{k}$ at $\boldsymbol{k}$ is similarly defined. The projection operator for the joint $\Delta\boldsymbol{k}$ at $\boldsymbol{k}$ probability of n photons in $\Delta\boldsymbol{k}$ at $\boldsymbol{k}$ with polarization σ and n_c photons in the complementary $\boldsymbol{k}$-space $\Delta_c\boldsymbol{k}$ with either polarization is

$$\hat{P}_\sigma^{(n,n_c)}(\boldsymbol{k}, \Delta_c\boldsymbol{k}) = \frac{1}{n!}\frac{1}{n_c!}\int_{\Delta \boldsymbol{k}} d\boldsymbol{k}_1 \cdots \int_{\Delta \boldsymbol{k}} d\boldsymbol{k}_n$$

$$\times \int_{\Delta_c \boldsymbol{k}} d\boldsymbol{k}'_1 \cdots \int_{\Delta_c \boldsymbol{k}} d\boldsymbol{k}'_n \sum_{\sigma'} \prod_{j'=1}^{n_c} [\hat{C}_{\sigma'}(\boldsymbol{k}'_{n+1-j'})^\dagger]$$

$$\times \prod_{j=1}^{n_c}[\hat{C}_\sigma(\boldsymbol{k}_{n+1-j})^\dagger]|\,0\rangle\langle 0\,| \prod_{l=1}^{n}[\hat{C}_\sigma(\boldsymbol{k}_l)] \prod_{l'=1}^{c}[\hat{C}_{\sigma'}(\boldsymbol{k}'_l)]. \tag{8.110}$$

The trace of the product of $\hat{P}_\sigma^{(n,n_c)}(\boldsymbol{k}, \Delta_c\boldsymbol{k})$ with $\hat{\rho}_c$ can be evaluated with the help of Eq. (8.109). The result is

$$\mathrm{Tr}[\hat{P}_\sigma^{(n,n_c)}(\boldsymbol{k}, \Delta_c\boldsymbol{k})\hat{\rho}_c]$$

$$= \frac{1}{n!}\left[\int_{\Delta \boldsymbol{k}} d\boldsymbol{k}\rho_\sigma(\boldsymbol{k})\right]^n \frac{1}{n_c!}\left[\int_{\Delta \boldsymbol{k}_c} d\boldsymbol{k}\rho_\sigma(\boldsymbol{k})\right]^{n_c} e^{-N_p}. \tag{8.111}$$

To find the probability $\mathcal{P}_\sigma^{(n)}(\boldsymbol{k}, \Delta\boldsymbol{k})$ of observing n photons in $\Delta\boldsymbol{k}$ at $\boldsymbol{k}$ with polarization σ regardless of the number of photons in $\Delta_c\boldsymbol{k}$ we sum Eq. (8.111) over n_c and get

$$\mathcal{P}_\sigma^{(n)}(\boldsymbol{k}, \Delta\boldsymbol{k}) = \sum_{n_c=0}^{\infty} \mathrm{Tr}[\hat{P}_\sigma^{(n,n_c)}(\boldsymbol{k}, \Delta_c\boldsymbol{k})\hat{\rho}_c]$$

$$= \frac{1}{n!}\left[\int_{\Delta \boldsymbol{k}} d\boldsymbol{k}\rho_\sigma(\boldsymbol{k})\right]^n \sum_{n_c=0}^{\infty} \frac{1}{n_c!}\left[\int_{\Delta_c \boldsymbol{k}'} d\boldsymbol{k}'\rho_{\sigma'}(\boldsymbol{k}')\right]^{n_c} e^{-N_p}$$

$$= \frac{1}{n!}\left[\int_{\Delta \boldsymbol{k}} d\boldsymbol{k}\rho_\sigma(\boldsymbol{k})\right]^n \exp\left[\int_{\Delta_c \boldsymbol{k}'} d\boldsymbol{k}'\rho_{\sigma'}(\boldsymbol{k}')\right] e^{-N_p}$$
$$= \frac{1}{n!}\left[\int_{\Delta \boldsymbol{k}} d\boldsymbol{k}\rho_\sigma(\boldsymbol{k})\right]^n \exp\left[\int_{\Delta_c \boldsymbol{k}'} d\boldsymbol{k}'\rho_{\sigma'}(\boldsymbol{k}') - N_p\right]. \quad (8.112)$$

But Eq. (8.104), modified for integrals extending over $\Delta\boldsymbol{k}$ and $\Delta_c\boldsymbol{k}$, shows that

$$\int_{\Delta \boldsymbol{k}} d\boldsymbol{k}\rho_\sigma(\boldsymbol{k}) = N_\sigma^{(\boldsymbol{k},\Delta\boldsymbol{k})}; \int_{\Delta_c \boldsymbol{k}'} \rho_{\sigma'}(\boldsymbol{k}') = N_p - N_\sigma^{(\boldsymbol{k},\Delta\boldsymbol{k})} \quad (8.113)$$

giving for Eq. (8.112)

$$\mathcal{P}_\sigma^{(n)}(\boldsymbol{k},\Delta\boldsymbol{k}) = \frac{1}{n!}[N_\sigma^{(\boldsymbol{k},\Delta\boldsymbol{k})}]^n \exp[-N_\sigma^{\boldsymbol{k}\cdot\Delta\boldsymbol{k}}], \quad (8.114)$$

which is the Poisson distribution.

Although some care must be exerted in defining the appropriate projection operators when $\boldsymbol{k}$ is allowed to vary continuously, the result for the distribution of coherent photons among the various modes is consistent with that for discrete modes.

8.4.3. *Average Fields for a Coherent Source*

To find appropriate forms for $\rho_\sigma(\boldsymbol{k})$ so that $\hat{\rho}_c$ represents a coherent laser pulse, we make use of the measurable field averages. We derive the average fields in terms of $\rho_\sigma(\boldsymbol{k})$. The average vector potential is

$$\mathrm{Tr}[\hat{\mathbf{A}}(\boldsymbol{r},t)\hat{\rho}_c] = \mathrm{Tr}\left\{\left(\frac{\hbar c}{16\pi^3}\right)^{\frac{1}{2}} \int \frac{d\boldsymbol{k}}{\sqrt{k}} \times \sum_{\sigma=1}^{2}[\hat{C}_\sigma(\boldsymbol{k})e^{i(\boldsymbol{k}\cdot\boldsymbol{r}-\omega t)} + \hat{C}_\sigma(\boldsymbol{k})^\dagger e^{-i(\boldsymbol{k}\cdot\boldsymbol{r}-\omega t)}]\boldsymbol{e}_\sigma\hat{\rho}_c\right\}$$
$$= \left(\frac{\hbar c}{4\pi^3}\right)^{\frac{1}{2}} \int \frac{d\boldsymbol{k}}{\sqrt{k}} \sum_{\sigma=1}^{2} \rho_\sigma^{\frac{1}{2}}(\boldsymbol{k}) \cos(\boldsymbol{k}\cdot\boldsymbol{r} - \omega t)\boldsymbol{e}\sigma. \quad (8.115)$$

Similarly, we have from Eqs. (8.42) and (8.43)

$$\mathrm{Tr}[\hat{\mathbf{E}}(\boldsymbol{r},t)\hat{\rho}_{\mathrm{c}}] = -\frac{c}{2\pi^{\frac{3}{2}}}\int d\boldsymbol{k}\sum_{\sigma}(\hbar ck)^{\frac{1}{2}}\rho_{\sigma}^{\frac{1}{2}}(\boldsymbol{k})\sin(\boldsymbol{k}\cdot r-\omega t)\boldsymbol{e}_{\sigma} \tag{8.116}$$

$$\mathrm{Tr}[\hat{\mathbf{B}}(\boldsymbol{r},t)\hat{\rho}_{\mathrm{c}}] = -\frac{1}{2\pi^{\frac{3}{2}}}\int d\boldsymbol{k}\sum_{\sigma}(\hbar ck)^{\frac{1}{2}}\rho_{\sigma}^{\frac{1}{2}}(\boldsymbol{k})\sin(\boldsymbol{k}\cdot r-\omega t)\boldsymbol{e}_{k}\times\boldsymbol{e}_{\sigma}. \tag{8.117}$$

For the energy density we need to calculate $\mathrm{Tr}[\hat{\mathbf{E}}^2\hat{\rho}_{\mathrm{c}}]$ and $\mathrm{Tr}[\hat{\mathbf{B}}^2\hat{\rho}_{\mathrm{c}}]$. In both cases, we will have to evaluate

$$[\hat{C}_{\sigma}(\boldsymbol{k})e^{i(\boldsymbol{k}\cdot\boldsymbol{r}-\omega t)}-\hat{C}_{\sigma}(\boldsymbol{k})^{\dagger}e^{-i(\boldsymbol{k}\cdot\boldsymbol{r}-\omega t)}] \times[\hat{C}_{\sigma'}(\boldsymbol{k}')e^{i(\boldsymbol{k}\cdot\boldsymbol{r}-\omega' t)}-\hat{C}_{\sigma'}(\boldsymbol{k}')^{\dagger}e^{-i(\boldsymbol{k}'\cdot\boldsymbol{r}-\omega' t)}].$$

After multiplying out this expression, we have

$$\begin{aligned}
&-[\hat{C}_{\sigma}(\boldsymbol{k})\hat{C}_{\sigma'}(\boldsymbol{k}')^{\dagger}e^{-i(\boldsymbol{k}\cdot\boldsymbol{r}-\omega' t)}e^{(\boldsymbol{k}\cdot\boldsymbol{r}-\omega t)}\\
&+\hat{C}(\boldsymbol{k}')^{\dagger}\hat{C}_{\sigma'}(\boldsymbol{k}')e^{i(\boldsymbol{k}'\cdot\boldsymbol{r}-\omega' t)}e^{-i(\boldsymbol{k}\cdot\boldsymbol{r}-\omega t)}\\
&-\hat{C}_{\sigma}(\boldsymbol{k})^{\dagger}\hat{C}_{\sigma'}(\boldsymbol{k}')^{\dagger}e^{i(\boldsymbol{k}\cdot\boldsymbol{r}-\omega t)}e^{i(\boldsymbol{k}'\cdot\boldsymbol{r}-\omega' t)}\\
&+\hat{C}_{\sigma}(\boldsymbol{k})\hat{C}_{\sigma'}(\boldsymbol{k}')e^{-i(\boldsymbol{k}\cdot\boldsymbol{r}-\omega t}e^{-i(\boldsymbol{k}'\cdot\boldsymbol{r}-\omega' t)}]\\
&-\delta(\boldsymbol{k}-\boldsymbol{k}')e^{-i(\boldsymbol{k}'\cdot\boldsymbol{r}-\omega' t)}\,e^{i(\boldsymbol{k}\cdot\boldsymbol{r}-\omega t)}.
\end{aligned} \tag{8.118}$$

With the use of Eq. (8.118), the continuous-$\boldsymbol{k}$ modification of Eq. (8.80), and the cyclic property of the trace, $\int d\boldsymbol{r}\,\mathrm{Tr}[\hat{\mathbf{E}}^2\hat{\rho}_{\mathrm{c}}]$ becomes

$$\begin{aligned}
\int d\boldsymbol{r}\mathrm{Tr}\,[\hat{\mathbf{E}}^2\hat{\rho}_{\mathrm{c}}] &= \frac{c^2}{16\pi^3}\int d\boldsymbol{k}\int d\boldsymbol{k}'\sum_{\sigma}(\hbar ck)^{\frac{1}{2}}(\hbar ck')^{\frac{1}{2}}\\
&\quad\times 2\left\{8\pi^3\rho_{\sigma}^{\frac{1}{2}}(\boldsymbol{k})\rho_{\sigma}^{\frac{1}{2}}(\boldsymbol{k}')[\delta(\boldsymbol{k}-\boldsymbol{k}')\right.\\
&\quad\left.-\frac{1}{2}\delta(\boldsymbol{k}+\boldsymbol{k}')(e^{2i\omega t}+e^{-2i\omega t)})]+\frac{1}{2}\delta(\boldsymbol{k}-\boldsymbol{k}')\right\}.
\end{aligned} \tag{8.119}$$

Then integration over $\boldsymbol{k}'$ gives the average total square electric field in a coherent photon ensemble (without the cosmic contribution from the vacuum):

$$\mathbf{E}^2 = c^2 \int d\boldsymbol{k} \sum_\sigma [\rho_\sigma^{\frac{1}{2}}(\boldsymbol{k})\rho_\sigma^{\frac{1}{2}}(\boldsymbol{k})$$
$$-\rho_\sigma^{\frac{1}{2}}(\boldsymbol{k})\rho_\sigma^{\frac{1}{2}}(-\boldsymbol{k})\cos(2i\omega t)]\hbar ck. \tag{8.120}$$

Now we must look at whether $\rho_\sigma^{\frac{1}{2}}(-\boldsymbol{k}) = \rho_\sigma^{\frac{1}{2}}(\boldsymbol{k})$ (*i.e.*, is an even function of $\boldsymbol{k}$) or whether $\rho_\sigma^{\frac{1}{2}}(-\boldsymbol{k}) = -\rho_\sigma^{\frac{1}{2}}(\boldsymbol{k})$ (*i.e.*, is an odd function of $\boldsymbol{k}$). If even, the average square of the field is

$$\mathbf{E}^2 = c^2 \int d\boldsymbol{k} \sum_\sigma \rho_\sigma(\boldsymbol{k})\hbar ck(1 - \cos 2\pi\omega t). \tag{8.121}$$

If odd,

$$\mathbf{E}^2 = c^2 \int d\boldsymbol{k} \sum_\sigma \rho_\sigma(\boldsymbol{k})\hbar ck(1 + \cos 2\pi\omega t). \tag{8.122}$$

In either case, the square of the electric field oscillates about its mean value at twice the frequency of the photons, the phase being dependent upon the symmetry properties of $\rho_\sigma^{\frac{1}{2}}(\boldsymbol{k})$.

Similarly for the average of the square of the magnetic field, so that the average energy is

$$E = \frac{1}{2}\left[\frac{1}{c^2}\mathbf{E}^2 + \mathbf{B}^2\right] = \int d\boldsymbol{k} \sum_\sigma \rho_\sigma(\boldsymbol{k})\hbar ck(1 \pm \cos 2\pi\omega t). \tag{8.123}$$

The average Poynting vector is also subject to this oscillation:

$$\langle \mathbf{S} \rangle = \frac{c}{\mu_0} \int d\boldsymbol{k} \sum_\sigma \rho_\sigma(\boldsymbol{k})\hbar ck(1 \pm \cos 2\pi\omega t)\boldsymbol{e}_k. \tag{8.124}$$

As a practical matter, it is rare that a coherent ensemble of photons has either an even or odd distribution about the $\boldsymbol{k}$-space origin. In an important application of coherent photon ensembles, a laser beam has a very narrow distribution about a wave vector $\boldsymbol{k}$, so that the $\boldsymbol{k}$-space origin doesn't come into play; we deal with this case in the following section.

8.4.4. *The k-Distribution for a Coherent Laser Pulse*

The simplest form for $\rho_\sigma(\boldsymbol{k})$ for a coherent laser pulse along the $\boldsymbol{k}$ axis is that of a Gaussian distribution of k_x and k_y about the origin and of k_z about k_0:

$$\rho_\sigma(\boldsymbol{k}) = N_\sigma(2\pi)^{-\frac{3}{2}} \exp[-(k_x^2 + k_y^2)/2(\Delta_\perp k)^2] \\ \times \exp[-k_z^2/2(\Delta_\| k)^2], \tag{8.125}$$

where N_σ is the number density of photons with $\boldsymbol{k}$ very close to $\boldsymbol{k}_0$ and with polarization σ in the collimated, near mono-chromatic beam. By inserting Eq. (8.126) into Eqs. (8.124) and (8.125) we can evaluate the average energy and Pointing vector, respectively. To simplify the integrals, we assume that $\Delta\boldsymbol{k}_\| \ll \Delta\boldsymbol{k}_\perp \ll k_0$; these assumptions are realistic, since the width of the cross section of a typical beam is much smaller than the length of the pulse along the beam axis and the band width is much smaller than the mean frequency of the photon beam. We may therefore set $\hbar ck = \hbar ck_0$ and $\omega = \omega_0 = \hbar ck_0$.

For a spatial distribution of the fields in a laser pulse, we need the Fourier transform of $\rho_\sigma{}^{\frac{1}{2}}(\boldsymbol{k})$; that is, the square root of Eq. (8.126) which is

$$\rho_\sigma{}^{\frac{1}{2}}(\boldsymbol{k}) = N_\sigma{}^{\frac{1}{2}}(2\pi)^{-\frac{3}{4}} \exp -(k_x^2 + k_y)/4(\Delta_\perp k)^2] \\ \times \exp[-(k_z - k_0)^2/4(\Delta_\| k)^2]. \tag{8.126}$$

Then its Fourier transform (which is *complex*) is

$$\tilde{\rho}_\sigma{}^{\frac{1}{2}}(\boldsymbol{r}) = (2\pi)^{-\frac{3}{2}} \int d\boldsymbol{k} e^{i\boldsymbol{k}\cdot\boldsymbol{r}} \rho_\sigma{}^{\frac{1}{2}}(\boldsymbol{k}) \\ = N_\sigma{}^{\frac{1}{2}} \left(\frac{2}{\pi}\right)^{\frac{3}{4}} \Delta_\perp \boldsymbol{k}(\Delta_\| \boldsymbol{k})^{\frac{1}{2}} e^{-(\Delta_\perp k)^2(x^2+y^2)} e^{-(\Delta_\| k)^2 z^2} e^{i\boldsymbol{k}_0 z}. \tag{8.127}$$

The trace of Eq. (8.118) with $\hat{\rho}_c$ is.

$$\begin{aligned}&-\rho_\sigma{}^{\frac{1}{2}}(\boldsymbol{k})\rho_{\sigma'}{}^{\frac{1}{2}}(\boldsymbol{k}')[e^{-i(\boldsymbol{k}'\cdot\boldsymbol{r}-\omega' t)}e^{i(\boldsymbol{k}\cdot\boldsymbol{r}-\omega' t)}\\&\quad+e^{i(\boldsymbol{k}'\cdot\boldsymbol{r}-\omega' t)}e^{-i(\boldsymbol{k}\cdot\boldsymbol{r}-\omega t)}\\&\quad-e^{i(\boldsymbol{k}\cdot\boldsymbol{r}-\omega t)}e^{i(\boldsymbol{k}'\cdot\boldsymbol{r}-\omega' t)}e^{-i(\boldsymbol{k}'\cdot\boldsymbol{r}-\omega t)}e^{-i(\boldsymbol{k}'\cdot\boldsymbol{r}-\omega' t)}]. \qquad (8.128)\end{aligned}$$

With Eq. (8.128) the energy density, ignoring the infinite contribution from the vacuum (and setting $\hbar ck = \hbar ck_0$ and $\omega = \omega_0 = \hbar ck_0$), becomes

$$\begin{aligned}E(\boldsymbol{r},t) = \frac{\hbar ck_0}{16\pi^3}\int d\boldsymbol{k}\int d\boldsymbol{k}'\sum_\sigma \rho_\sigma{}^{\frac{1}{2}}(\boldsymbol{k})\rho_\sigma{}^{\frac{1}{2}}(\boldsymbol{k}')\\ \times[e^{i\boldsymbol{k}\cdot\boldsymbol{r}}e^{-i\boldsymbol{k}'\cdot\boldsymbol{r}}+e^{i\boldsymbol{k}'\cdot\boldsymbol{r}}e^{-i\boldsymbol{k}\cdot\boldsymbol{r}}\\ -e^{i\boldsymbol{k}\cdot\boldsymbol{r}}e^{-\boldsymbol{k}'\cdot\boldsymbol{r}}e^{2i\omega_0 t}-e^{i\boldsymbol{k}\cdot\boldsymbol{r}}e^{i\boldsymbol{k}'\cdot\boldsymbol{r}}e^{-2i\omega_0 t}. \qquad (8.129)\end{aligned}$$

Now that the only functions of $\boldsymbol{k}$ and $\boldsymbol{k}'$ in Eq. (8.129) are $\rho_\sigma{}^{\frac{1}{2}}(\boldsymbol{k})$ and $\rho_\sigma{}^{\frac{1}{2}}(\boldsymbol{k}')$, the integrals will give their Fourier transforms:

$$\begin{aligned}E(\boldsymbol{r},t) = \frac{1}{2}\hbar ck_0\sum_\sigma[2\tilde{\rho}_\sigma{}^{\frac{1}{2}}(\boldsymbol{r})^*\tilde{\rho}_\sigma{}^{\frac{1}{2}}(\boldsymbol{r})\\ -\tilde{\rho}_\sigma{}^{\frac{1}{2}}(\boldsymbol{r})^*\tilde{\rho}_\sigma{}^{\frac{1}{2}}(\boldsymbol{r})^*e^{2i\omega_0 t}-\tilde{\rho}_\sigma{}^{\frac{1}{2}}(\boldsymbol{r})\tilde{\rho}_\sigma{}^{\frac{1}{2}}(\boldsymbol{r})e^{-2i\omega_0 t}]. \qquad (8.130)\end{aligned}$$

Insertion of the expression for $\tilde{\rho}_\sigma^{\frac{1}{2}}(\boldsymbol{r})$ given in Eq. (8.128) and its complex conjugate into Eq. (8.131) gives the energy density:

$$\begin{aligned}E(\boldsymbol{r},t) = 2\hbar ck_0\sum_0 N_\sigma\left(\frac{2}{\pi}\right)^{\frac{3}{2}}(\Delta_\perp k)^2(\Delta_\parallel k)\\ \times e^{-2(\Delta_\perp k)^2(x^2+y^2)}e^{-2(\Delta_\parallel k)^2 z^2}\sin^2(k_0 z-\omega_0 t). \qquad (8.131)\end{aligned}$$

Expressed in terms bandwidth of the beam, $\Delta\omega = c\Delta_\parallel k$, and its cross-sectional width, $d = (2\Delta_\perp k)^{-1}$, Eq. (8.131) can be written

$$\begin{aligned}E(\boldsymbol{r},t) = \left(\frac{2}{\pi}\right)^{\frac{3}{2}}\hbar ck_0\sum_\sigma N_\sigma\frac{\Delta\omega}{2cd^2}\exp[-(x^2+y^2)/(2d^2)]\\ \times\exp[-2(\Delta\omega)^2(z/c-t)^2]\sin^2[k_0(z-ct)]. \qquad (8.132)\end{aligned}$$

Integration over $\boldsymbol{r}$ gives the total energy of the pulse;

$$E = \int d\boldsymbol{r} E(\boldsymbol{r}, t) = \sum_{\sigma} N_{\sigma}\, \hbar c k_0 [1 - e^{-\omega_0^2/2(\Delta\omega)^2}]. \quad (8.133)$$

This result shows the reduction of the total average energy due to the finite bandwidth. This is usually not a problem, since the band center ω_0 is much larger than its width $\Delta\omega$.

Since the average Poynting vector is given by

$$\langle \mathbf{S} \rangle = \frac{1}{\mu^0} \langle \mathbf{E} \times \mathbf{B} \rangle = c \langle E^2 \rangle \boldsymbol{e}_k, \quad (8.134)$$

the transport of energy by the pulse is likewise reduced by the exponential in Eq. (8.133).

8.4.5. *Focused Laser Pulse*

For a focused laser pulse, the focal point z_0 is introduced as an additional parameter in the expression for the $\boldsymbol{k}$-space distribution

$$\rho_{\sigma}(\boldsymbol{k}) = \left(\frac{4cN_{\sigma} d^2}{(2\pi)^{\frac{3}{2}}} \right) \exp\{-2d^2(k_x^2 + k_y^2) - \frac{c^2}{2(\Delta\omega)^2} [(k_z - k_0) - \frac{d^2}{z_0}(k_x^2 + k_y^2)]^2\}. \quad (8.135)$$

As $z_0 \to \infty$ the $\boldsymbol{k}$-space distribution approaches that of an unfocused beam.

The Fourier transform of $\rho_{\sigma}{}^{\frac{1}{2}}(\boldsymbol{k})$ is

$$\tilde{\rho}_{\sigma}{}^{\frac{1}{2}}(\boldsymbol{r}) = N_{\sigma} \left(\frac{2}{\pi} \right)^{\frac{3}{4}} \left(\frac{\Delta\omega}{2cd^2} \right)^{\frac{1}{2}} (1 - iz/z_0)^{-1} \times \exp[-(x^2 + y^2)/4d^2(1 - iz/z_0)] \times \exp[-(\Delta\omega)^2 z^2/c^2] \exp(ik_0 z). \quad (8.136)$$

From Eqs. (8.130) and (8.136) we obtain the energy density of a focused laser:

$$E(\boldsymbol{r}) = \left(\frac{2}{\pi}\right)^{\frac{3}{2}} \sum_{\sigma} N_{\sigma}\hbar c k_0 \left(\frac{\Delta\omega}{2cd^2}\right) (1 + z^2/z_0^2)^{-1}$$
$$\times \exp[-(x^2 + y^2)/2d^2(1 + z^2/z_0^2)]$$
$$\times \exp[-2(\Delta\omega^2 z^2/c^2] \sin^2 k_0 z. \tag{8.137}$$

This result agrees with the classical result given by Yariv,[6] except for the $\sin^2$ oscillating factor. We see from Eq. (8.138) that the beam cross section takes its minimum value at the focal point $z = z_0$.

8.5. Ensemble Operator and Fields for Incoherent Light

Now we seek the grand canonical ensemble operator for incoherent light, as we did for the perfect gas in Chapter 6. Incoherent light need not be in thermal equilibrium with matter, but is in any case characterized by randomness of phase. Specifically, we take incoherent light to be emitted by multiple light sources that have no definite phase relation with one another. Thus there are degrees of coherence and incoherence. In Section 8.4 we treated one extreme point on this scale, namely light emitted from a single coherent source. In this section, we treat the other extreme: the superposition of signals from such a large number of unphased sources that their phases can be taken to be completely random.

8.5.1. *Randomization of the Phases*

We begin by returning to Eq. (8.97) for the ensemble operator for a single coherent signal, but this time explicitly including the phases $\varphi_0\boldsymbol{k}$ of the distribution functions; that is, we write

$$\rho_{\sigma}{}^{\frac{1}{2}}(\boldsymbol{k}) = \beta_{\sigma}(\boldsymbol{k}) e^{i\varphi_{\sigma}(\boldsymbol{k})}, \tag{8.138}$$

[6] Yariv, A., *Quantum Electronics*, Wiley (1975, 2nd edition).

where $\beta_\sigma(\boldsymbol{k})$ and $\varphi_0(\boldsymbol{k})$ are real-valued functions of the mode $\boldsymbol{k}\sigma$. Expanding the exponential operators in Eq. (8.97) and omitting the normalizing factor for now, we have

$$\hat{\rho} = \sum_{n=0}^{\infty}\sum_{m=0}^{\infty}\frac{1}{n!m!}\prod_{j=1}^{n}\int d\boldsymbol{k}_j \sum_{\sigma_j}\beta_{\sigma_j}(\boldsymbol{k}_j)e^{i\varphi_j(\boldsymbol{k}_j)}$$
$$\times \prod_{l=1}^{m}\int d\boldsymbol{k}_l \sum_{\sigma_l}\beta_{\sigma_l}(\boldsymbol{k}_l)e^{-i\varphi_l(\boldsymbol{k}_l)}$$
$$\times \prod_{l=1}^{m}[\hat{C}_{\sigma_l}(\boldsymbol{k}_l)^\dagger]|\,0\rangle\langle 0\,|\prod_{j=1}^{n}[\hat{C}_{\sigma_j}(\boldsymbol{k}_j)]. \qquad (8.139)$$

According to our definition of complete incoherence, we take the phases to be random; this means that $\varphi_0(\boldsymbol{k})$ is a multi-valued function of $\boldsymbol{k}\sigma$. Thus Eq. (8.140) is to be thought of as representing a superposition of sources with random phases; that is, we must sum Eq. (8.140) over all possible values of the phases. Since we have not yet normalized $\hat{\rho}$, we may replace the sum over phases by an average over phases. To find the effect of random phases on the form of $\hat{\rho}$, we average the phases in the following way: first pair each mode $k_j\sigma_j$ with a mode $k_l\sigma_l$. Consider one such pair, $\boldsymbol{k}\sigma$ and $\boldsymbol{k}'\sigma'$; the double integral and sums over these variables may be represented as

$$\hat{I} = \int d\boldsymbol{k}\sum_{\sigma}\int d\boldsymbol{k}'\sum_{\sigma'}\hat{\mathbf{F}}_{\sigma'}(\boldsymbol{k}')\hat{\mathbf{G}}_{\sigma}(\boldsymbol{k})e^{i[\varphi_\sigma(\boldsymbol{k})-\varphi_{\sigma'}(\boldsymbol{k}')]}. \quad (8.140)$$

Now replace the integrals and sums by sums over cubes of volume $\Delta\boldsymbol{k}$ in $\boldsymbol{k}$-space, enumerated by the index q:

$$\hat{\mathbf{I}} = (\Delta\boldsymbol{k}^2)\sum_{q}\sum_{q'}\hat{\mathbf{F}}_{q'}\hat{\mathbf{G}}_q e^{i(\varphi_q-\varphi'_q)}. \qquad (8.141)$$

Averaging the random multiple values of φ_q and $\varphi_{q'}$ within the cubes q and q' gives

$$\langle e^{i(\varphi_q-\varphi_{q'})}\rangle = \frac{1}{(2\pi)^2}\int_0^{2\pi}d\varphi\int_0^{2\pi}d\varphi' e^{i(\varphi_q-\varphi_{q'})} = \delta_{qq'}, \quad (8.142)$$

where upon the phase-averaged expression for $\hat{\mathbf{I}}_{\text{sum}}$ becomes

$$\langle \hat{\mathbf{I}}_{\text{sum}} \rangle = (\Delta \boldsymbol{k})^2 \sum_q \hat{\mathbf{F}}_q \hat{\mathbf{G}}_q. \tag{8.143}$$

Returning to the integral form, we have

$$\langle \hat{\mathbf{I}} \rangle = \Delta \boldsymbol{k} \int d\boldsymbol{k} \sum_\sigma \hat{\mathbf{F}}_\sigma(\boldsymbol{k}) \hat{\mathbf{G}}_\sigma(\boldsymbol{k}). \tag{8.144}$$

Similar results are obtained for each pair of modes in Eq. (8.140). For terms with $m \neq n$ there is at least one unpaired mode; the result of phase averaging in this case is

$$\frac{1}{2\pi} \int_0^{2\pi} d\varphi e^{i\varphi} = 0. \tag{8.145}$$

Applying these results to Eq. (8.140), including the $n!$ ways of pairing modes, produces the form

$$\hat{\rho}_{\text{h}} = |\,0\rangle\langle 0\,| + \sum_{n=1}^{\infty} \frac{1}{n!} \prod_{j=1}^{n} \int d\boldsymbol{k}_j \sum_{\sigma_j} \Phi_{\sigma_j}(\boldsymbol{k}_j)$$
$$\times \prod_{j=1}^{n} [\hat{C}_{\sigma_{(n+1-j)}}(\boldsymbol{k}_{(n+1-j)})]|0\rangle\langle 0\,| \prod_{l=1}^{n} [\hat{C}_{\sigma_l}(\boldsymbol{k}_l)], \tag{8.146}$$

where we have defined $\Phi_\sigma(\boldsymbol{k})$ as the dimensionless function $\Delta \boldsymbol{k} \beta_\sigma(\boldsymbol{k})^2$ and adopt the rule of nesting the operators that were used in defining the projection operator in Eq. (8.108).[7] For reasons made clear in Section 8.5.3, we describe the form of $\hat{\rho}_{\text{h}}$ as homogeneous.

[7] See also the discussion following Eq. (8.108).

8.5.2. *Properties of the Random-Phase Ensemble Operator*

The ensemble operator in Eq. (8.146) no longer obeys the eigenvalue relations obeyed by coherent ensembles. Instead, we have

$$\hat{C}_\sigma(\boldsymbol{k})\hat{\rho}_{\rm h} = \sum_{n=1}^{\infty}\frac{1}{n!}\prod_{j=1}^{n}\int d\boldsymbol{k}_j\sum_{\sigma_j}\Phi_{\sigma_j}(\boldsymbol{k}_j)$$
$$\times\hat{C}_\sigma(\boldsymbol{k})\prod_{j=1}^{n}[\hat{C}_{\sigma_{(n+1-j)}}(\boldsymbol{k}_{n+1-j})^\dagger]|0\rangle\langle 0|\prod_{l=1}^{n}[\hat{C}_{\sigma_l}(\boldsymbol{k}_l)]. \tag{8.147}$$

Using Wick's theorem, we normal-order the operators on the left of the vacuum projection to get

$$\hat{C}_\sigma(\boldsymbol{k})\prod_{j=1}^{n}[\hat{C}_{\sigma_{(n+1-j)}}(\boldsymbol{k}_{(n+1-j)})^\dagger]|\,0\rangle$$
$$=\sum_{q=1}^{n}\delta_{\sigma_q\sigma}\delta(\boldsymbol{k}_q-\boldsymbol{k})\prod_{\substack{j=1\\ j\neq+1-q}}[\hat{C}_{\sigma_{(n+1-j)}}(\boldsymbol{k}_{(n+1-j)})^\dagger]|0\rangle. \tag{8.148}$$

After insertion of Eq. (8.148) into Eq. (8.147), integration over $\boldsymbol{k}_q$ and summation over σ_q yield the factor $\Phi_\sigma\boldsymbol{k}$ and the operator $\hat{C}_\sigma(\boldsymbol{k})$ in the operator product on the right side of the vacuum projection. Renumbering the remaining modes other than $\boldsymbol{k}_q\sigma_q$ we rewrite Eq. (8.147) with the aid of the n terms in Eq. (8.148) as follows:

$$\hat{C}_\sigma(\boldsymbol{k})\hat{\rho}_{\rm h} = \Phi_\sigma(\boldsymbol{k})\sum_{n=1}^{\infty}\frac{n}{n!}\prod_{j=1}^{n}\int d\boldsymbol{k}_j\sum_{\sigma_j}\Phi_{\sigma_j}(\boldsymbol{k}_j)$$
$$\times\prod_{j=1}^{n-1}[\hat{C}_{\sigma_{(n+1-j)}}(\boldsymbol{k}_{n+1-j})^\dagger]|0\rangle\langle 0|\prod_{l=1}^{n-1}[\hat{C}_{\sigma_l}(\boldsymbol{k}_l)]\hat{C}_\sigma(\boldsymbol{k}). \tag{8.149}$$

Since the sum sandwiched between $\Phi_\sigma(\boldsymbol{k})$ and $\hat{C}_\sigma(\boldsymbol{k})$ in Eq. (8.149) is seen to be $\hat{\rho}_{\rm h}$, the relations we seek are

$$\hat{C}_\sigma(\boldsymbol{k})\hat{\rho}_{\rm h} = \Phi_\sigma(\boldsymbol{k})\hat{\rho}_{\rm h}\hat{C}_\sigma(\boldsymbol{k}) \tag{8.150a}$$

and its Hermitian conjugate

$$\hat{\rho}_{\rm h}\hat{C}_\sigma(\boldsymbol{k})^\dagger = \hat{C}_\sigma(\boldsymbol{k})^\dagger\hat{\rho}_{\rm h}\Phi_\sigma(\boldsymbol{k}). \tag{8.150b}$$

These equations therefore display the distinctive property of a homogeneous ensemble of photons.

First, we use Eqs. (8.150) to show that the number operator commutes with $\hat{\rho}_{\rm h}$. Dividing both sides of Eq. (8.150b) by $\Phi_\sigma(\boldsymbol{k})$ gives the useful result

$$\hat{C}_\sigma(\boldsymbol{k})^\dagger\hat{\rho}_{\rm h} = \Phi_\sigma^{-1}(\boldsymbol{k})\hat{\rho}_{\rm h}\hat{C}_\sigma(\boldsymbol{k})^\dagger. \tag{8.151}$$

Then from Eqs. (8.150a) and (8.151) we have

$$\begin{aligned}\hat{C}(\boldsymbol{k})^\dagger\hat{C}_\sigma(\boldsymbol{k})\hat{\rho}_{\rm h} &= \Phi_\sigma(\boldsymbol{k})\hat{C}_\sigma(\boldsymbol{k})^\dagger\hat{\rho}_{\rm h}\hat{C}_\sigma(\boldsymbol{k})\\ &= \Phi_\sigma(\boldsymbol{k})\Phi_\sigma^{-1}(\boldsymbol{k})\hat{\rho}_{\rm h}\hat{C}_\sigma(\boldsymbol{k})^\dagger\hat{C}_\sigma(\boldsymbol{k}) = \hat{\rho}_{\rm h}\hat{C}_\sigma(\boldsymbol{k})^\dagger\hat{C}_\sigma(\boldsymbol{k}).\end{aligned} \tag{8.152}$$

Although $\hat{\rho}_{\rm h}$ commutes with $\hat{C}_\sigma(\boldsymbol{k})^\dagger\hat{C}_\sigma(\boldsymbol{k})$, it is obviously *not* an eigenstate of the number operator in view of Eqs. (8.150), but rather a superposition of such eigenstates; accordingly, the average values of **A**, **E**, and **B** are all zero for an incoherent ensemble.

A more remarkable and useful result is the following: from Eq. (8.150a), the boson commutation rule, and the cyclic property of the trace,

$$\begin{aligned}\mathrm{Tr}[\hat{C}_{\sigma'}(\boldsymbol{k}')^\dagger\hat{C}_\sigma(\boldsymbol{k})\hat{\rho}_{\rm h}] &= \Phi_\sigma(\boldsymbol{k})\mathrm{Tr}[\hat{C}_{\sigma'}(\boldsymbol{k}')^\dagger\hat{\rho}_{\rm h}\hat{C}_\sigma(\boldsymbol{k})]\\ &= \Phi_\sigma(\boldsymbol{k})\mathrm{Tr}[\hat{C}_\sigma(\boldsymbol{k})\hat{C}_{\sigma'}(\boldsymbol{k}')^\dagger\hat{\rho}_{\rm h}]\\ &= \Phi_\sigma(\boldsymbol{k})\mathrm{Tr}[\hat{C}_{\sigma'}(\boldsymbol{k}')^\dagger\hat{C}_{\sigma'}(\boldsymbol{k})\hat{\rho}_{\rm h}]\\ &\quad + \Phi_\sigma(\boldsymbol{k})\delta_{\sigma'\sigma}\delta(\boldsymbol{k}-\boldsymbol{k}')\mathrm{Tr}\hat{\rho}_{\rm h}.\end{aligned} \tag{8.153}$$

Collecting the terms in $\mathrm{Tr}[\hat{C}_{\sigma'}(\boldsymbol{k}')^\dagger \hat{C}_\sigma(\boldsymbol{k})\hat{\rho}_\mathrm{h}]$, we have

$$\mathrm{Tr}[\hat{C}'_\sigma(\boldsymbol{k}')^\dagger \hat{C}_\sigma(\boldsymbol{k})\hat{\rho}_\mathrm{h}] = \frac{\Phi_\sigma(\boldsymbol{k})}{1-\Phi_\sigma(\boldsymbol{k})}\delta_{\sigma'\sigma}\delta(\boldsymbol{k}-\boldsymbol{k}')\mathrm{Tr}\hat{\rho}_\mathrm{h}, \tag{8.154}$$

which is the Bose-Einstein distribution.

8.5.3. *Homogeneous Light Beams and Black Body Radiation*

When we trace Eq. (8.118) with the incoherent ensemble operator to calculate the energy, we find that

$$\mathrm{Tr}[\hat{C}_{\sigma'}(\boldsymbol{k}')\hat{C}_\sigma(\boldsymbol{k})\hat{\rho}_\mathrm{h}] = \mathrm{Tr}[\hat{C}_{\sigma'}(\boldsymbol{k}')^\dagger \hat{C}_{\sigma'}(\boldsymbol{k})^\dagger \hat{C}_\sigma(\boldsymbol{k})^\dagger\hat{\rho}_\mathrm{h}] = 0, \tag{8.155}$$

because each term of $\hat{\rho}_\mathrm{h}$ has an equal number of $\hat{C}_\sigma(\boldsymbol{k})^\dagger$'s and $\hat{C}_\sigma(\boldsymbol{k})$'s; so these terms leading to oscillations in the coherent case are absent. Using the methods leading to Eqs. (8.121)–(8.125), we calculate average densities of the fields in $\boldsymbol{r}$-space:

$$\begin{aligned}\mathbf{E}^2(\boldsymbol{r}) &= \mathrm{Tr}[\hat{\mathbf{E}}(\boldsymbol{r})\cdot\hat{\mathbf{E}}(\boldsymbol{r})]/\mathrm{Tr}\hat{\rho}_\mathrm{h}\\ &= \frac{c^2}{16\pi^3}\int d\boldsymbol{k}\sum_\sigma\int d\boldsymbol{k}'\sum_{\sigma'}\hbar c(kk')^{\frac{1}{2}}e^{i(\boldsymbol{k}-\boldsymbol{k}')\cdot\boldsymbol{r}}\\ &\quad\times\mathrm{Tr}\{[\hat{C}_{\sigma'}(\boldsymbol{k}')^\dagger\hat{C}_\sigma(\boldsymbol{k}) + \hat{C}_{\sigma'}(-\boldsymbol{k}')\hat{C}_\sigma(-\boldsymbol{k})^\dagger\\ &\quad-\hat{C}_{\sigma'}(\boldsymbol{k}')^\dagger\hat{C}_\sigma(-\boldsymbol{k})^\dagger - \hat{C}_{\sigma'}(-\boldsymbol{k}')\hat{C}_\sigma(\boldsymbol{k})]\hat{\rho}_\mathrm{h}\}/\mathrm{Tr}\hat{\rho}_\mathrm{h}\\ &= \frac{c^2}{16\pi^3}\int d\boldsymbol{k}\sum_\sigma\hbar ck\frac{\Phi_\sigma(\boldsymbol{k})}{1-\Phi_\sigma(\boldsymbol{k})}\end{aligned} \tag{8.156}$$

$$\mathbf{B}^2(\boldsymbol{r}) = \frac{1}{16\pi^3}\int d\boldsymbol{k}\sum_\sigma\hbar ck\frac{\Phi_\sigma(\boldsymbol{k})}{1-\Phi_\sigma(\boldsymbol{k})} \tag{8.157}$$

The energy density is therefore

$$E(\boldsymbol{r}) = \frac{1}{(2\pi)^3}\int d\boldsymbol{k}\sum_\sigma\hbar ck\frac{\Phi_\sigma(\boldsymbol{k})}{1-\Phi_\sigma(\boldsymbol{k})}. \tag{8.158}$$

These fields are seen to be uniform in space. The form of $\hat{\rho}_\mathrm{h}$ given in Eq. (8.146) obviously does not correspond to a pulse, but to a model for a spatially homogeneous region of an incoherent light beam or the isotropic radiation in a black body, for example. In a black body, thermal equilibrium between the walls and the field requires the dimensionless function $\Phi_\sigma(\boldsymbol{k})$ to be equal to e^{-hck/k_BT}, whereupon Eq. (8.158) becomes the Planck radiation law for the energy density in a black body,

$$E(\boldsymbol{r}) = \frac{8\pi}{c^3}\int_0^\infty v^2 dv \frac{hv}{e^{hv/k_BT} - 1}. \tag{8.159}$$

The normalization of $\hat{\rho}_h$ requires the calculation of $\mathrm{Tr}\,\hat{\rho}_h$, which is somewhat intricate, but since it is essentially the same as the derivation of Eq. (6.40) for the bosonic case, we shall give only the result here:

$$\mathrm{Tr}\,\hat{\rho}_h = \exp\left\{-\frac{V}{(2\pi)^3}\sum_\sigma \int d\boldsymbol{k}\ln[1 - \boldsymbol{\Phi}_\sigma(\boldsymbol{k})]\right\}. \tag{8.160}$$

Before going on to the major purpose of this section, namely the development of an ensemble for an incoherent pulse, we note an important point: the total electromagnetic energy in the volume V for an ensemble of photons represented by Eq. (8.146) is just the expression for $\mathrm{E}(\boldsymbol{r})$ given in Eq. (8.158) multiplied by V. Alternatively, we can calculate the total energy from the trace of the Hamiltonian given in Eq. (8.54) with $\hat{\rho}_h$ as follows:

$$\begin{aligned}
\int d\boldsymbol{r}E(\boldsymbol{r}) &= \mathrm{Tr}[\hat{\mathbf{H}}^0\hat{\rho}_h] = \int d\boldsymbol{k}\hbar ck\sum_\sigma Tr[\hat{C}_\sigma(\boldsymbol{k})^\dagger\hat{C}_\sigma(\boldsymbol{k})\hat{\rho}_h] \\
&= \delta(\boldsymbol{k}-\boldsymbol{k})\int d\boldsymbol{k}\sum_\sigma \hbar ck\frac{\Phi_\sigma \boldsymbol{k}}{1-\Phi_\sigma(\boldsymbol{k})} \\
&= \frac{V}{(2\pi)^3}\int d\boldsymbol{k}\sum_\sigma \hbar ck\frac{\Phi_\sigma(\boldsymbol{k})}{1-\phi_\sigma(\boldsymbol{k})},
\end{aligned} \tag{8.161}$$

in agreement with Eq. (8.158).

8.5.4. *An Incoherent Inhomogeneous Light Pulse*

To obtain a pulse that is localized in space, the photons must interact with collimating and chopping devices; that is, they must be absorbed and re-emitted by matter. But propagation of the pulse out of this photon-matter system precludes the establishment of thermal equilibrium. The phases of $\Phi_\sigma(\boldsymbol{k})$ in $\hat{\rho}_{\rm h}$ are random under such circumstances, but are essentially independent of $\boldsymbol{k}$ over the range contributing to the pulse; that is, the phases in a small cube q in $\boldsymbol{k}$-space will no longer be random relative to cube q'. In this case, phase averaging does not cause collapse of $\hat{\rho}$ into the form of $\hat{\rho}_{\rm h}$ by the mode-pairing used to obtain the homogeneous ensemble operator. This means that the $\boldsymbol{k}$ and σ variables for the string of $\hat{C}^\dagger$'s on the left of the vacuum projection will be independent of those for the string of $\hat{C}$'s to the right, but for a term in which $m \neq n$ the term vanishes. The result is

$$\hat{\rho}_{\rm i} = |0\rangle\langle 0| + \sum_{n=1}^{\infty} \frac{1}{(n!)^2} \prod_{j=1}^{n} \left[\int d\boldsymbol{k}_j \sum_{\sigma_j} \rho_{\sigma_j}{}^{\frac{1}{2}}(\boldsymbol{k}_j) \right]$$

$$\times \prod_{l=1} \left[\int d\boldsymbol{k}'_l \sum_{\sigma'_l} \rho_{\sigma'_l}{}^{\frac{1}{2}}(\boldsymbol{k}'_j) \right]$$

$$\times \prod_{j'=1}^{n} [\hat{C}_{\sigma_{(n+1-j')}}(\boldsymbol{k}_{n+1-j})^\dagger]|\,0\rangle\langle 0\,| \prod_{l'=1}^{n} [\hat{C}_{\sigma'_l}(\boldsymbol{k}'_{l'}], \quad (8.162)$$

where $\rho_\sigma{}^{\frac{1}{2}}\,(\boldsymbol{k})$ is a real-valued function. The distinguishing property of the ensemble operator of Eq. (8.162) is easily found by left-multiplying by $\hat{C}_\sigma(\boldsymbol{k})$. With the help of Eq. (8.148) we have

$$\hat{C}_\sigma(\boldsymbol{k})\hat{\rho}_i = \rho_\sigma{}^{\frac{1}{2}}(\boldsymbol{k}) \sum_{n=1}^{\infty} \frac{n}{(n!)^2}$$

$$\times \prod_{j=1}^{n} \left[\sum_{\sigma_j} \int d\boldsymbol{k}_j \rho_{\sigma_j}{}^{\frac{1}{2}}(\boldsymbol{k}_j) \right] \left[\prod_{l=1}^{n} \sum_{\sigma_l} \int d\boldsymbol{k}_l \rho_{\sigma_l}{}^{\frac{1}{2}}(\boldsymbol{k}_l) \right]$$

$$\times \prod_{j'=1}^{n} [\hat{C}_{\sigma_{(n+1-j')}}(\boldsymbol{k}_{n+1-j'})^\dagger]|0\rangle\langle 0| \prod_{l'=1}^{n} [\hat{C}_{\sigma'_l}(\boldsymbol{k}'_l)]. \quad (8.163)$$

Multiplication of Eq. (8.163) on the right by $\hat{C}_{\sigma'}(\boldsymbol{k}')^\dagger$ similarly gives

$$\hat{C}_\sigma(\boldsymbol{k})\hat{\rho}\hat{C}_{\sigma'}(\boldsymbol{k}')^\dagger = \rho_\sigma{}^{\frac{1}{2}}(\boldsymbol{k})\rho_{\sigma'}{}^{\frac{1}{2}}(\boldsymbol{k}')\sum_{n=1}^{\infty}\frac{n^2}{(n!)^2} \times \prod_{j=1}^{n}\left[\sum_{\sigma_j}\int d\boldsymbol{k}_j\rho_{\sigma_j}{}^{\frac{1}{2}}(\boldsymbol{k})_j\right]\left[\prod_{l=1}^{n}\sum_{\sigma_l}\int d\boldsymbol{k}_l\rho_{\sigma_l}{}^{\frac{1}{2}}(\boldsymbol{k}_l)\right] \times \prod_{j'=1}^{n-1}[\hat{C}_{\sigma_{(n+1-j')}}(\boldsymbol{k})_{(n+1-j')})^\dagger]|0\rangle\langle 0|\prod_{l'=1}^{n-1}[\hat{C}_{\sigma'_l}(\boldsymbol{k})'_l], \tag{8.164a}$$

which is seen to be

$$\hat{C}_\sigma(\boldsymbol{k})\hat{\rho}_{\rm i}\hat{C}_{\sigma'}(\boldsymbol{k}')^\dagger = \rho_{\sigma'}{}^{\frac{1}{2}}(\boldsymbol{k})\rho_{\sigma'}{}^{\frac{1}{2}}(\boldsymbol{k}')\hat{\rho}_{\rm i}. \tag{8.164b}$$

From Eq. (8.164b) and the cyclic property of the trace, we obtain

$$\mathrm{Tr}[\hat{C}_\sigma(\boldsymbol{k})\hat{C}_{\sigma'}(\boldsymbol{k}')^\dagger\hat{\rho}_{\rm i}] = \rho_\sigma{}^{\frac{1}{2}}(\boldsymbol{k})\rho_{\sigma'}{}^{\frac{1}{2}}(\boldsymbol{k}')\mathrm{Tr}\hat{\rho}_i. \tag{8.165}$$

Since there is no restriction on the values of $\boldsymbol{k}\sigma$ and $\boldsymbol{k}'\sigma'$ we also have

$$\mathrm{Tr}[\hat{C}_\sigma(\boldsymbol{k})\hat{C}_\sigma(\boldsymbol{k})^\dagger\hat{\rho}_{\rm i}] = \rho_\sigma(\boldsymbol{k})\mathrm{Tr}\hat{\rho}_{\rm i}, \tag{8.166}$$

in contrast to the average photon number in $(\boldsymbol{k})$-space for a homogeneous ensemble given in Eq. (8.154); that is, the photons represented by $\hat{\rho}_{\rm i}$ do not obey the Bose-Einstein distribution. On the other hand, since $\hat{\rho}_i$ is similar to $\rho_{\rm h}$ in that the same number of $\hat{C}^\dagger$'s and $\hat{C}$'s are contained in each term, Eq. (8.155) is obeyed by $\hat{\rho}_{\rm i}$. But in common with $\hat{\rho}_{\rm c}$, because of the *inhomogeneous* form of $\hat{\rho}_{\rm i}$ (that is, the integration and summation variables in the operators on each side of the vacuum projection are independent of one another), the normalization of $\hat{\rho}_i$ is

$$\mathrm{Tr}\hat{\rho}_{\mathrm{i}} = e^{N_p} \tag{8.167}$$

and the photons obey the Poisson distribution given in Eq. (8.114).

Turning now to the calculation of $E(\boldsymbol{r}, t)$, from $\hat{\rho}_{\mathrm{i}}$, we find from Eqs. (8.155) and (8.156) that

$$E(\boldsymbol{r}, t) = \frac{1}{16\pi^3}\hbar c k_0 \int d\boldsymbol{k} \int d\boldsymbol{k}' \sum_\sigma \rho_\sigma{}^{\frac{1}{2}}(\boldsymbol{k})\rho_\sigma{}^{\frac{1}{2}}(\boldsymbol{k}')$$
$$\times \left[e^{i(\boldsymbol{k})-\boldsymbol{k}')\cdot\boldsymbol{r}-(\omega-\omega')t} - e^{-i(\boldsymbol{k}-\boldsymbol{k}')\cdot\boldsymbol{r}+(\omega-\omega')t}\right], \tag{8.168}$$

where we have assumed that $\Delta_\| k \ll \Delta_\perp k \ll k_0$ [see the discussion following Eq. (8.126)] and have used the approximations of Eqs. (8.127) and (8.128). This result for $E(\boldsymbol{r}, t)$ shows that $\hat{\rho}_{\mathrm{i}}$ can be used as the ensemble operator for an incoherent, inhomogeneous light pulse, for if we take $\rho_\sigma(\boldsymbol{k})$ to be given by Eq. (8.126), we obtain from Eq. (8.168)

$$E(\boldsymbol{r}, t) = \frac{1}{2}\left(\frac{2}{\pi}\right)^{\frac{3}{2}} \hbar c k_0 \sum_\sigma N_\sigma \frac{\Delta\omega}{2cd^2} \exp[-(x^2+y^2)/(2d^2)]$$
$$\times \exp[-2(\Delta\omega)^2(z/c-t)^2], \tag{8.169}$$

which is the same result as Eq. (8.132), except that the oscillating factor $\sin^2[k_0(z-ct)]$ is replaced by its average value $1/2$. The average total photon energy E_{i} for the inhomogeneous, incoherent pulse is given by integration of Eq. (8.167) over $\boldsymbol{r}$:

$$E_{\mathrm{i}} = \int d\boldsymbol{r} E(\boldsymbol{r}, t) = \sum_\sigma N_\sigma \hbar c k_0. \tag{8.170}$$

Comparison of Eq. (8.170) with the average total photon energy E_{c} for the coherent pulse given by Eq. (8.134) we see that

$$E_{\mathrm{c}} = E_{\mathrm{i}}[1 - e^{-\omega_0^2/2(\Delta\omega)^2}]. \tag{8.171}$$

The incoherent pulse energy is the maximum attainable by a coherent pulse, and reducing the bandwidth of a coherent pulse enhances its energy output. In addition, high-frequency lasers will be more energy-efficient than lower-frequency lasers. For an optical dye laser (high frequency, low bandwidth) the coherent pulse energy is essentially that of an incoherent, inhomogeneous pulse. On the contrary, for a coherent infrared laser, there will be energy loss due to the coherence effect.

8.6. Summary and Discussion for Chapter 8

Three different forms for a second-quantized ensemble operator for photons have been identified and their distinguishing physical and mathematical properties have been demonstrated. We started with the extreme case of a coherent ensemble $\hat{\rho}_{\mathrm{c}}$ as given in Eq. (8.97) for which the wave vector is continuously variable and showed that by properly choosing the wave-vector distribution, $\hat{\rho}_{\mathrm{c}}$ can be used to describe a coherent, focused Gaussian light pulse of the type produced by a pulsed laser.

By randomizing the phases associated with different regions of $\boldsymbol{k}$-space, we showed that $\hat{\rho}_{\mathrm{c}}$ takes on the homogeneous form $\hat{\rho}_{\mathrm{h}}$ as given in Eq. (8.146), which describes isotropic black body radiation and the homogeneous interior of an incoherent light beam. If the phases are randomized, but are taken to be independent of $\boldsymbol{k}$ over the range that contributes to the pulse, $\hat{\rho}_{\mathrm{c}}$ takes on the inhomogeneous form $\hat{\rho}_{\mathrm{i}}$ as given in Eq. (8.162), which can be used to describe an incoherent, inhomogeneous Gaussian light pulse by proper choice of the $\boldsymbol{k}$ distribution.

The derivations of these three ensemble operators is admittedly difficult for readers not well-versed in second-quantization through the use of Fock operators, but the takeaways for this chapter, in addition to the explicit expressions $\hat{\rho}_{\mathrm{c}}$, $\hat{\rho}_{\mathrm{h}}$, and $\hat{\rho}_{\mathrm{i}}$, are mainly the following distinguishing mathematical properties:

$$\begin{aligned}
\hat{C}_{\sigma}(\boldsymbol{k})\hat{\rho}_{\mathrm{c}} &= \rho_{\sigma}(\boldsymbol{k})^{\frac{1}{2}}\hat{\rho}_{\mathrm{c}} \\
\hat{C}_{\sigma}(\boldsymbol{k})\hat{\rho}_{\mathrm{h}} &= \Phi_{\sigma}(\boldsymbol{k})\hat{\rho}_{\mathrm{h}}\hat{C}_{\sigma}(\boldsymbol{k}) \\
\hat{C}_{\sigma}(\boldsymbol{k})\hat{\rho}_{\mathrm{i}}\hat{C}_{\sigma'}(\boldsymbol{k})^{\dagger} &= \rho_{\sigma}(\boldsymbol{k})^{\frac{1}{2}}\rho_{\sigma'}(\boldsymbol{k}')^{\frac{1}{2}}\hat{\rho}_{\mathrm{i}} \\
\hat{C}_{\sigma}(\boldsymbol{k})\hat{\rho}_{\mathrm{i}} &\neq \rho_{\sigma}(\boldsymbol{k})^{\frac{1}{2}}\hat{\rho}_{\mathrm{i}}
\end{aligned} \tag{S8.1}$$

where $\Phi_\sigma(\boldsymbol{k}) = \Delta(\boldsymbol{k})\rho_\sigma(\boldsymbol{k})$. The property for $\hat{\rho}_{\rm h}$ given in Eq. (S8.1) leads to the Bose-Einstein distribution of photons among the modes, while the properties for $\hat{\rho}_{\rm c}$ and $\hat{\rho}_{\rm i}$ give rise to Poisson distributions for the number of photons in a given region of $\boldsymbol{k}$-space. The corresponding energy densities $E_{\rm c}$, $E_{\rm h}$, and $E_{\rm i}$ are all different: $E_{\rm h}(\boldsymbol{r})$ is independent of $\boldsymbol{r}$ and is identical to the Planck radiation law when $\Phi_\sigma(\boldsymbol{k})$ takes the thermal equilibrium form $e^{-hck/k_{\rm B}T}$, while $E_{\rm c}$ and $E_{\rm i}$ are $\boldsymbol{r}$-independent and differ only by an oscillating factor in the coherent case. For Gaussian pulses, integration of $E_{\rm c}$ and $E_{\rm i}$ over $\boldsymbol{r}$ gives total energies related by Eq.(8.169).

Because of their different mathematical and physical properties, care must be taken to choose the ensemble operator appropriate to the experimental conditions in any theoretical analysis of phenomena resulting from the interaction of radiation and matter. The importance of the manner in which the phases of particles that compose a grand canonical ensemble is demonstrated by the results analyzed in this chapter on photons.

In the next chapter we use a general molecular current density to obtain a second-quantized coupling of radiation and matter so that the results of such interaction can be treated consistently in a grand canonical ensemble of photons and molecules.

Chapter 9

Light-Molecule Interaction

9.1. Introduction

In exploring some of the phenomena that can be observed when light interacts with molecules, we first must determine the contribution to the Hamiltonian from the interaction; that is, $\mathrm{H_I}$. For guidance, we look back at Chapter 5, in which we saw that invariance of the Lagrangian to local phase changes led to the result, described in Sections 5.2.3, 5.3.1, and 5.2.2.5 of that chapter, in which the momentum $\boldsymbol{p}$ of a charged particle is replaced by $\boldsymbol{p} - (q/c)\mathbf{A}$, where q is the charge, c the speed of light, and $\mathbf{A}$ the vector potential. In classical mechanics, the kinetic energy becomes

$$\frac{(\boldsymbol{p} - \frac{q}{c}\mathbf{A})^2}{2m} = \frac{1}{2m}\left[p^2 - 2\frac{q}{c}\boldsymbol{p}\cdot\mathbf{A} + \left(\frac{q}{c}\right)^2 \mathbf{A}^2\right], \tag{9.1}$$

from which we designate

$$\mathrm{H} = \mathrm{H_0} + \mathrm{H_I}; \quad \mathrm{H_0} = \frac{p^2}{2m}; \quad \mathrm{H_I} = -\frac{q}{mc}\boldsymbol{p}\cdot\mathbf{A} + \left(\frac{q}{c}\right)^2 \mathrm{A}^2. \tag{9.2}$$

Our next task is to express this interaction Hamiltonian in second-quantized form.

From there we go on to a generalization to a multipolar interaction, treat both stimulated and spontaneous photon emission, photon absorption, and the Planck radiation law in the present context.

9.2. Quantization of the Interaction

Quantization of the vector potential was derived in Chapter 8; it has the form

$$\hat{\mathbf{A}}(\boldsymbol{r}) = \left(\frac{\hbar c}{16\pi^3}\right)^{\frac{1}{2}} \int \frac{d\boldsymbol{k}}{\sqrt{k}} \sum_{\sigma=1}^{2} e^{i\boldsymbol{k}\cdot\boldsymbol{r}}[\hat{C}_\sigma(\boldsymbol{k}) + \hat{C}_\sigma(-\boldsymbol{k})^\dagger]\boldsymbol{e}_k, \quad (9.3)$$

where we have changed the sign of $\boldsymbol{k}$ in the second term so as to have a single exponential in the expression. To quantize $\frac{q}{mc}\boldsymbol{p}$, we begin with the quantized charge density in $\boldsymbol{r}$ space, $\hat{\rho}_q(\boldsymbol{r})$:

$$\hat{\rho}_q(\boldsymbol{r}) = q\hat{\psi}(\boldsymbol{r})^\dagger\hat{\psi}(\boldsymbol{r}). \quad (9.4)$$

Next, we use the Fourier transforms

$$\hat{\psi}(\boldsymbol{r})^\dagger = (2\pi\hbar)^{-\frac{3}{2}} \int d\boldsymbol{p}\, e^{\frac{i}{\hbar}\boldsymbol{p}\cdot\boldsymbol{r}}\hat{a}(\boldsymbol{p})^\dagger,$$

$$\hat{\psi}(\boldsymbol{r}) = (2\pi\hbar)^{-\frac{3}{2}} \int d\boldsymbol{p}\, e^{-\frac{i}{\hbar}\boldsymbol{p}\cdot\boldsymbol{r}}\hat{a}(\boldsymbol{p}) \quad (9.5a,b)$$

to get

$$\hat{\rho}_q(\boldsymbol{r}) = \frac{q}{(2\pi\hbar)^3} \int d\boldsymbol{p} \int d\boldsymbol{p}' e^{\frac{i}{\hbar}(\boldsymbol{p}-\boldsymbol{p}')\cdot\boldsymbol{r}}\hat{a}(\boldsymbol{p}')^\dagger\hat{a}(\boldsymbol{p}). \quad (9.6)$$

The momentum operator is

$$\hat{\boldsymbol{p}} = \int d\boldsymbol{p}\hat{a}(\boldsymbol{p})^\dagger\boldsymbol{p}\hat{a}(\boldsymbol{p}). \quad (9.7)$$

9.2.1. *The Charge Current*

The classical charge current density is $\boldsymbol{j}_q(\boldsymbol{r}) = \rho_q v$; we can insert $v = \boldsymbol{p}/m$ into Eq. (9.6) by "splitting" between $\boldsymbol{p}$ and $\boldsymbol{p}'$ to get the quantized form

$$\hat{\boldsymbol{j}}_q(\boldsymbol{r}) = \frac{q}{(2\pi\hbar)^3} \int d\boldsymbol{p} \int d\boldsymbol{p}' e^{\frac{i}{\hbar}(\boldsymbol{p}-\boldsymbol{p}')\cdot\boldsymbol{r}}\hat{a}(\boldsymbol{p}')^\dagger\frac{1}{2m}(\boldsymbol{p}+\boldsymbol{p}')\hat{a}(\boldsymbol{p}). \quad (9.8)$$

Now we dot $\hat{\boldsymbol{j}}_q(\boldsymbol{r})$ with $\hat{\mathbf{A}}(\boldsymbol{r})$ and integrate over $\boldsymbol{r}$ to obtain

$$\begin{aligned}&\int d\boldsymbol{r}\hat{\boldsymbol{j}}_q(\boldsymbol{r})\cdot\hat{\mathbf{A}}(\boldsymbol{r})\\&\quad=\frac{q}{m}\int d\boldsymbol{p}\int\frac{d\boldsymbol{k}}{\sqrt{k}}\,\hat{a}(\boldsymbol{p}+\hbar\boldsymbol{k})^\dagger\left(\boldsymbol{p}+\frac{1}{2}\hbar\boldsymbol{k}\right)\hat{a}(\boldsymbol{p})\\&\qquad\times\left(\frac{\hbar c}{16\pi^3}\right)^{\frac{1}{2}}\sum_{\sigma=1}^{2}[\hat{C}_\sigma(\boldsymbol{k})+\hat{C}_\sigma(-\boldsymbol{k})^\dagger]\boldsymbol{e}_k.\end{aligned}\tag{9.9}$$

At this point we will ignore the $\mathbf{A}^2$ term in Eq. (9.2) and return to it later. Then from Eq. (9.9) and with Eq. (9.2) as a guide, we find the interaction Hamiltonian to be

$$\begin{aligned}\hat{\mathrm{H}}_{\mathrm{I}}&=-\frac{1}{c}\int d\boldsymbol{r}\hat{\boldsymbol{j}}_q(\boldsymbol{r})\cdot\hat{\mathbf{A}}(\boldsymbol{r})\\&=\frac{q\hbar}{m}\int d\boldsymbol{r}\int d\boldsymbol{k}\frac{(\boldsymbol{p}+\frac{1}{2}\hbar\boldsymbol{k})\cdot\boldsymbol{e}_k}{(16\pi^3\hbar ck)^{\frac{1}{2}}}\\&\quad\times\hat{a}^\dagger(\boldsymbol{k}+\hbar\boldsymbol{k})\sum_\sigma[\hat{C}_\sigma(\boldsymbol{k})+\hat{C}_\sigma(-\boldsymbol{k})]\hat{a}(\boldsymbol{k}).\end{aligned}\tag{9.10}$$

So far, we have assumed the molecules to be point charges. The Hamiltonian in Eq. (9.10) is for the monopole interaction. To treat neutral molecules with internal structure, we must expand the notion of current density as follows: let

$$\hat{\boldsymbol{j}}(\boldsymbol{k})=\int d\boldsymbol{p}\sum_{\alpha\beta}\hat{a}_\beta(\boldsymbol{p}+\hbar\boldsymbol{k})^\dagger\boldsymbol{j}_{\alpha\beta}(\boldsymbol{p},\boldsymbol{k})\hat{a}_\alpha(p).\tag{9.11}$$

The matrix $\boldsymbol{j}_{\alpha\beta}(\boldsymbol{p},\boldsymbol{k})+$ is C-valued and its indices α and β label the internal states of the molecules: electronic, vibrational, rotational, angular momentum, electron spin, and nuclear spin. These matrices are sums of the various interactions between the molecules and the photon field, notably the monopole we have just identified in Eq. (9.10), the electric dipole, the magnetic dipole, the electric quadrupole, and the $\mathbf{A}^2$ term. Here we give these terms without proof, deferring their derivation until later:

$$\begin{aligned}\boldsymbol{j}_{\alpha\beta}(\boldsymbol{p},\boldsymbol{k}) &= \frac{q}{m}\boldsymbol{p}\delta_{\alpha\beta} + \frac{i}{\hbar}(\varepsilon_\beta - \varepsilon_\alpha)\boldsymbol{\mu}_{\beta\alpha} \\ &\quad + ic\left[\boldsymbol{M}_{\beta\alpha} + \frac{i}{2mc}\left(\boldsymbol{p} + \frac{1}{2}\hbar\boldsymbol{k}\right) \times \boldsymbol{\mu}_{\beta\alpha}\right] \\ &\quad \times \boldsymbol{k}\frac{1}{2\hbar}(\varepsilon_\beta - \varepsilon_\alpha)\boldsymbol{k}\cdot\mathbf{Q}_{\beta\alpha} \\ &\quad + \frac{i}{2m}\boldsymbol{k}\cdot\left[\boldsymbol{\mu}_{\beta\alpha}\left(\boldsymbol{p} + \frac{1}{2}\hbar\boldsymbol{k}\right) + \left(\boldsymbol{p} + \frac{1}{2}\hbar\boldsymbol{k}\right)\boldsymbol{\mu}_{\beta\alpha}\right]. \qquad (9.12)\end{aligned}$$

In Eq. (9.12),

$$\boldsymbol{\mu}_{\beta\alpha} = \langle\beta|\sum_{j=1}^{s} q_j(\hat{\boldsymbol{x}}_j - \hat{\boldsymbol{x}}_j^e)|\alpha\rangle \quad \text{electric dipole moment} \qquad (9.13)$$

$$\boldsymbol{M}_{\beta\alpha} = \langle\beta|\sum_{j=1}^{s} \frac{q_j}{2m_j c}(\hat{\mathbf{L}}_j + 2\hat{\mathbf{S}}_j)|\alpha\rangle \quad \text{magnetic dipole moment} \qquad (9.14)$$

$$\mathbf{Q}_{\beta\alpha} = \langle\beta|\sum_{j=1}^{s} q_j(\hat{\boldsymbol{x}}_j - \hat{\boldsymbol{x}}_j^e)(\hat{\boldsymbol{x}}_j - \hat{\boldsymbol{x}}_j^e)|\alpha\rangle \ \text{electric quadrupole tensor} \qquad (9.15)$$

The j summations enumerate the atoms in the molecule; $\hat{\boldsymbol{x}}_j$ is the position of atom j and $\hat{\boldsymbol{x}}_j^e$, its equilibrium position; $\hat{\boldsymbol{L}}_j$ and $\hat{\boldsymbol{S}}_j$ are its orbital and spin angular momentum, respectively; α and β are its initial and final internal states.

9.2.2. *The Electric Dipole Interaction*

The electric dipole Hamitonian is

$$\begin{aligned}\hat{\mathrm{H}}_{\mathrm{I}}^{\mathrm{ed}} &= -\frac{i}{\sqrt{16\pi^3}}\sum_{\alpha\beta}\boldsymbol{\mu}_{\beta\alpha}\cdot\boldsymbol{e}_k\int d\boldsymbol{p}\int d\boldsymbol{k}\hat{a}_\beta(\boldsymbol{p}+\hbar\boldsymbol{k})^\dagger \\ &\quad \times \frac{(\varepsilon_\beta - \varepsilon_\alpha)}{(\hbar ck)^{\frac{1}{2}}}\hat{a}_\alpha(\boldsymbol{p})\sum_{\sigma=1}^{2}[\hat{C}_\sigma(\boldsymbol{k}) + \hat{C}_\sigma(-\boldsymbol{k})^\dagger]. \qquad (9.16)\end{aligned}$$

As a first step in deriving the electric dipole interaction of radiation and molecules, consider a molecule as a cluster of s particles (electrons, nuclei), each of charge q_j and position r_j; the internal states of the molecule (electronic, vibrational, rotational, and electron and nuclear spin) are designated by lower-case Greek letters, α, β, γ, *etc.*, with *kets* and *bras* $|\alpha\rangle$ and $\langle\beta|$. The momentum states of the center of mass of the molecules are created and annihilated by $\hat{a}(\boldsymbol{p})^\dagger$ and $\hat{a}(\boldsymbol{p})$, respectively. Creators and annihilators of combined momentum and internal states are designated by $\hat{a}_\alpha(\boldsymbol{p})^\dagger$ and $\hat{a}_\beta(\boldsymbol{p})$, where

$$\hat{a}_\alpha(\boldsymbol{p})^\dagger = \hat{a}(\boldsymbol{p})^\dagger|\alpha\rangle, \quad \hat{a}_\alpha(\boldsymbol{p}) = \langle\alpha|\hat{a}(\boldsymbol{p}). \tag{9.17}$$

Then, because of the completeness of the internal states, $\sum_\alpha |\alpha\rangle\langle\alpha| = 1$, we have

$$\hat{a}(\boldsymbol{p})^\dagger = \sum_\alpha \hat{a}_\alpha(\boldsymbol{p})^\dagger|\alpha\rangle, \quad \hat{a}(\boldsymbol{p}) = \sum_\alpha \langle\alpha|\hat{a}_\alpha(\boldsymbol{p}). \tag{9.18}$$

Similarly,

$$\begin{aligned} \hat{\psi}_\alpha(\boldsymbol{r})^\dagger &= \hat{\psi}(\boldsymbol{r})^\dagger|\alpha\rangle, \quad \hat{\psi}_\alpha(\boldsymbol{r}) = \langle\alpha|\hat{\psi}(\boldsymbol{r}), \\ \hat{\psi}(\boldsymbol{r})^\dagger &= \sum_\alpha \hat{\psi}_\alpha(\boldsymbol{r})^\dagger\langle\alpha|, \quad \hat{\psi}(\boldsymbol{r}) = \sum_\alpha |\alpha\rangle\hat{\psi}_\alpha(\boldsymbol{r}). \end{aligned} \tag{9.19}$$

We assume that the vector potential is relatively uniform over the region occupied by the particles in a given molecule. Accordingly, the vector potential operator is written as a function of the center of mass $\boldsymbol{r}$ of the molecule:

$$\hat{\mathbf{A}}(\boldsymbol{r}) = \left(\frac{\hbar c}{16\pi^3}\right)^{\frac{1}{2}} \int \frac{d\boldsymbol{k}}{\sqrt{k}} \sum_{\sigma=1}^{2} e^{i\boldsymbol{k}\cdot\boldsymbol{r}}[\hat{C}_\sigma(\boldsymbol{k}) + \hat{C}_\sigma(-\boldsymbol{k})^\dagger]\boldsymbol{e}_k, \tag{9.20}$$

$$\hat{\rho}(\boldsymbol{r}) = \hat{\psi}(\boldsymbol{r})^\dagger\hat{\psi}(\boldsymbol{r}). \tag{9.21}$$

Now we turn to the probability current density operator. We start with the probability density $\hat{\rho}(\boldsymbol{r})$ for molecules at position $\boldsymbol{r}$. To express $\hat{\rho}(\boldsymbol{r})$ in terms of the molecular momentum states, we use the Fourier transforms

$$\hat{\psi}(\boldsymbol{r})^\dagger = \frac{1}{(2\pi\hbar)^{\frac{3}{2}}}\int d\boldsymbol{r} e^{-\frac{i}{\hbar}\boldsymbol{p}\cdot\boldsymbol{r}}, \quad \hat{\psi}(\boldsymbol{r}) = \frac{1}{(2\pi\hbar)^{\frac{3}{2}}}\int d\boldsymbol{r} e^{\frac{i}{\hbar}\boldsymbol{p}\cdot\boldsymbol{r}} \tag{9.22}$$

to obtain

$$\hat{\rho}(\boldsymbol{r}) = \frac{1}{(2\pi\hbar)^3}\int d\boldsymbol{p}\int d\boldsymbol{p}' e^{\frac{i}{\hbar}(\boldsymbol{p}-\boldsymbol{p}')\cdot\boldsymbol{r}}\hat{a}(\boldsymbol{p}')^\dagger \hat{a}(\boldsymbol{p}). \tag{9.23}$$

It is convenient at this point to transfer the factor $e^{i\boldsymbol{k}\cdot\boldsymbol{r}}$ from $\hat{\mathbf{A}}(\boldsymbol{r})$ to $\hat{\rho}(\boldsymbol{r})$ so that the $\boldsymbol{r}$ dependence is all in the probability density:

$$\hat{\rho}(\boldsymbol{r}) = \frac{1}{(2\pi\hbar)^3}\int d\boldsymbol{p}\int d\boldsymbol{p}' e^{\frac{i}{\hbar}(\boldsymbol{p}-\boldsymbol{p}+\hbar\boldsymbol{k})\cdot\boldsymbol{r}}\hat{a}(\boldsymbol{p}')^\dagger \hat{a}(\boldsymbol{p}). \tag{9.24}$$

The first step in turning Eq. (9.24) into a charge current is to multiply $\hat{\rho}(\boldsymbol{r})$ by the total charge on the molecule and the velocity of its center of mass, $q\dot{\boldsymbol{r}} = q\boldsymbol{p}/m$; this gives the electric monopole approximation. Next, we consider the particles in the molecule and add a term consisting of the charge on each particle, q_j times the velocity of the particle relative to that of the center of mass, $\frac{\partial}{\partial t}(\boldsymbol{r}_j - \boldsymbol{r})$, and summing over all particles j. These insertions give:

$$\hat{\boldsymbol{j}}(\boldsymbol{r},\boldsymbol{k}) = \frac{1}{(2\pi\hbar)^3}\int d\boldsymbol{p}\int d\boldsymbol{p}' e^{\frac{i}{\hbar}(\boldsymbol{p}-\boldsymbol{p}'+\hbar\boldsymbol{k})\cdot\boldsymbol{r}}\hat{a}(\boldsymbol{p})^\dagger \times\left[\frac{q}{m}\boldsymbol{p} + \sum_{j=1}^{s} q_j\frac{\partial}{\partial t}(\boldsymbol{r}_j - \boldsymbol{r})\right]\hat{a}(\boldsymbol{p}). \tag{9.25}$$

Since

$$\int d\boldsymbol{r} e^{\frac{i}{\hbar}(\boldsymbol{p}-\boldsymbol{p}'-\hbar\boldsymbol{k})\cdot\boldsymbol{r}} = (2\pi\hbar)^3\delta(\boldsymbol{p}-\boldsymbol{p}'+\hbar\boldsymbol{k}), \tag{9.26}$$

integration of Eq. (9.25) over $\boldsymbol{r}$ gives

$$\hat{\boldsymbol{j}}(\boldsymbol{r}) = \int d\boldsymbol{p}\int d\boldsymbol{p}'\delta(\boldsymbol{p}-\boldsymbol{p}'+\hbar\boldsymbol{k})\hat{a}(\boldsymbol{p}')^\dagger\left[\frac{q}{m}\boldsymbol{p} + \sum_{j=1}^{s} q_j\frac{\partial}{\partial t}(\boldsymbol{r}_j - \boldsymbol{r})\right] \times\hat{a}(\boldsymbol{p}). \tag{9.27}$$

Then integration of Eq. (9.27) over $\boldsymbol{p}'$ gives

$$\hat{\boldsymbol{j}}(\boldsymbol{k}) = \int d\boldsymbol{p}\hat{a}(\boldsymbol{p}+\hbar\boldsymbol{k})^{\dagger}\left[\frac{q}{m}\boldsymbol{p} + \sum_{j=1}^{s} q_j \frac{\partial}{\partial t}(\boldsymbol{r}_j - \boldsymbol{r})\right]\hat{a}(\boldsymbol{p}). \tag{9.28}$$

Use of Eq. (9.18) allows Eq. (9.28) to be written as

$$\hat{\boldsymbol{j}}(\boldsymbol{k}) = \int d\boldsymbol{p}\sum_{\beta\alpha}\hat{a}_{\beta}(\boldsymbol{p}+\hbar\boldsymbol{k})^{\dagger}\langle\beta|\left[\frac{q}{m}\boldsymbol{p} + \sum_{j=1}^{s} q_j \frac{\partial}{\partial t}(\boldsymbol{r}_j - \boldsymbol{r})\right]|\alpha\rangle\hat{a}_{\alpha}(\boldsymbol{p}). \tag{9.29}$$

The derivative w.r.t. t, inside the brackets, is

$$\langle\beta|\frac{\partial}{\partial t}\sum_{j}^{s} q_j(\boldsymbol{r}_j - \boldsymbol{r})|\alpha\rangle = \langle\beta|\frac{i}{\hbar}[\hat{H}^0, \sum_{j}^{s} q_j(\boldsymbol{r}_j - \boldsymbol{r})]_{-}\,|\alpha\rangle, \tag{9.30}$$

where $\hat{H}^0$ is the molecular Hamiltonian, $\hat{H}^0 = \sum_{\gamma}|\gamma\rangle\varepsilon_{\gamma}\langle\gamma|$; thus, we have

$$\langle\beta|\frac{\partial}{\partial t}\sum_{j}^{s} q_j(\boldsymbol{r}_j - \boldsymbol{r})|\alpha\rangle = \frac{i}{\hbar}(\varepsilon_{\beta} - \varepsilon_{\alpha})\langle\beta|\sum_{j}^{s} q_j(\boldsymbol{r}_j - \boldsymbol{r})|\alpha\rangle \tag{9.31}$$

and Eq. (9.29) becomes

$$\hat{\boldsymbol{j}}(\boldsymbol{k}) = \int d\boldsymbol{p}\sum_{\beta\alpha}\hat{a}_{\beta}(\boldsymbol{p}+\hbar\boldsymbol{k})^{\dagger} \times \langle\beta|\left[\frac{q}{m}\boldsymbol{p} + \frac{i}{\hbar}(\varepsilon_{\beta} - \varepsilon_{\alpha})\sum_{j=1}^{s} q_j \frac{\partial}{\partial t}(\boldsymbol{r}_j - \boldsymbol{r})\right]|\alpha\rangle\hat{a}_{\alpha}(\boldsymbol{p}). \tag{9.32}$$

The first term in the brackets is

$$\langle\beta|\,\frac{q}{m}\boldsymbol{p}\,|\alpha\rangle = \frac{q}{m}\boldsymbol{p}\delta_{\beta\alpha}, \tag{9.33}$$

since $q\boldsymbol{p}/m$ is unaffected by the internal state and $\langle\beta|\alpha\rangle = \delta_{\beta\alpha}$. Thus we have

$$\hat{\boldsymbol{j}}(\boldsymbol{k}) = \int d\boldsymbol{p} \sum_{\beta\alpha} \hat{a}_\beta(\boldsymbol{p} + \hbar\boldsymbol{k})^\dagger \left[\frac{q}{m}\boldsymbol{p}\delta_{\beta\alpha} + \frac{i}{\hbar}(\varepsilon_\beta - \varepsilon_\alpha)\boldsymbol{\mu}_{\beta\alpha}\right] \hat{a}_\alpha(\boldsymbol{p}), \tag{9.34}$$

where $\boldsymbol{\mu}_{\beta\alpha}$ is the *transition electric dipole moment,*

$$\boldsymbol{\mu}_{\beta\alpha} = \langle\beta| \sum_j^s q_j(\boldsymbol{r}_j - \boldsymbol{r})|\alpha\rangle. \tag{9.35}$$

It remains to find the interaction Hamiltonian operator by dotting $\hat{\boldsymbol{j}}(\boldsymbol{k})$ with $\hat{\mathbf{A}}(\boldsymbol{k})$, multiplying by $-1/c$ and integrating over $\boldsymbol{k}$:

$$\begin{aligned}\hat{\mathrm{H}}_{\mathrm{I}} &= -\frac{1}{c}\int d\boldsymbol{k}\hat{\boldsymbol{j}}(\boldsymbol{k}) \cdot \hat{\boldsymbol{A}}(\boldsymbol{k}) \\ &= -\frac{1}{c}\int d\boldsymbol{k}\int d\boldsymbol{p}\sum_{\beta\alpha} \hat{a}_\beta(\boldsymbol{p} + \hbar\boldsymbol{k})^\dagger \\ &\quad\times \left[\frac{q}{m}\boldsymbol{p}\delta_{\beta\alpha} + \frac{i}{\hbar}(\varepsilon_\beta - \varepsilon_\alpha)\boldsymbol{\mu}_{\beta\alpha}\right] \hat{a}_\alpha(\boldsymbol{p}) \\ &\quad\times \left(\frac{hc}{16\pi^3 k}\right)^{\frac{1}{2}} \sum_{\sigma=1}^{2} [\hat{C}_\sigma(\boldsymbol{k}) + \hat{C}_\sigma(-\boldsymbol{k})^\dagger]\boldsymbol{e}_k.\end{aligned} \tag{9.36}$$

Rearrangement of Eq. (9.36) gives

$$\begin{aligned}\hat{\mathrm{H}}_{\mathrm{I}}^{\mathrm{em}} &= -\frac{\hbar c}{4\pi^{3/2}}\int d\boldsymbol{k}\int d\boldsymbol{p}\sum_{\alpha} \hat{a}_\alpha(\boldsymbol{p} + \hbar\boldsymbol{k})^\dagger \frac{q}{m}\boldsymbol{p}\hat{a}_\alpha(\boldsymbol{p}) \\ &\quad\times (\hbar ck)^{-\frac{1}{2}} \sum_{\sigma=1}^{2} [\hat{C}_\sigma(\boldsymbol{k}) + \hat{C}_\sigma(-\boldsymbol{k})^\dagger]\boldsymbol{e}_k\end{aligned} \tag{9.37}$$

and

$$\begin{aligned}\hat{\mathrm{H}}_{\mathrm{I}}^{\mathrm{ed}} &= \frac{i}{4\pi^{3/2}}\sum_{\alpha\beta} \boldsymbol{\mu}_{\beta\alpha} \cdot \boldsymbol{e}_k \int d\boldsymbol{p}\int d\boldsymbol{k} \\ &\quad\times \hat{a}_\beta(\boldsymbol{p} + \hbar\boldsymbol{k})^\dagger \frac{(\varepsilon_\beta - \varepsilon_\alpha)}{(\hbar ck)^{1/2}} \hat{a}_\alpha(\boldsymbol{p}) \sum_{\sigma=1}^{2} [\hat{C}_\sigma(\boldsymbol{k}) + \hat{C}_\sigma(-\boldsymbol{k})^\dagger]\end{aligned} \tag{9.38}$$

for the electric monopole and electric dipole interaction Hamiltonians, respectively. To use the interaction Hamiltonian to find rates of photon absorption and emission by molecules, we need the ensemble operators for both molecules and photons.

9.3. Homogeneous Grand Canonical Ensemble Operators

In this section we borrow heavily from Chapters 6 and 8, which the reader may wish to review.

9.3.1. *The Molecular Ensemble*

From Chapter 6 we saw that the molecular homogeneous grand canonical (isodasic) ensemble operator is

$$\hat{\rho}_{\mathrm{m}} = |0\rangle\langle 0| + \sum_{n=1}^{\infty} \frac{1}{n!} \prod_{l=1}^{n} \left[\int d\boldsymbol{p} \sum_{\alpha_l} \boldsymbol{\Phi}_{\alpha_l}(\boldsymbol{p}_l) \right]$$

$$\times \prod_{j=1}^{n} [\hat{a}_{\alpha(n-j+1)}(\boldsymbol{p}'_{(n-j+1)})^{\dagger}] |0\rangle\langle 0| \prod_{k=1}^{n} [\hat{a}_{\alpha_k}(\boldsymbol{p}_k)], \quad (9.39)$$

where $\hat{a}_{\alpha}(\boldsymbol{p})^{\dagger}$ and $\hat{a}_{\alpha}(\boldsymbol{p})$ are the creator and annihilator, respectively, of a molecule in internal state α with momentum $\boldsymbol{p}$, and $\Phi_{\alpha}(\boldsymbol{p})$ is a function that is essentially the relative number of molecules in that state. Although the form of Eq. (9.39) is complicated looking, it was shown in Chapter 6 that the essential property of $\hat{\rho}_{\mathrm{m}}$ is

$$\hat{a}_{\alpha}(\boldsymbol{p})\hat{\rho}_{\mathrm{m}} = \Phi_{\alpha}(\boldsymbol{p})\hat{\rho}_{\mathrm{m}}\hat{a}_{\alpha}(\boldsymbol{p}) \quad \text{and} \quad \hat{\rho}_{\mathrm{m}}\hat{a}_{\alpha}(\boldsymbol{p})^{\dagger} = \Phi_{\alpha}(\boldsymbol{p})\hat{a}_{\alpha}(\boldsymbol{p})^{\dagger}\hat{\rho}_{\mathrm{m}}. \quad (9.40)$$

Average or expectation values for molecular properties represented by the operator $\hat{O}$ are given by

$$\langle O\rangle = \mathrm{Tr}[\hat{O}\hat{\rho}_{\mathrm{m}}]/\mathrm{Tr}[\hat{\rho}_{\mathrm{m}}], \quad (9.41)$$

where we have used the symbol Tr to indicate *trace* of the operator string in the following brackets. The trace has the cyclic property

$$\mathrm{Tr}[\hat{A}\hat{B}\hat{C}] = \mathrm{Tr}[\hat{C}\hat{A}\hat{B}] = \mathrm{Tr}[\hat{B}\hat{C}\hat{A}]. \quad (9.42)$$

9.3.2. *The Photon Ensemble*

From Chapter 8 we saw that the ensemble operator for homogeneous photons has the same form as Eq. (9.39):

$$\hat{\rho}_{\mathrm{r}} = |0\rangle\langle 0| + \sum_{n=1}^{n} \frac{1}{n!} \prod_{l=1}^{n} \left[\int d\boldsymbol{k} \sum_{\sigma_l=1}^{2} \chi_{\sigma_l}(\boldsymbol{k}_l) \right]$$

$$\times \prod_{j=1}^{n} [\hat{C}_{\sigma(n-j+1)}(\boldsymbol{k}_{(n-j+1)})^{\dagger}] |0\rangle\langle 0| \prod_{k=1}^{n} [\hat{C}_{\sigma_k}(\boldsymbol{k}_k)]. \quad (9.43)$$

We now integrate over the wave vector $\boldsymbol{k}$ and sum over the polarity σ. The subscript r on $\hat{\rho}$ stands for radiation. This radiation ensemble operator has the same property as that for molecules, namely

$$\hat{C}_\sigma(\boldsymbol{k})\hat{\rho}_{\mathrm{r}} = \chi_\sigma(\boldsymbol{k})\hat{\rho}_{\mathrm{r}}\hat{C}_\sigma(\boldsymbol{k}) \quad \text{and} \quad \hat{\rho}_{\mathrm{r}}\hat{C}_\sigma(\boldsymbol{k})^{\dagger} = \chi_\sigma(\boldsymbol{k})\hat{C}_\sigma(\boldsymbol{k})^{\dagger}\hat{\rho}_{\mathrm{r}}, \quad (9.44)$$

and for a photon property O

$$\langle O \rangle = \mathrm{Tr}[\hat{O}\hat{\rho}_{\mathrm{r}}]/\mathrm{Tr}[\hat{\rho}_{\mathrm{r}}]. \quad (9.45)$$

9.4. Dynamics

If we wish to calculate the rate of photon absorption and emission, we will need the time dependence of the number operators. In Chapter 3, time-dependent perturbation theory was developed in the Schrödinger, Heisenberg, and Dirac (or interaction) pictures. Here it is most convenient to use the interaction picture, which we review briefly.

The time dependence of an operator $\hat{O}$ is given by

$$\hat{O}(t) = e^{\frac{i}{\hbar}\hat{\mathrm{H}}t}\hat{O}e^{-\frac{i}{\hbar}\hat{\mathrm{H}}t}, \quad (9.46)$$

where $\hat{\mathrm{H}}$ is the total Hamiltonian for two interacting systems. This total Hamiltonian consists of two parts: the Hamiltonians for each system without interaction, $\hat{\mathrm{H}}_0 = \hat{\mathrm{H}}_{01} + \hat{\mathrm{H}}_{02}$, plus the Hamiltonian for the interaction, $\hat{\mathrm{H}}' = \hat{\mathrm{H}} - \hat{\mathrm{H}}_0$. Accordingly, we may write Eq. (9.46) as

$$\hat{O}(t) = e^{\frac{i}{\hbar}\hat{\mathrm{H}}t}e^{-\frac{i}{\hbar}\hat{\mathrm{H}}_0 t}e^{\frac{i}{\hbar}\hat{\mathrm{H}}_0 t}\hat{O}e^{-\frac{i}{\hbar}\hat{\mathrm{H}}_0 t}e^{\frac{i}{\hbar}\hat{\mathrm{H}}_0 t}e^{-\frac{i}{\hbar}\hat{\mathrm{H}}t} = \hat{U}^\dagger(t)\hat{O}_{\mathrm{I}}(t)\hat{U}(t), \tag{9.47}$$

where

$$\hat{O}_{\mathrm{I}}(t) = e^{\frac{i}{\hbar}\hat{\mathrm{H}}_0 t}\hat{O}e^{-\frac{i}{\hbar}\hat{\mathrm{H}}_0 t} \tag{9.48}$$

and

$$\hat{U}(t) = e^{\frac{i}{\hbar}\hat{\mathrm{H}}_0 t}e^{-\frac{i}{\hbar}\hat{\mathrm{H}}t}. \tag{9.49}$$

To develop a perturbation series for $\hat{U}(t)$, we take its time derivative:

$$\begin{aligned}\frac{\partial}{\partial t}\hat{U}(t) &= \frac{i}{\hbar}e^{\frac{i}{\hbar}\hat{\mathrm{H}}_0 t}(\hat{\mathrm{H}}_0 - \hat{\mathrm{H}})e^{-\frac{i}{\hbar}\hat{\mathrm{H}}t}\\ &= -\frac{i}{\hbar}\hat{\mathrm{H}}_{\mathrm{I}}(t)e^{\frac{i}{\hbar}\hat{\mathrm{H}}_0 t}e^{-\frac{i}{\hbar}\hat{\mathrm{H}}t}\\ &= -\frac{i}{\hbar}\hat{\mathrm{H}}_{\mathrm{I}}(t)\hat{U}(t).\end{aligned} \tag{9.50}$$

Integration of Eq. (9.50) gives

$$\hat{\mathrm{H}}_{\mathrm{I}}(t) \equiv e^{\frac{i}{\hbar}\hat{\mathrm{H}}_0 t}(\hat{\mathrm{H}} - \hat{\mathrm{H}}_0)e^{-\frac{i}{\hbar}\hat{\mathrm{H}}_0 t}. \tag{9.51}$$

To first order, Eq. (9.51) gives

$$\hat{U}(t) = 1 - \frac{i}{\hbar}\int_0^t dt'\hat{\mathrm{H}}_{\mathrm{I}}(t')\hat{U}(t'). \tag{9.52}$$

Then from Eq. (9.47) we have, to first order

$$\hat{U}^{(1)}(t) = -\frac{i}{\hbar}\int_0^t dt'\hat{\mathrm{H}}_{\mathrm{I}}(t') \tag{9.53}$$

and

$$\hat{O}_{\mathrm{I}}^{(1)}(t) = \frac{i}{\hbar}\int_0^t dt'[\hat{\mathrm{H}}_{\mathrm{I}}(t'), \hat{O}_{\mathrm{I}}(t)]_-. \tag{9.54}$$

After writing out the commutator bracket and using the trace cyclic property, Eq. (9.54) becomes

$$\mathrm{Tr}[\hat{O}^{(1)}(t)\hat{\rho}] = \frac{i}{\hbar}\int_0^t dt'\mathrm{Tr}\{\hat{\mathrm{H}}_{\mathrm{I}}(t')[\hat{O}_{\mathrm{I}}(t),\hat{\rho}]_{-}\}. \tag{9.56}$$

Then the time derivative of the average value of O to first order is

$$\begin{aligned}\langle\dot{O}^{(1)}(t)\rangle &= \frac{\partial}{\partial t}\mathrm{Tr}[\hat{O}^{(1)}(t)\hat{\rho}] = \frac{i}{\hbar}\mathrm{Tr}\{\hat{\mathrm{H}}_{\mathrm{I}}(t)[\hat{O}_{I}(t),\hat{\rho}]_{-}\}\\ &\quad + \frac{i}{\hbar}\int_0^t dt'\mathrm{Tr}\left\{\hat{\mathrm{H}}_{\mathrm{I}}(t')\left[\frac{\partial}{\partial t}\hat{O}_{\mathrm{I}}(t),\hat{\rho}\right]_{-}\right\}.\end{aligned} \tag{9.57}$$

The commutation bracket in the first term in Eq. (9.56) is

$$[\hat{O}_{\mathrm{I}}(t),\hat{\rho}]_{-} = [e^{\frac{i}{\hbar}H_0 t}\hat{O}e^{-\frac{i}{\hbar}H_0 t},\hat{\rho}]_{-}. \tag{9.58}$$

But if $\hat{\rho} = \hat{\rho}_{\mathrm{m}}\hat{\rho}_{\mathrm{r}}$ is a product of the ensemble operators for homogeneous molecules and photons, respectively, and if $\hat{\mathrm{H}}_0 = \hat{\mathrm{H}}_{\mathrm{m}0} + \hat{\mathrm{H}}_{\mathrm{r}0}$; *i.e.*, if

$$\begin{aligned}\hat{\mathrm{H}}_{\mathrm{m}0} &= \int d\boldsymbol{p}\sum_{\alpha}\hat{a}_{\alpha}(\boldsymbol{p})^{\dagger}\hat{a}_{\alpha}(\boldsymbol{p})\varepsilon_{\alpha}(p)\\ \hat{\mathrm{H}}_{\mathrm{r}0} &= \int d\boldsymbol{k}\sum_{\sigma}\hat{C}_{\sigma}(\boldsymbol{k})^{\dagger}\hat{C}_{\sigma}(\boldsymbol{k})\hbar ck\end{aligned} \tag{9.59a,b}$$

then in view of Eqs. (9.40) and (9.44) it is seen that $[\hat{O}_{\mathrm{I}}(t),\hat{\rho}]_{-} = 0$ and therefore that the first term in Eq. (9.57) vanishes. Similarly, since

$$\frac{\partial}{\partial t}\hat{O}_{\mathrm{I}}(t) = \frac{i}{\hbar}[\hat{\mathrm{H}}_0\hat{O}_{\mathrm{I}}(t) - \hat{O}_{\mathrm{I}}(t)\hat{\mathrm{H}}_0], \tag{9.60}$$

we have

$$\left[\frac{\partial}{\partial t}\hat{O}_{\mathrm{I}}(t),\hat{\rho}\right]_{-} = \left\{\frac{i}{\hbar}\left[\hat{\mathrm{H}}_0\hat{O}_{\mathrm{I}}(t) - \hat{O}_{\mathrm{I}}(t)\hat{\mathrm{H}}_0\right],\hat{\rho}\right\}_{-}. \tag{9.61}$$

In our present application, $\hat{O}$ will be a number operator for molecules and for photons; in this case $\hat{O}$ and $\hat{\mathrm{H}}_0$ commute so that Eq. (9.61) will vanish; therefore there is no first-order contribution.

In second order, we have from Eq. (9.52)

$$\begin{aligned}\hat{O}^2(t) &= \hat{U}^{(1)}(t)^\dagger \hat{O}_{\rm I}(t)\hat{U}^{(1)}(t) \\ &\quad + \hat{U}^{(2)}(t)^\dagger \hat{O}_{\rm I}(t)\hat{U}^{(0)}(t) + \hat{U}^{(0)}(t)^\dagger \hat{O}_{\rm I}(t)\hat{U}^2(t) \\ &= \hat{U}^{(1)}(t)^\dagger \hat{O}_{\rm I}(t)\hat{U}^{(1)}(t) + \hat{U}^{(2)}(t)^\dagger \hat{O}_{\rm I}(t) + \hat{O}_{\rm I}(t)\hat{U}^{(2)}(t).\end{aligned} \tag{9.62}$$

We take the trace of Eq. (9.62) with $\hat{\rho}$ term-by-term, using Eq. (9.52); the first term is

$$\begin{aligned}\langle O(t)\rangle &= \mathrm{Tr}[\hat{U}^{(1)}(t)^\dagger \hat{O}_{\rm I}(t)\hat{U}^{(1)}(t)\hat{\rho}] \\ &= \frac{1}{\hbar^2}\int_0^t dt' \int_0^t dt'' \mathrm{Tr}[\hat{\mathrm{H}}_{\rm I}(t')^\dagger \hat{O}_{\rm I}(t)\hat{\mathrm{H}}_{\rm I}(t'')\hat{\rho}],\end{aligned} \tag{9.63}$$

the time derivative of which is

$$\langle \dot{O}(t)\rangle = \frac{1}{\hbar^2}\int_0^t dt' \mathrm{Tr}[\hat{\mathrm{H}}_{\rm I}(t)^\dagger \hat{O}_{\rm I}(t)\hat{\mathrm{H}}_{\rm I}(t')\hat{\rho} + \hat{\mathrm{H}}_{\rm I}(t')^\dagger \hat{O}_{\rm I}(t)\hat{\mathrm{H}}_{\rm I}(t)\hat{\rho}]. \tag{9.64}$$

Since in the present application $\hat{O}_{\rm I}(t) = \hat{O}$ and $[\hat{\mathrm{H}}_0, \hat{\rho}]_- = 0$, with use of the cyclic property of the trace, the first term in Eq. (9.64) can be written

$$\begin{aligned}&\frac{1}{\hbar^2}\int_0^t dt' \mathrm{Tr}[\hat{\mathrm{H}}_{\rm I}(t)^\dagger \hat{O}\hat{\mathrm{H}}_{\rm I}(t)\hat{\rho}] \\ &= \frac{1}{\hbar^2}\int_0^t dt' \mathrm{Tr}[e^{\frac{i}{\hbar}\hat{\mathrm{H}}_0 t}\hat{\mathrm{H}}_{\rm I}^\dagger e^{-\frac{i}{\hbar}\hat{\mathrm{H}}_0 t}\hat{O} e^{\frac{i}{\hbar}\hat{\mathrm{H}}_0 t'}\hat{\mathrm{H}}_{\rm I} e^{-\frac{i}{\hbar}\hat{\mathrm{H}}_0 t'}\hat{\rho}] \\ &= \frac{1}{\hbar^2}\int_0^t dt' \mathrm{Tr}[e^{-\frac{i}{\hbar}\hat{\mathrm{H}}_0 t'}\hat{\rho} e^{\frac{i}{\hbar}\hat{\mathrm{H}}_0 t}\hat{\mathrm{H}}_{\rm I}^\dagger e^{-\frac{i}{\hbar}\hat{\mathrm{H}}_0 (t-t')}\hat{O}\hat{\mathrm{H}}_{\rm I}] \\ &= \frac{1}{\hbar^2}\int_0^t dt' \mathrm{Tr}[e^{\frac{i}{\hbar}\hat{\mathrm{H}}_0 (t-t')}\hat{\mathrm{H}}_{\rm I}^\dagger e^{-\frac{i}{\hbar}\hat{\mathrm{H}}_0 (t-t')}\hat{O}\hat{\mathrm{H}}_{\rm I}\hat{\rho}] \\ &= \frac{1}{\hbar^2}\int_0^t dt' \mathrm{Tr}[\hat{\mathrm{H}}_{\rm I}(t-t')^\dagger \hat{O}\hat{\mathrm{H}}_{\rm I}\hat{\rho}].\end{aligned} \tag{9.65}$$

We now change the integration variable to $\tau = t - t'$; $dt' = -d\tau$; $t' = 0 \to \tau = t$; $t' = t \to \tau = 0$, so that Eq. (9.65) becomes

$$\frac{1}{\hbar^2}\int_0^t dt' \mathrm{Tr}[\hat{\mathrm{H}}_\mathrm{I}(t)^\dagger \hat{O}\hat{\mathrm{H}}_\mathrm{I}(t)\hat{\rho}] = \frac{1}{\hbar^2}\int_0^t d\tau \mathrm{Tr}[\hat{\mathrm{H}}_\mathrm{I}(\tau)^\dagger \hat{O}\hat{\mathrm{H}}_\mathrm{I}\hat{\rho}]. \quad (9.66)$$

For the second term in Eq. (9.64), we have

$$\begin{aligned}
&\frac{1}{\hbar^2}\int_0^t dt' \mathrm{Tr}[\hat{\mathrm{H}}_\mathrm{I}(t')^\dagger \hat{O}\hat{\mathrm{H}}_\mathrm{I}(t)\hat{\rho}]\\
&= \frac{1}{\hbar^2}\int_0^t dt' \mathrm{Tr}[e^{\frac{i}{\hbar}\hat{\mathrm{H}}_0 t'}\hat{\mathrm{H}}_\mathrm{I}^\dagger e^{-\frac{i}{\hbar}\hat{\mathrm{H}}_0 t'}\hat{O}e^{\frac{i}{\hbar}\hat{\mathrm{H}}_0 t}\hat{\mathrm{H}}_\mathrm{I}e^{-\frac{i}{\hbar}\hat{\mathrm{H}}_0 t}\hat{\rho}]\\
&= \frac{1}{\hbar^2}\int_0^t dt' \mathrm{Tr}[e^{-\frac{i}{\hbar}\hat{\mathrm{H}}_0 t}\hat{\rho}e^{\frac{i}{\hbar}\hat{\mathrm{H}}_0 t'}\hat{\mathrm{H}}_\mathrm{I}^\dagger e^{-\frac{i}{\hbar}\hat{\mathrm{H}}_0 (t'-t)}\hat{O}\hat{\mathrm{H}}_\mathrm{I}]\\
&= \frac{1}{\hbar^2}\int_0^t dt' \mathrm{Tr}[e^{\frac{i}{\hbar}\hat{\mathrm{H}}_0 (t'-t)}\hat{\mathrm{H}}_\mathrm{I}^\dagger e^{-\frac{i}{\hbar}\hat{\mathrm{H}}_0 (t'-t)}\hat{O}\hat{\mathrm{H}}_\mathrm{I}\hat{\rho}]\\
&= \frac{1}{\hbar^2}\int_0^t dt' \mathrm{Tr}[\hat{\mathrm{H}}_\mathrm{I}(t'-t)^\dagger \hat{O}\hat{H}_\mathrm{I}\hat{\rho}]. \qquad (9.67)
\end{aligned}$$

Then when we change the integration variable to $\tau = t' - t$; $dt' = d\tau$ and $t' = 0 \to \tau = -t$; $t' = t \to \tau = 0$, Eq. (9.67) becomes

$$\frac{1}{\hbar^2}\int_0^t dt' \mathrm{Tr}[\hat{\mathrm{H}}_\mathrm{I}(t')^\dagger \hat{O}\hat{\mathrm{H}}_\mathrm{I}(t)\hat{\rho}] = \frac{1}{\hbar^2}\int_{-t}^0 dt' \mathrm{Tr}[\hat{\mathrm{H}}_\mathrm{I}(\tau)^\dagger \hat{O}\hat{\mathrm{H}}_\mathrm{I}\hat{\rho}]. \quad (9.68)$$

Adding Eqs. (9.67) and (9.68) gives

$$\langle \dot{O}(t)\rangle = \frac{1}{\hbar^2}\int_{-t}^t d\tau \mathrm{Tr}[\hat{\mathrm{H}}_\mathrm{I}(\tau)^\dagger \hat{O}\hat{\mathrm{H}}_\mathrm{I}\hat{\rho}]. \quad (9.69)$$

Before inserting expressions for $\hat{\mathrm{H}}_\mathrm{I}(\tau)^\dagger$ and $\hat{\mathrm{H}}_\mathrm{I}$ into Eq. (9.69), look at $\hat{\mathrm{H}}_\mathrm{I}$ as given in Eq. (9.38): the initial molecular state is α and the final molecular state β; the sum in the second line allows us to choose the photon process, whether emission of a photon with the choice $\hat{C}_\sigma{}^\dagger$ or absorption of a photon with the choice $\hat{C}_\sigma$. We will treat emission first by writing for $\hat{\mathrm{H}}_{\mathrm{I}}$

$$\hat{\mathrm{H}}_{\mathrm{I}}^{\mathrm{ed}} = -\frac{i}{4\pi^{3/2}} \sum_{\alpha\beta} \boldsymbol{\mu}_{\beta\alpha} \cdot \boldsymbol{e}_k \int d\boldsymbol{p} \int d\boldsymbol{k} \frac{(\varepsilon_\beta - \varepsilon_\alpha)}{(\hbar c k)^{1/2}} \times \hat{a}_\beta(\boldsymbol{p} - \hbar\boldsymbol{k})^\dagger \hat{a}_\alpha(\boldsymbol{p}) \sum_{\sigma=1}^{2} \hat{C}_\sigma(\boldsymbol{k})^\dagger, \tag{9.70}$$

where we have changed the sign of the symmetric integration variable $\boldsymbol{k}$ to reflect the fact that the emitted photon has momentum $\boldsymbol{k}$ at the expense of the momentum of the final state of the molecule. Then we have

$$\hat{\mathrm{H}}_{\mathrm{I}}^{\mathrm{ed}}(\tau)^\dagger = \frac{i}{4\pi^{3/2}} \sum_{\alpha''\beta''} \boldsymbol{\mu}_{\beta''\alpha''} \cdot \boldsymbol{e}_{k''} \int d\boldsymbol{p}'' \int d\boldsymbol{k}'' \frac{(\varepsilon_{\beta''} - \varepsilon_{\alpha''})}{(\hbar c k'')^{1/2}} \times \exp\left[-\frac{i}{\hbar}\left(\varepsilon_{\beta''} - \varepsilon_{\alpha''} + \hbar c k'' - \frac{\hbar}{m}\boldsymbol{p}'' \cdot \boldsymbol{k}'' + \frac{\hbar^2 k''^2}{2m}\right)\tau\right] \hat{a}_{\alpha''}(\boldsymbol{p}'')^\dagger \hat{a}_{\beta''}(\boldsymbol{p}'' - \hbar\boldsymbol{k}'') \times \sum_{\sigma''=1}^{2} \hat{C}_{\sigma''}(\boldsymbol{k}'')^\dagger. \tag{9.71}$$

Now we insert Eqs. (9.70) and (9.71) into Eq. (9.69) to get the rate of photon emission at $t = \infty$.

$$\langle \dot{N}_{\mathrm{emission}}(\infty) \rangle = \frac{1}{\hbar^2} \frac{1}{16\pi^3} \int d\boldsymbol{p} \sum_{\alpha\beta} \int d\boldsymbol{p}'' \sum_{\alpha''\beta''} \int d\boldsymbol{p}' \times \sum_{\gamma} \int d\boldsymbol{k} \int d\boldsymbol{k}'' \int d\boldsymbol{k}' \frac{(\varepsilon_\beta - \varepsilon_\alpha)(\varepsilon_{\beta''} - \varepsilon_{\alpha''})}{(\hbar c k)^{1/2}(\hbar c k'')^{1/2}} \boldsymbol{\mu}_{\beta''\alpha''} \cdot \boldsymbol{e}_{k''}\, \boldsymbol{e}_k \cdot \boldsymbol{\mu}_{\beta\alpha} \int_{-\infty}^{\infty} d\tau \exp\left[-\frac{i}{\hbar}\left(\varepsilon_{\beta''} - \varepsilon_{\alpha''} + \hbar c k'' - \frac{\hbar}{m}\boldsymbol{p}'' \cdot \boldsymbol{k}'' + \frac{\hbar^2 k''^2}{2m}\right)\tau\right] \mathrm{Tr}[\hat{a}_{\alpha''}(\boldsymbol{p}'')^\dagger \hat{a}_{\beta''}(\boldsymbol{p}'' - \hbar\boldsymbol{k}'')$$

$$\times \sum_{\sigma''} \hat{C}_{\sigma''}(\boldsymbol{k}')\hat{a}_\gamma(\boldsymbol{p}')^\dagger \hat{a}_\gamma(\boldsymbol{p}') \sum_{\sigma'} \hat{C}_{\sigma'}(\boldsymbol{k}')^\dagger \hat{C}_{\sigma'}(\boldsymbol{k}')\hat{a}_\beta(\boldsymbol{p} - \hbar\boldsymbol{k})^\dagger$$
$$\times \hat{a}_\alpha(\boldsymbol{p}) \sum_{\sigma} \hat{C}_\sigma(\boldsymbol{k})^\dagger \hat{\rho}]. \tag{9.72}$$

When the strings of operators are put into normal order and selection is made for single-photon, single-molecule interactions which result in the emission of a photon of mode $\boldsymbol{k}\sigma$ the result is[1]

$$\langle \dot{N}_{\text{emission}}(\infty)\rangle = \frac{1}{\hbar}\frac{1}{8\pi^2}\int d\boldsymbol{p} \sum_{\alpha\beta} \int d\boldsymbol{k} \frac{(\varepsilon_\beta - \varepsilon_\alpha)^2}{(\hbar ck)} \boldsymbol{\mu}_{\beta\alpha} \cdot \boldsymbol{e}_k \boldsymbol{e}_k \cdot \boldsymbol{\mu}_{\beta\alpha}$$
$$\times \frac{V}{(2\pi\hbar)^3} \delta\left(\varepsilon_\beta - \varepsilon_\alpha + \hbar ck - \frac{\hbar}{m}\boldsymbol{p}\cdot\boldsymbol{k} + \frac{\hbar^2 k^2}{2m}\right)$$
$$\times \Phi_\alpha(\boldsymbol{p}) \left[\frac{\chi_\sigma(\boldsymbol{k})}{1 - \chi_\sigma(\boldsymbol{k})} + 1\right]. \tag{9.73}$$

We have shown in the steps leading to Eq. (6.50) that for a perfect gas with an average of N_{m} molecules at thermal equilibrium,

$$\Phi_\alpha(\boldsymbol{p}) = N_{\text{m}}\zeta_\alpha e^{-p^2/2mk_{\text{B}}T}, \tag{9.74}$$

where ζ_α is the fraction of the molecules in the state α which at thermal equilibrium is $e^{-\varepsilon_\alpha/k_{\text{B}}T}/\sum_\gamma e^{-\varepsilon_\gamma/k_{\text{B}}T}$. After insertion of Eq. (9.74) into Eq. (9.73) the integration over $\boldsymbol{p}$ becomes

$$\int d\boldsymbol{p}\delta\left(\varepsilon_\beta - \varepsilon_\alpha + \hbar ck - \frac{\hbar}{m}\boldsymbol{p}\cdot\boldsymbol{k} + \frac{\hbar^2k^2}{2m}\right) e^{-p^2/2mk_{\text{B}}T}. \tag{9.75}$$

We accommodate the $\boldsymbol{p}\cdot\boldsymbol{k}$ term in the delta function by breaking the vector $\boldsymbol{p}$ into the component parallel to $\boldsymbol{k}$, namely $p_\parallel$, and the

[1] For the derivation of Eq. (9.73), see Appendix A.

perpendicular 2-vector $p_\perp$:

$$\begin{aligned}
&\int d\boldsymbol{p}\delta\left(\varepsilon_\beta - \varepsilon_\alpha + \hbar ck - \frac{\hbar}{m}\boldsymbol{p}\cdot\boldsymbol{k} + \frac{\hbar^2k^2}{2m}\right)e^{-p^2/2mk_\mathrm{B}T}\\
&\quad = \int d\boldsymbol{p}_\perp e^{-p_\perp^2/2mk_\mathrm{B}T}\frac{m}{\hbar k}\int dp_\parallel e^{-p_\parallel^2/2mk_\mathrm{B}T}\\
&\quad \times \delta[p_\parallel - \frac{m}{\hbar k}(\hbar ck + \varepsilon_\beta - \varepsilon_\alpha + \hbar^2c^2k^2)^2/2m]\\
&\quad = \frac{m}{\hbar k}(2\pi mk_\mathrm{B}T)\exp\left[-\left(\frac{m^2}{\hbar^2k^2}\right)\frac{(\hbar ck + \varepsilon_\beta - \varepsilon_\alpha + \hbar^2k^2/2m)^2}{2mk_\mathrm{B}T}\right].
\end{aligned} \tag{9.76}$$

With the help of Eqs. (9.74)–(9.76), Eq. (9.73) becomes

$$\begin{aligned}
&\langle\dot{N}_\mathrm{emission}(\infty)\rangle\\
&\quad = \frac{1}{\hbar}\frac{mc}{8\pi^2}\sum_{\alpha\beta}N_\mathrm{m}\zeta_\alpha\int d\boldsymbol{k}\frac{(\varepsilon_\beta - \varepsilon_\alpha)^2}{(\hbar ck)^2}\boldsymbol{\mu}_{\beta\alpha}\cdot\boldsymbol{e}_k\boldsymbol{e}_k\cdot\boldsymbol{\mu}_{\beta\alpha}(2pmk_\mathrm{B}T)\\
&\quad \times \frac{V}{(2\pi\hbar)^3}\exp\left[-\left(\frac{m^2}{\hbar^2k^2}\right)\frac{(\hbar ck + \varepsilon_\beta - \varepsilon_\alpha + \hbar^2k^2/2m)^2}{2mk_\mathrm{B}T}\right]\\
&\quad \times \left[\frac{\chi_\sigma(\boldsymbol{k})}{1 - \chi_\sigma(\boldsymbol{k})} + 1\right].
\end{aligned} \tag{9.77}$$

For a single-photon absorption term contained in Eq. (9.72), we can contract the photon operator string as follows:

$$\overbrace{\hat{C}_{\sigma''}(\boldsymbol{k}'')\hat{C}_{\sigma'}(\boldsymbol{k}')^\dagger\hat{C}_{\sigma'}(\boldsymbol{k}')\hat{C}_\sigma(\boldsymbol{k})^\dagger} = \delta_{\sigma''\sigma}\delta(\boldsymbol{k}'' - \boldsymbol{k})\hat{C}_{\sigma'}(\boldsymbol{k}')^\dagger\hat{C}_{\sigma'}(\boldsymbol{k}'). \tag{9.78}$$

The trace of the expression in Eq. (9.78) with the radiation component of the ensemble operator is

$$\begin{aligned}
&\delta_{\sigma''\sigma}\delta(\boldsymbol{k}'' - \boldsymbol{k})\mathrm{Tr}[\hat{C}_{\sigma'}(\boldsymbol{k}')^\dagger\hat{C}_{\sigma'}(\boldsymbol{k}')\hat{\rho}_l]\\
&\quad = \delta_{\sigma''\sigma}\delta(\boldsymbol{k}'' - \boldsymbol{k})\frac{V}{(2\pi)^3}\frac{\chi\sigma'(\boldsymbol{k})}{1 - \chi_{\sigma'}(\boldsymbol{k})}\mathrm{Tr}\hat{\rho}_l.
\end{aligned} \tag{9.79}$$

For consistency with the emission terms, we take $\varepsilon_\beta < \varepsilon_\alpha$. This means that the initial state for the absorption term is β and the final state α will have the momentum of the absorbed photon added to its original momentum. Therefore, the molecular operator string for this absorption term is

$$\mathrm{Tr}[\hat{a}_{\beta''}(\boldsymbol{p}'')^\dagger \hat{a}_{\alpha''}(\boldsymbol{p}'' + \hbar\mathbf{k}'')\hat{a}_\gamma(\boldsymbol{p}')^\dagger \hat{a}_\gamma(\boldsymbol{p}')\hat{a}_\alpha(\boldsymbol{p} + \hbar\boldsymbol{k})^\dagger \hat{a}_\beta(\boldsymbol{p})\hat{\rho}_{\mathrm{m}}]. \tag{9.80}$$

The derivation of the expression for the rate of photon absorption is similar to that for induced emission given in Appendix A, followed by the derivation of Eq. (9.77); the result is

$$\begin{aligned}\langle \dot{N}_{\mathrm{absorption}}(\infty)\rangle &= \frac{1}{\hbar}\frac{mc}{8\pi^2}\sum_{\alpha,\beta} N\zeta_\beta \int d\boldsymbol{k} \\ &\times \frac{(\varepsilon_\beta - \varepsilon_\alpha)^2}{(\hbar ck)^2}\boldsymbol{\mu}_{\beta\alpha}\cdot \boldsymbol{e}_k\boldsymbol{e}_k \cdot \boldsymbol{\mu}_{\beta\alpha}(2\pi m k_{\mathrm{B}}T) \\ &\times \frac{V}{(2\pi\hbar)^3}\exp\left[-\left(\frac{m^2}{\hbar^2 k^2}\right)\frac{(\hbar ck + \varepsilon_\beta - \varepsilon_\alpha + \hbar^2k^2/2m)^2}{2mk_{\mathrm{B}}T}\right] \\ &\times \frac{V}{(2\pi)^3}\frac{\chi_\sigma(\boldsymbol{k})}{1-\chi_\sigma(\boldsymbol{k})}.\end{aligned} \tag{9.81}$$

Indeed, the only difference between Eqs. (9.77) and (9.81) is the subscript on ζ and the factor $\frac{V}{(2\pi)^3}$ on the last line in the absorption expression. As for the last two terms in the second line of Eq. (9.62), the one-photon terms after normal ordering give $\boldsymbol{k} = 0$ for the absorbed photon, so those second-order terms can be ignored.

9.5. The Einstein Coefficients and Planck's Black Body Formula

At the steady state (dynamic equilibrium), the rate of emission equals the rate of absorption for each frequency $\nu_{\alpha\beta} = 2\pi(\varepsilon_\alpha - \varepsilon_\beta)/\hbar$; that is,

$$\dot{N}^{\mathrm{emiss}}_{\beta\leftarrow\alpha}(\infty) = \dot{N}^{\mathrm{ste}}_{\beta\leftarrow\alpha}(\infty) + \dot{N}^{\mathrm{spe}}_{\beta\leftarrow\alpha}(\infty) = \dot{N}^{\mathrm{abs}}_{\alpha\leftarrow\beta}(\infty), \tag{9.82}$$

where the superscripts "emiss", "ste", "spe", and "abs" indicate emission, stimulated emission, spontaneous emission, and absorption, respectively. From Eqs. (9.77), (9.82), and (9.83), we can write

$$\begin{aligned}\dot{N}^{\mathrm{ste}}_{\beta\leftarrow\alpha}(\infty) &= \mathrm{B}_{\alpha\beta}N_{\mathrm{m}}\zeta_\alpha \\ \dot{N}^{\mathrm{abs}}_{\alpha\leftarrow\beta}(\infty) &= \mathrm{B}_{\alpha\beta}N_{\mathrm{m}}\zeta_\beta \\ \dot{N}^{\mathrm{spe}}_{\beta\leftarrow\alpha}(\infty) &= \mathrm{A}_{\alpha\beta}N_{\mathrm{m}}\zeta_\alpha,\end{aligned} \tag{9.83a,b,c}$$

where

$$\mathrm{B}_{\alpha\beta} = \mathrm{A}_{\alpha\beta}\sum_\sigma N_\sigma(\boldsymbol{k}). \tag{9.84}$$

Note that $N_\sigma(\boldsymbol{k})$ is the *total* number of photons in mode $\boldsymbol{k}\sigma$ in the volume V, *not* a number density. Inserting Eq. (9.84) into Eq. (9.83) gives

$$\mathrm{B}_{\alpha\beta}N_{\mathrm{m}}\zeta_\alpha + \mathrm{A}_{\alpha\beta}N_{\mathrm{m}}\zeta_\alpha = \mathrm{B}_{\alpha\beta}N_{\mathrm{m}}\zeta_\beta. \tag{9.85}$$

Now divide each term in Eq. (9.84) by $\mathrm{B}_{\alpha\beta}N_{\mathrm{m}}\zeta_\alpha$, make use of Eq. (9.85), and rearrange:

$$\sum_\sigma N_\sigma(\boldsymbol{k}) = (\zeta_\beta/\zeta_\alpha - 1)^{-1}. \tag{9.86}$$

At thermal equilibrium,

$$\zeta_\beta/\zeta_\alpha = e^{(\varepsilon_\alpha-\varepsilon_\beta)/k_{\mathrm{B}}T} = e^{h\nu_{\alpha\beta}/k_{\mathrm{B}}T}, \tag{9.87}$$

so that Eq. (9.86) becomes

$$\sum_\sigma N_\sigma(\boldsymbol{k}) = \frac{1}{e^{h\nu_{\alpha\beta}/k_{\mathrm{B}}T} - 1}. \tag{9.88}$$

Our treatment of a system of molecules and photons in thermal equilibrium applies to black body radiation, where often the observed quantities are frequency and radiation energy density per unit volume. Accordingly, we need to convert $N_\sigma(\boldsymbol{k})$ to $N_\sigma(\nu)$. We begin by noting that $\boldsymbol{k}$ is a 3-vector and ν is a scalar. Since the radiation in the black body is essentially isotropic, $d\boldsymbol{k} = 4\pi k^2 dk$; thus

$$\sum_\sigma N_\sigma(\boldsymbol{k})d\boldsymbol{k} = 4\pi k^2 \sum_\sigma N_\sigma(\boldsymbol{k})dk = \sum_\sigma N_\sigma(\nu)d\nu \qquad (9.89)$$

so that

$$\sum_\sigma N_\sigma(\boldsymbol{k}) = \frac{\sum_\sigma N_\sigma(\nu)}{4\pi k^2}\frac{d\nu}{dk} = \frac{\sum_\sigma N_\sigma(\nu)}{4\pi k^2}\frac{c}{2\pi}$$

$$= \frac{c^3}{32\pi^4 h\nu^3}\sum_\sigma N_\sigma(\nu)h\nu. \qquad (9.90)$$

From Eqs. (9.89) and (9.90), we have

$$\sum_\sigma N_\sigma(\nu)h\nu = \frac{32\pi^4 h\nu^3}{c^3}\frac{1}{e^{h\nu_{\alpha\beta}/k_\mathrm{B}T} - 1}, \qquad (9.91)$$

a form of the Planck radiation law.[2]

9.6. Multiphoton Absorption

Multiphoton absorption processes in intense focused laser beams have become important tools for the spectroscopy of molecules, in particular for obtaining the ro-vibrational structure of electronically excited states. An example of the photon Fock operator chain for 2-photon absorption by a molecule is given in Eq. (9A.2a) below. If this photon chain is coupled with the molecule chain

$$\hat{a}_\alpha(\boldsymbol{p} + \hbar\boldsymbol{k} + \hbar\boldsymbol{k}')^\dagger \hat{a}_\gamma(\boldsymbol{p}' + \hbar\boldsymbol{k})\hat{a}_\gamma(\boldsymbol{p}' + \hbar\boldsymbol{k})^\dagger \hat{a}_\beta(\boldsymbol{p}')$$

(and left-multiplied by its complex conjugate if we are deriving the analog of Eq. (9.82) for 2-photon absorption), we have an operator for the absorption of a photon by a molecule in internal state β, exciting it to an intermediate state γ, which absorbs another photon and is excited to the final state α.

[2]For a thermodynamics-based derivation of the Planck distribution, see Appendix B.

Both the theory and experimental design for multiphoton ionization are highly specialized and are beyond the scope of this introduction to the quantum field theory of photon-molecule interaction. Readers interested in this field should see reviews of multiphoton processes.[3]

9.7. Appendix A: Derivation of Equation (9.73)

The string of photon operators in Eq. (9.72) for emission of a $\boldsymbol{k}\sigma$ photon is

$$\hat{C}_{\sigma''}(\boldsymbol{k}'')\hat{C}_{\sigma'}(\boldsymbol{k}')^\dagger\hat{C}_{\sigma'}(\boldsymbol{k}')\hat{C}_{\sigma'}(\boldsymbol{k})^\dagger, \tag{9A.1}$$

the normal order of which is

$$\hat{C}_{\sigma''}(\boldsymbol{k}'')^\dagger\hat{C}_{\sigma'}(\boldsymbol{k}')^\dagger\hat{C}_{\sigma'}(\boldsymbol{k}')\hat{C}_{\sigma}(\boldsymbol{k}) \quad \text{2 photons absorbed} \tag{9A.2a}$$

$$+\overbrace{\hat{C}_{\sigma''}(\boldsymbol{k}'')\hat{C}_{\sigma'}(\boldsymbol{k}')^\dagger}\hat{C}_{\sigma'}(\boldsymbol{k}')\hat{C}_{\sigma}(\boldsymbol{k})^\dagger \quad \text{1 photon emitted} \tag{9A.2b}$$

$$+\overbrace{\hat{C}_{\sigma}(\boldsymbol{k})\hat{C}^\dagger_{\sigma'}(\boldsymbol{k}')\hat{C}_{\sigma'}(\boldsymbol{k}')\hat{C}_{\sigma^\dagger}(\boldsymbol{k})} \quad \text{1 photon absorbed} \tag{9A.2c}$$

$$+\overbrace{\hat{C}_{\sigma''}(\boldsymbol{k}'')\hat{C}_{\sigma'}(\boldsymbol{k}')^\dagger}\,\overbrace{\hat{C}_{\sigma'}(\boldsymbol{k}')\hat{C}_{\sigma}(\boldsymbol{k})^\dagger} \quad \text{spontaneous emission.} \tag{9A.2d}$$

Carrying out the contractions for the 1-photon emissions, we have

$$\overbrace{\hat{C}_{\sigma''}(\boldsymbol{k}'')\hat{C}_{\sigma'}(\boldsymbol{k}')^\dagger}\hat{C}_{\sigma'}(\boldsymbol{k}')\hat{C}_{\sigma}(\boldsymbol{k})^\dagger = \delta_{\sigma\sigma'}\delta(\boldsymbol{k}-\boldsymbol{k}')\hat{C}_{\sigma}(\boldsymbol{k})^\dagger\hat{C}_{\sigma'}(\boldsymbol{k}') \tag{9A.3a}$$

and

$$\hat{C}_{\sigma''}(\boldsymbol{k}'')\hat{C}_{\sigma'}(\boldsymbol{k}')^\dagger\,\overbrace{\hat{C}_{\sigma'}(\boldsymbol{k}')\hat{C}_{\sigma}(\boldsymbol{k})^\dagger} = \delta_{\sigma'\sigma}\delta(\boldsymbol{k}'-\boldsymbol{k})\hat{C}_{\sigma''}(\boldsymbol{k}'')\hat{C}_{\sigma'}(\boldsymbol{k}')^\dagger. \tag{9A.3b}$$

---

[3]Lambropoulos, P. and Smith, S. J. (eds), *Multiphoton Processes*, Springer-Verlag (1984).

The trace of the radiation factor of the ensemble operator with the remaining normal-ordered pair of operators in Eq. (9A.3) is

$$\begin{aligned}\mathrm{Tr}[\hat{C}_\sigma(\boldsymbol{k})^\dagger \hat{C}_{\sigma'}(\boldsymbol{k}')\hat{\rho}_l] &= \mathrm{Tr}[\hat{C}_{\sigma'}(\boldsymbol{k}')\hat{\rho}_l \hat{C}_\sigma(\boldsymbol{k})^\dagger] \\ &= \chi_\sigma(\boldsymbol{k})\mathrm{Tr}[\hat{\rho}_l \hat{C}_{\sigma'}(\boldsymbol{k}')\hat{C}_\sigma(\boldsymbol{k})^\dagger] = \chi_\sigma(\boldsymbol{k})\mathrm{Tr}[\hat{C}_{\sigma'}(\boldsymbol{k}')\hat{C}_\sigma(\boldsymbol{k})^\dagger \hat{\rho}_l] \\ &= \chi_{\sigma'}(\boldsymbol{k}')\mathrm{Tr}[\hat{C}_\sigma(\boldsymbol{k})^\dagger \hat{C}_{\sigma'}(\boldsymbol{k}')\hat{\rho}_l] + \chi_{\sigma'}(\boldsymbol{k}')\delta_{\sigma\sigma'}\delta(\boldsymbol{k}-\boldsymbol{k}')\mathrm{Tr}\hat{\rho}_l. \qquad (9A.4)\end{aligned}$$

Rearrangement of Eq. (9A.4) gives

$$\mathrm{Tr}[\hat{C}_\sigma(\boldsymbol{k})^\dagger \hat{C}_{\sigma'}(\boldsymbol{k}')\hat{\rho}_l] = \frac{\chi_\sigma(\boldsymbol{k})}{1-\chi_\sigma(\boldsymbol{k})}\delta_{\sigma\sigma'}\delta(\boldsymbol{k}-\boldsymbol{k}')\mathrm{Tr}\hat{\rho}_l. \qquad (9A.5)$$

Similarly, for Eq. (9A.3b) we have

$$\mathrm{Tr}[\hat{C}_{\sigma''}(\boldsymbol{k}'')^\dagger \hat{C}_{\sigma'}(\boldsymbol{k}')\hat{\rho}_l] = \frac{\chi_{\sigma'}(\boldsymbol{k}')}{1-\chi_{\sigma'}(\boldsymbol{k}')}\delta_{\sigma''\sigma'}\delta(\boldsymbol{k}''-\boldsymbol{k}')\mathrm{Tr}\hat{\rho}_l. \qquad (9A.6)$$

For the spontaneous emission term we have

$$\begin{aligned}&\mathrm{Tr}[\hat{C}_{\sigma''}(\boldsymbol{k}'')\hat{C}_{\sigma'}(\boldsymbol{k}')^\dagger \hat{C}_{\sigma'}(\boldsymbol{k}')\hat{C}_\sigma(\boldsymbol{k})^\dagger \rho_l] \\ &\quad = \delta_{\sigma''\sigma'}\delta(\boldsymbol{k}''-\boldsymbol{k}')\delta_{\sigma\sigma'}\delta(\boldsymbol{k}-\boldsymbol{k}')\mathrm{Tr}\hat{\rho}_l. \qquad (9A.7)\end{aligned}$$

Similarly for the trace of the molecular factor of the ensemble operator with the string of molecular operators in Eq. (9.72), keeping only the single-molecule term:

$$\begin{aligned}&\mathrm{Tr}[\hat{a}_{\alpha''}(\boldsymbol{p}'')^\dagger \hat{a}_{\beta''}(\boldsymbol{p}''-\hbar\boldsymbol{k}'')\hat{a}_\gamma(\boldsymbol{p}')^\dagger \hat{a}_\gamma(\boldsymbol{p}')\hat{a}_\beta(\boldsymbol{p}-\hbar\boldsymbol{k})^\dagger \hat{a}_\alpha(\boldsymbol{p})\hat{\rho}_{\mathrm{m}}] \\ &\quad = \delta_{\beta''\gamma}\delta_{\gamma\beta}\delta_{\alpha''\alpha}\delta(\boldsymbol{p}''-\boldsymbol{p})\delta(\boldsymbol{p}'-\boldsymbol{p}+\hbar\boldsymbol{k})\frac{V}{(2\pi\hbar)^3}\frac{\Phi_\alpha(\boldsymbol{p})}{1\mp\Phi_\alpha(\boldsymbol{p})}\mathrm{Tr}\hat{\rho}_{\mathrm{m}}. \qquad (9A.8)\end{aligned}$$

The factor $\frac{V}{(2\pi\hbar)^3}$ results from the redundancy in the product $\delta[\hbar(\boldsymbol{k}-\boldsymbol{k}')]\delta(\boldsymbol{k}-\boldsymbol{k}')$.

In the high-temperature (Boltzmann) limit, which applies in most cases of interest to us,

$$\frac{\Phi_\alpha(\boldsymbol{p})}{1\mp\Phi_\alpha(\boldsymbol{p})} \to \Phi_\alpha(\boldsymbol{p}). \qquad (9A.9)$$

We also see from Eqs. (9A.5) and (9A.6) that only $\boldsymbol{p}, \boldsymbol{k}, \alpha, \beta$, and σ survive the integration over $\boldsymbol{k}', \boldsymbol{k}''$, and sums over σ, σ', and σ''. After putting Eqs. (9A.5)–(9A.9) into Eq. (9.72), we obtain Eq. (9.73).

9.8. Appendix B: A Thermodynamic Derivation of the Planck Distribution

9.8.1. *The Empirical Observations*

Rayleigh[4] found that at low frequency ν the radiation from a black body as a function of temperature T has the energy density ρ_ν (*i.e.*, the radiation energy divided by volume V of the black body) is well represented by the empirical formula

$$\rho_\nu = Ak_{\mathrm{B}}T, \tag{9B.1}$$

where k_{B} is the Boltzmann constant and the coefficient A is proportional to ν^2. Wien,[5] on the other hand, found that at high frequencies ρ_ν follows the exponential law

$$\rho_\nu = Ce^{-B\nu/T}, \tag{9B.2}$$

with C proportional to ν^3.

Planck[6] was successful at finding a formula that united these two limiting laws and as a bonus, to discover a new (*i.e.*, Bose-Einstein) statistics for oscillators that ushered in quantum theory. We will not deal here with the statistics, but only with the thermodynamic reasoning leading to the identification of the Planck constant.

9.8.2. *Thermodynamic Temperature*

Since the black body is assumed to be in equilibrium, the laws of thermodynamics apply. The First Law states that the increased energy of a system is the sum of the increase in heat δQ and the work done on the system δW and that, while both δQ and δW depend

---

[4]Strutt, J., Lord Rayleigh, *Philos. Mag.* **49**, 539 (1900).
[5]Wien, W., *Annalen der Physik und Chemie* **294**, 662 (1896).
[6]Planck, M., *Annalen der Physik* **306**, 69 (1900).

upon the details of the process (*i.e.*, are path functions), their sum dE does not (*i.e.*, E is a state function). This is expressed as

$$dE = \delta Q + \delta W. \tag{9B.3}$$

The Second Law defines the state function entropy S by

$$dS = dQ_{\mathrm{r}}/T, \tag{9B.4}$$

where the subscript r (for *reversible*) indicates that the system is heated so gradually that it remains in virtual equilibrium, and that therefore the process by which heat is added is reversible.

Then if Eq. (9B.3) is applied to a reversible process in which the work done is restricted to a decrease in volume V under the application of an external pressure p, the combination of Eqs. (9B.3) and (9B.4) give

$$dE = TdS + pdV. \tag{9B.5}$$

Solving Eq. (9B.5) for dS gives

$$dS = \frac{1}{T}(dE - pdV), \tag{9B.6}$$

from which we obtain

$$\left(\frac{\partial S}{\partial E}\right)_V = \frac{1}{T}. \tag{9B.7}$$

Eq. (9B.7) applies to the present problem when we divide both S and E in the partial derivative by the volume V. Then we have

$$\left(\frac{\partial S_V}{\partial \rho_V}\right)_V = \frac{1}{T}, \tag{9B.8}$$

where S_V is the entropy density. Solving the two limiting laws, Eqs. (9B.1) and (9B.2), for $1/T$ we have

$$\left(\frac{\partial S_V}{\partial \rho_V}\right)_V = \frac{1}{T} = \begin{Bmatrix} \dfrac{Ak_{\mathrm{B}}}{\rho_V} & (\text{small } V) \\ -\dfrac{1}{BV}\ln\left(\frac{\rho_V}{C}\right) & (\text{large } V) \end{Bmatrix}. \tag{9B.9}$$

To see more clearly the required bridge between the two limiting expressions, we take the second derivative with respect to ρ_V:

$$\left(\frac{\partial^2 S_V}{\partial \rho_V^2}\right)_V = -\frac{1}{T^2}\left(\frac{\partial T}{\partial \rho_V}\right) = \begin{Bmatrix} -\dfrac{Ak_B}{\rho_V^2} & (\text{small}\, V) \\ -\dfrac{1}{BV\rho_V} & (\text{large}\, V) \end{Bmatrix}. \quad (9B.10)$$

9.8.3. *Planck's Empirical Formula*

A simple common formula that has the limiting forms of Eq. (9.B.10) is

$$\left(\frac{\partial^2 S_V}{\partial \rho_V^2}\right)_V = -\frac{Ak_B}{\rho_V(\rho_V + Ak_B BV)}. \quad (9B.11)$$

So now we need to find ρ_V from this formula. By integrating Eq. (9B.8) once by ρ_V we will get $1/T$:

$$\left(\frac{\partial S_V}{\partial \rho_V}\right)_V = \int \left(\frac{\partial^2 S_V}{\partial \rho_V^2}\right)_V d\rho_V = -Ak_B \int \frac{d\rho_V}{\rho_V(\rho_V + Ak_B BV)} = \frac{1}{T}. \quad (9B.12)$$

The integration can be accomplished with the substitution

$$\frac{1}{\rho_V(\rho_V + Ak_B BV)} = \frac{1}{Ak_B BV}\left\{\frac{1}{\rho_V} - \frac{1}{(\rho_V + Ak_B BV)}\right\}. \quad (9B.13)$$

Eq. (9B.12) then becomes

$$\begin{aligned} \frac{1}{T} &= -\frac{1}{BV}\int \left\{\frac{1}{\rho_V} - \frac{1}{(\rho_V + Ak_B BV)}\right\} d\rho_V \\ &= \frac{1}{BV}\ln\left(\frac{\rho_V + Ak_B BV}{\rho_V}\right) = \frac{1}{BV}\ln\left(1 + \frac{Ak_B BV}{\rho_V}\right). \end{aligned} \quad (9B.14)$$

From Eq. (9B.14) we see that ρ_V is given by

$$\rho_V = \frac{Ak_B BV}{e^{BV/T} - 1}. \quad (9B.15)$$

This expression for ρ_V has the correct limits. As V becomes small: $\rho_V \sim Ak_B T$. As V becomes large: $\rho_V \sim Ak_B TBVe^{BV/T}$. Since A is observed to be $\propto V^2$, then the coefficient C of Eq. (9B.2) is seen to be $\propto V^3$, as observed.

Classical oscillators in equilibrium with surroundings at temperature T have average energy per oscillator $\langle\varepsilon\rangle = k_B T$, but according to Planck's formula the average energy of the oscillators in a black body is

$$\langle\varepsilon\rangle = \frac{k_B B V}{e^{BV/T} - 1}. \tag{9B.16}$$

9.8.4. *Interpretation of Equation (9B.16)*

Planck concluded that a black body takes on integral multiples of a factor proportional to the frequency ν. This assumption gives an average energy

$$\langle\varepsilon\rangle = \frac{\sum_{n=0}^{\infty} nh\nu e^{-nh\nu/k_B T}}{\sum_{n=0}^{\infty} e^{-nh\nu/k_B T}} = \frac{h\nu}{e^{h\nu/k_B T} - 1}. \tag{9B.17}$$

Comparison of Eqs. (9B.16) and (9B.17) shows that Planck's constant is $h = Bk_B$. An evaluation in 1900 of the empirical constant B gave the estimate $h \sim 6.55 \times 10^{-34}$ J-s. The official value is now fixed at exactly $6.62607015 \times 10^{-34}$ J-s.

Planck's thorough knowledge of thermodynamics and statistical mechanics was essential to his discovery, though he remained a skeptic of the quantum theory as subsequently developed by Einstein, Schrödinger, Heisenberg, Bohr, and others, as did some of these founders of the theory. But in its latest version, the subject of this book, its predictive reliability is unmatched by any theory of nature and no experiment has thus far proved it wrong.

Chapter 10

Conclusions, Acknowledgements, and References

10.1. Conclusions

The development of the isodasic ensemble operator for perfect gases in Chapter 6, for which the defining equation is Eq. (6.28):

$$\hat{a}_\alpha(\boldsymbol{p})\hat{\rho}_0(\beta) = \boldsymbol{\Phi}_\alpha(\beta)\hat{\rho}_0(\beta)\hat{a}_\alpha(\boldsymbol{p})$$

and the treatment of imperfect gases and liquids in Chapter 7 are the main reasons for this book. We hope some readers will be inspired by these chapters and by Chapters 8 and 9 on photons and their interactions with molecules to extend the ideas beyond what has been presented there.

10.2. Acknowledgements

By way of acknowledgments, Phil Johnson and his student Le-ping Li first turned my attention to the question of the quantum-theoretical representation of gases; the result is contained in Chapter 6. But I could not have accomplished our contribution to the answer without the tutelage on the basic points of atomic and molecular quantum theory by Martin Karplus and the patient hearing and encouragement we received from him and his Harvard colleagues while we were working out the quantum field theory of molecular ensembles. I thank Lionel Raff for reviewing early drafts of Chapters 2–4, Rudy Marcus and Dudley Hershbach for their patient counseling. Both the Departments of Chemistry at Stony Brook and of Chemistry and

Chemical Biology at Harvard were generous in facilitating sabbaticals during which the ideas leading to this volume were developed. I am grateful to my family, wife Martha, daughter Cynthia, son John, and my grandchildren Elisabeth, Constance, Jessica, and Trevor. I would have had more interaction with them were it not for my spending so much time at this endeavor. And certainly not least is my huge debt to my editor at World Scientific, Shaun Tan, who guided me through the protocols of modern publishing; his professionalism is much admired. His typesetter also handled the challenges of the equations quite admirably.

10.3. Notes

The following is a listing of the references in the footnotes appearing in the main text:

1. Tong, D., *Lectures on Quantum Field Theory*, Cambridge University (2006). Page 1.
2. Avery, J., *Creation and Annihilation Operators*, McGraw-Hill (1976). Page 9.
3. Wick, G. C., *Phys. Rev.* **80**, 268 (1950). Pages 56 and 209.
4. Glauber, R. J., *Phys. Rev.* **131**, 2766 (1963). Page 64.
5. Noether, E., *Nachrichten von der Gesellschaft der Wissenschaften zu Göttingen, Mathematisch-Physikalische Klasse* **1918**, 235 (1918). Page 95.
6. Schwinger, J., in *Quantum Theory of Angular Momentum*, Academic Press (1965), p. 229. Page 148.
7. Porter, R. N., *AIP Conf. Proc.* **1102**, 219 (2009). Pages 150 and 181.
8. Yasuda, M. and Shimizu, F., *Phys. Rev. Lett.* **77**, 3090 (1996); Fölling, S., *et al.*, *Nature* **434**, 481 (2005); Greiner, M., *et al.*, *Phys. Rev. Lett.* **94**, 110401 (2005); Schellekins, M., *et al.*, *Science* **310**, 648 (2005). Page 179.
9. Debye, P. and Hückel, E., *Phys. Zeits.* **24**, 185 (1928). Page 187.
10. Glauber, R. J., *Phys. Rev.* **130**, 2529 (1963); Cahill, K. E. and Glauber, R. J., *Phys. Rev.* **177**, 1857 (1969). Page 211.

11. Louisell, W. H., *Quantum Statistical Properties of Radiation*, Wiley (1973). Page 211.
12. Yariv, A., *Quantum Electronics*, Wiley (1975, 2nd edition). Page 228.
13. Lambropoulos, P. and Smith, S. J. (eds), *Multiphoton Processes*, Springer-Verlag (1984). Page 260.
14. Strutt, J., Lord Rayleigh, *Philos. Mag.* **49**, 539 (1900). Page 262.
15. Wien, W., *Annalen der Physik und Chemie* **294**, 662 (1896). Page 262.
16. Planck, M., *Annalen der Physik* **306**, 69 (1900). Page 262.